NEUROMETHODS

For other titles published in this series, go to
www.springer.com/series/7657

NEUROMETHODS

Animal Models of Epilepsy

Methods and Innovations

Edited by

Scott C. Baraban Ph.D.

University of California, San Francisco, CA, USA

Humana Press

Editor
Scott C. Baraban
Department of Neurological Surgery
Box 0520
University of California, San Francisco
513 Parnassus Avenue
San Francisco, CA 94143
USA
scott.baraban@ucsf.edu

Series Editor
Wolfgang Walz
University of Saskatchewan
Saskatoon, SK, Canada

ISSN 0893-2336 e-ISSN 1940-6045
ISBN 978-1-60327-262-9 e-ISBN 978-1-60327-263-6
DOI 10.1007/978-1-60327-263-6
Springer Dordrecht Heidelberg London New York

Library of Congress Control Number: 2009920531

Printed on acid-free paper

Springer is part of Springer Science+Business Media (www.springer.com)

Preface to the Series

Under the guidance of its founders Alan Boulton and Glen Baker, the Neuromethods series by Humana Press has been very successful since the first volume appeared in 1985. In about 17 years, 37 volumes have been published. In 2006, Springer Science + Business Media made a renewed commitment to this series. The new program will focus on methods that are either unique to the nervous system and excitable cells or which need special consideration to be applied to the neurosciences. The program will strike a balance between recent and exciting developments like those concerning new animal models of disease, imaging, in vivo methods, and more established techniques. These include immunocytochemistry and electrophysiological technologies. New trainees in neurosciences still need a sound footing in these older methods in order to apply a critical approach to their results. The careful application of methods is probably the most important step in the process of scientific inquiry. In the past, new methodologies led the way in developing new disciplines in the biological and medical sciences. For example, Physiology emerged out of Anatomy in the 19th century by harnessing new methods based on the newly discovered phenomenon of electricity. Nowadays, the relationships between disciplines and methods are more complex. Methods are now widely shared between disciplines and research areas. New developments in electronic publishing also make it possible for scientists to download chapters or protocols selectively within a very short time of encountering them. This new approach has been taken into account in the design of individual volumes and chapters in this series.

Wolfgang Walz

Preface

Epilepsy is fairly unique among the various neurological disorders as it provides the neuroscientist with almost boundless opportunities to examine basic neurobiological mechanisms. Not surprisingly, advances in epilepsy research are closely tied to development of innovative neurobiological methodologies. In many cases the practical application of these innovations – especially in the context of a neurological disorder with anatomical, molecular, electrophysiological, and behavioral components such as epilepsy – can be found in the development of new animal models. In turn, our understanding of the pathogenesis of epilepsy (and new therapy development) greatly benefits from these models. Taking advantage of transgenic and homologous recombination techniques, laboratories have recently moved beyond the standard convulsant or stimulation models in rat to develop novel mouse models of epilepsy. This forward thinking approach has recently been applied to genetically tractable "simple" species such as *Drosophila melanogaster* (fruit flies), *Caenorhabditis elegans* (worms), *Xenopus laevis* (tadpoles), and *Danio rerio* (zebrafish).

With contributions from prominent investigators in this field, this book provides a review of these emerging animal models of epilepsy. Prior textbooks devoted to models of seizure and epilepsy almost exclusively categorized rat models with little attention paid to these more innovative approaches. Here we attempt to diverge from the conventional epilepsy literature and focus on animal models that attempt to incorporate the latest technological advancements in neurobiology. While some of these models and approaches are, admittedly, at very early stages of development and may ultimately fall short of widespread utilization, it is through the consideration and presentation of these models that the authors' hope to advance and challenge the field of epilepsy research. Here we also attempt to move beyond a strict review of animal models and also include innovative approaches to epilepsy research that are just now appearing in the literature. These range from modeling seizure activity in silica to advanced strategies for seizure detection and gene therapy. These latter innovations are important as some may lead to better therapeutic treatments for patients suffering from intractable forms of epilepsy.

As a *Neuromethods* book, our overall objective is to provide an accessible and thought-provoking review of recent advancements in epilepsy research. Contributing authors encompass a wide spectrum of expertise from basic neurobiology, through sophisticated electrophysiology and genetics, to practicing epilepsy clinicians. Authors were carefully selected who not only offer a broad perspective on epilepsy, but, in each case, have pioneered the innovative approaches described in this book.

Scott C. Baraban

Contents

Contributors

SCOTT C. BARABAN • *Department of Neurological Surgery, University of California, San Francisco, CA, USA*
SUZANNE B. BAUSCH • *Department of Pharmacology and Graduate Program in Neuroscience, Uniformed Services University, Bethesda, MD, USA*
ROBERT BRENNER • *Department of Physiology, University of Texas Health Science Center at San Antonio, San Antonio, TX, USA*
AMY R. BROOKS-KAYAL • *Division of Neurology, Pediatric Regional Epilepsy Program, Children's Hospital of Philadelphia, Philadelphia, PA, USA*
MARIA ELISA CALCAGNOTTO • *Department of Physiology, Universidade Federal de São Paulo - UNIFESP, São Paulo, Brazil*
GUY A. CALDWELL • *Departments of Biological Sciences, Neurology and Neuroscience, Center for Neurodegeneration and Experimental Therapeutics, University of Alabama at Birmingham, Birmingham, AL, USA*
KIM A. CALDWELL • *Departments of Biological Sciences, Neurology and Neuroscience, Center for Neurodegeneration and Experimental Therapeutics, University of Alabama at Birmingham, Birmingham, AL, USA*
QING H. CHEN • *Department of Physiology, University of Texas Health Science Center at San Antonio, San Antonio, TX, USA*
JAVIER ECHAUZ, J.E. • *Research, Atlanta, GA, USA*
KEVIN C. ESS • *Departments of Neurology and Pediatrics, Vanderbilt Kennedy Center, Vanderbilt University, Nashville, TN, USA*
STACEY B. FOTI • *Program in Neurobiology, University of North Carolina at Chapel Hill, Chapel Hill, NC, USA*
KURT HAAS • *Department of Cellular and Physiological Sciences and the Brain Research Centre, University of British Columbia, Vancouver, BC, Canada*
D. SESATH HEWAPATHIRANE • *Department of Cellular and Physiological Sciences and the Brain Research Centre, University of British Columbia, Vancouver, BC, Canada*
DOROTHY JONES-DAVIS • *Department of Neurological Surgery, University of California, San Francisco, CA, USA*
MICHELLE M. KRON • *Department of Neurology, University of Michigan, Ann Arbor, MI, USA*
MARK F. LEPPERT • *Department of Human Genetics, University of Utah, Salt Lake City, UT, USA*
BRIAN LITT • *Departments of Neurology and Bioengineering, University of Pennsylvania, Philadelphia, PA, USA*
CODY J. LOCKE • *Department of Biological Sciences, The University of Alabama, Tuscaloosa, AL, USA*
THOMAS J. MCOWN • *University of North Carolina Gene Therapy Center, University of North Carolina School of Medicine, Chapel Hill, NC, USA*

Robert J. Morgan • *Department of Anatomy and Neurobiology, University of California, Irvine, CA, USA*
James F. Otto • *Anticonvulsant Drug Development Program, University of Utah, Salt Lake City, UT, USA*
Jack M. Parent • *Department of Neurology, University of Michigan, Ann Arbor, MI, USA*
David Petrik • *Department of Physiology, University of Texas Health Science Center at San Antonio, San Antonio, TX, USA*
Shelley J. Russek • *Department of Pharmacology and Experimental Therapeutics, Boston University School of Medicine, Boston, MA, USA*
Joy Y. Sebe • *Department of Neurological Surgery, University of California, San Francisco, CA, USA*
Nanda A. Singh • *Department of Human Genetics, University of Utah, Salt Lake City, UT, USA*
Ivan Soltesz • *Department of Anatomy and Neurobiology, University of California, Irvine, CA, USA*
Juan Song • *Departments of Environmental Sciences, Policy and Management, Division of Organisms and Environment, Molecular and Cellular Biology, Division of Neurobiology, University of California, Berkeley, Berkeley, CA, USA*
Mark A. Tanouye • *Departments of Environmental Sciences, Policy and Management, Division of Organisms and Environment, Molecular and Cellular Biology, Division of Neurobiology, University of California, Berkeley, Berkeley, CA, USA*
Andrew J. Trevelyan • *Institute of Neuroscience, Newcastle University, Medical School, Framlington Place, Newcastle upon Tyne, UK*
H. Steve White • *Anticonvulsant Drug Development Program, University of Utah, Salt Lake City, UT, USA*
Karen Wilcox • *Department of Pharmacology and Toxicology, University of Utah, Salt Lake City, UT, USA*
Stephen Wong • *Departments of Neurology and Bioengineering, University of Pennsylvania, Philadelphia, PA, USA*
Rafael Yuste • *HHMI, Department of Biological Sciences, Columbia University, New York, NY, USA*

Color Plates

Color Plate 1 **In vivo imaging of dendritic arbor growth and synapse formation in *X. laevis* tadpoles.** (**A**) Single-cell electroporation allowing targeted neuronal labeling and transfection in vivo. Scale bar = 2 mm. (Chapter 3, Fig. 4; *see* discussion on p. 53 and complete caption on p. 54)

Color Plate 2 **Tscl morphant zebrafish.** (*Anatomy*) Histological analysis of control and *tsc1a* morphants. Immunostaining with the neuronal marker HuC/HuD of control-injected (**A**) and *tsc1a* morphant (**B**) larvae, and semi-thin plastic sections of control-injected (**C,E,G,I**) and *tsc1a* morphant (**D,F,H,J**) larvae are shown. (Chapter 4, Fig. 3; *see* discussion on p. 69 and complete caption on p. 70)

Color Plate 3 Confocal images of brain sections from $Tsc1^{Synapsin}$CKO mice immunostained for SMI311 (red, neuronal marker) and phospho-S6 (green). This staining reveals large, dysplastic appearing neurons from white matter **(A)** or the cortex **(B)**. (Chapter 5, Fig. 4; *see* discussion on p. 81)

Color Plate 4 **Diagram of BK channel α (pore-forming) subunit and accessory *β* subunits.** Annotations indicate relevant functional domains and sites containing epilepsy mutations ($\beta 3$ delA750 and gain-of-function mutation D434G for α subunit). (Chapter 6, Fig. 1; *see* discussion on p. 88)

Color Plate 5 **BK and *β*4 effect on action potential waveform in dentate gyrus granule cells.** Action potential was selected from 10th spike during a 300 pA current injection. Average data are plotted at right. Red trace is $\beta 4$ knockout, blue trace is wild type plus SK channel blocker UCL1684, black trace is wild type. (Chapter 6, Fig. 3; *see* discussion on p. 97)

Color Plate 6 **Cortical interneuron loss in selected GABAergic interneuron-deficient mice.** Wild-type mice display normal connectivity in the cortex. (Chapter 8, Fig. 3; *see* discussion on p. 134 and complete caption on p. 135)

Color Plate 7 **Somatic Ca^{2+} transients reveal episodic localized recruitment. (A)** The derivation of the cellular Ca^{2+}signal. (Chapter 9, Fig. 4; *see* discussion on p. 156 and complete caption on p. 157)

Chapter 1

The Nematode, *Caenorhabditis elegans*, as an Emerging Model for Investigating Epilepsy

Cody J. Locke, Kim A. Caldwell, and Guy A. Caldwell

Abstract

Acquiring a holistic understanding of epilepsy, an intricate and often multifactorial disorder, is certain to necessitate considerable effort from a diverse and global assortment of scientists and clinicians. For most of its history, a combination of basic research with rodent models and clinical research with patients has dominated the approach of the epilepsy research community. These complementary approaches will undoubtedly remain as the ultimate direction in which research should move before new treatments for epilepsy are established, as mammalian nervous systems are exemplified by a degree of complexity, which sets them apart from those of invertebrates. Nonetheless, epilepsy researchers are inundated with inexplicable genetic, anatomical, and physiological defects, which underlie the scores of epilepsies and epilepsy syndromes. Likewise, the techniques required for effectively characterizing the underlying defects associated with epilepsy via mammalian systems are not only laborious and time consuming but also costly. Because epilepsy represents a significant disease burden worldwide *(1)*, there is a serious need to accelerate research by considering non-traditional approaches. The subject of this chapter is one such promising approach that employs what is often described as the "most understood animal on the planet", a nematode, known as *Caenorhabditis elegans*.

Key words: nematode, *C. elegans*, PTZ, lissencephaly.

1. *Caenorhabditis elegans* as a Model for General Human Biology

With the aim of generating a model to understand the nervous system in its entirety, Nobel laureate Sydney Brenner chose *C. elegans* in the early 1960s. *C. elegans* is a transparent nematode, which grows to roughly a millimeter in length. Laboratory "worms" commonly feed on *Escherichia coli* in agar-filled Petri dishes. As a predominately hermaphroditic species with

Scott C. Baraban (ed.), *Animal Models of Epilepsy: Methods and Innovations, vol. 40*

DOI 10.1007/978-1-60327-263-6_1 Springerprotocols.com

only a rare male population, *C. elegans* is also a prolific self-fertilizer and can generate approximately 300 offspring per parent. A typical worm is capable of rapid development, metamorphosing from a fertilized egg to a full-sized adult in less than 3 full days. Overall, these attributes ensure that worms could be cheaply and copiously available for analysis at any time.

Regardless of how rapidly and inexpensively it can be studied, choosing an appropriate model for research into a human disorder also depends upon the relevance of that model to the disorder in question. One might assume that use of a "simple" animal model like *C. elegans* for epilepsy research would be nonsensical. However, despite its size, *C. elegans* is actually a multifaceted organism. *C. elegans* adult hermaphrodites only have 959 somatic cells. Yet, these somatic cells include a reproductive system, glands, muscle, hypodermis, intestines, and a nervous system *(2)*. In addition to these anatomical and associated physiological characteristics, which share unmistakable homology with those of mammals, *C. elegans* also shares substantial evolutionary conservation of cellular and molecular pathways with mammals, including humans. In fact, a comparison of the *C. elegans* genome with the human genome revealed that approximately 70% of human disease genes and pathways are found in *C. elegans (3)*.

Accordingly, *C. elegans* has become the primary model organism for thousands of scientists interested in studying the molecular and cellular basis of assorted biological phenomena, ranging from cell signaling, the cell cycle, and regulation of gene expression to metabolism, sex determination, and aging *(4)*. A particularly applicable result of studying the genetics of aging in *C. elegans* has been the identification of DAF-16, a negative regulator of insulin-like signaling *(5)*. A mouse homolog of DAF-16, the forkhead transcription factor, FOXO, was later found to play a comparable role in insulin signaling, establishing a new paradigm for diabetes research *(6)*. Similarly, the molecular nature of polycystic kidney disorder was first demonstrated following the identification of mating-deficient *C. elegans* mutants for *lov-1*, a polycystin-encoding gene important for regulation of fluid sensing in ciliated cells *(7, 8)*. While illuminating much about the aforementioned processes, *C. elegans* researchers have also twice been awarded the Nobel Prize in Physiology or Medicine for discovering apoptosis in 2002 and RNA interference in 2006, validating the use of *C. elegans* in human disease research. In accordance with Brenner's vision to create a model organism capable of shedding light on the mysteries of a nervous system, *C. elegans* has also become a valuable system for the study of synaptic function and development.

2. Overview of the *C. elegans* Nervous System

Because the multifaceted nature of epilepsy inherently requires study of disparate regions of the mammalian brain at molecular, cellular, and physiological levels, epilepsy researchers have an outstanding opportunity to accelerate discoveries in neuroscience. From the perspective of the *C. elegans* research community, understanding how a relatively simple nervous system operates will lead to a more efficient elucidation of more complex systems. Fittingly, epilepsy research may benefit from this reductionist approach, if clinically pertinent neurobiological phenomena are truly conserved between humans and invertebrates. The following is designed to reveal how our current understanding of the structure and function of the *C. elegans* nervous system can be exploited to further characterize epilepsy genes, susceptibility factors, and various drugs that target them (**Table 1.1**).

Table 1.1
Description of *C. elegans* synaptic transmission genes included in the text

Gene name	Sequence name	Protein identity	Human homolog	BLAST E-value[1]	Confirmed function in *C. elegans*
Acetylcholine transmission					
unc-4	*F26C11.2*	Paired-class (homeodomain) transcription factor	*UNCX*	2.3E-30	Acetylcholine transmission
GABA transmission					
exp-1	*H35N03.1*	Cationic-selective GABA receptor	*GBRB3*	6.3E-50	Excitatory GABA transmission
unc-25	*Y37D8A.23*	Glutamic acid decarboxylase (GAD)	*DCE1*	1.0E-159	General GABA transmission
unc-46	*C04F5.3*	Novel sorting protein with similarity to LAMP family members	*C20orf103*	2.1E-03	General GABA transmission
unc-47	*T20G5.6*	Vesicular inhibitory amino acid transporter (hVIAAT)	*SLC32A1*	1.5E-86	General GABA transmission
unc-30	*B0564.10*	PTX (homeodomain) transcription factor	*PITX1*	9.0E-29	Inhibitory GABA transmission

(continued)

Table 1.1(continued)

Gene name	Sequence name	Protein identity	Human homolog	BLAST E-value[1]	Confirmed function in *C. elegans*
unc-49	*T21C12.1*	Anionic-selective heterotrimeric GABA(A) receptor	*GABRA1*	2.0E-63	Inhibitory GABA transmission
General synaptic function					
dgk-1	*C09E10.2*	Diacylglycerol kinase	*DGKQ*	8.2E-162	Negative regulation of synaptic transmission
eat-16	*C16C2.2*	Regulator of G-protein signaling (RGS)	*RGS7*	2.5E-83	Negative regulation of synaptic transmission
goa-1	*C26C6.2*	G-protein alpha subunit (G(o)alpha)	*GNAO2*	6.4E-162	Negative regulation of synaptic transmission
slo-1	*Y51A2D.19*	Ca[2+]-activated K+ channel slowpoke alpha subunit	*KCNMA1*	0	Negative regulation of synaptic transmission
unc-43	*K11E8.1*	Type II calcium/ calmodulin-dependent protein kinase (CaMKII)	*CAMK2A*	0	Negative regulation of synaptic transmission
snb-1	*T10H9.4*	Synaptobrevin/vesicle-associated membrane protein	*VAMP1*	1.5E-33	Positive regulation of synaptic transmission
unc-104	*C52E12.2*	Kinesin-like motor protein	*KIF1B*	0	Positive regulation of synaptic transmission
unc-13	*ZK524.2*	Syntaxin-binding protein UNC-13	*UNC13B*	0	Positive regulation of synaptic transmission
unc-2	*T02C5.5*	Voltage-gated Ca[2+] channel alpha subunit	*CACNA1A*	0	Positive regulation of synaptic transmission

(continued)

Table 1.1 (continued)

Gene name	Sequence name	Protein identity	Human homolog	BLAST E-value[1]	Confirmed function in *C. elegans*
Muscle contraction					
myo-3	*K12F2.1*	Myosin class II heavy chain (MHC A)	*MYH13*	0	Muscle contraction
tax-6	*C02F4.2*	Calmodulin-dependent calcineurin A subunit alpha	*PPP3CA*	2.8E-199	Muscle contraction
unc-27	*ZK721.2*	Troponin I	*TNNI3*	6.1E-17	Muscle contraction
unc-54	*F11C3.3*	Myosin class II heavy chain (MHC B)	*MYH7*	0	Muscle contraction
Nerve cell cytoskeleton					
cdka-1	*T23F11.3*	Cyclin-dependent kinase 5 (CDK5) activator p35/Nck5a	*CDK5R1*	2.0E-42	Inhibitory GABA transmission
dhc-1	*T21E12.4*	Cytoplasmic dynein heavy chain	*DNAH9*	0	Inhibitory GABA transmission
lis-1	*T03F6.5*	Platelet-activating factor acetylhydrolase IB subunit alpha/LIS1	*PAFAH1B1*	6.3E-138	Inhibitory GABA transmission
nud-1	*F53A2.4*	Nuclear distribution protein NUDC	*NUDC*	2.3E-83	Inhibitory GABA transmission

[1]BLAST E-values obtained from WormBase (www.wormbase.org)

3. General Discussion of *C. elegans* Features

3.1. Gross Organization of the C. elegans *Nervous System*

Of the 959 somatic cells comprising the adult *C. elegans* body plan, only 302 are neurons *(9)*. Unlike other animal models, the anatomy and connectivity of the *C. elegans* nervous system has been characterized in exquisite detail following reconstruction of serial section electron micrographs *(10)*. Exemplifying the underlying complexity of the *C. elegans* nervous system, worm neurons were originally grouped into 118 distinct classes based on differences in morphology, connectivity, and function, including mechanosensation, chemosensation, and thermosensation. *C. elegans* neurons are organized into three major regions of neuropil, including the nerve ring, the ventral nerve cord, and the dorsal nerve cord, as well as several associated ganglia. It is estimated that nearly all of the approximately 7000 synapses in *C. elegans* are concentrated in the three major neuropils. The majority of neural processes in *C. elegans* normally serve dual roles as both axons and dendrites, as most synapses in *C. elegans* are *en passant*, instead of *terminaux*, like mammalian synapses. The nerve ring, often thought of as the worm equivalent of a brain, is a bundle of approximately 160 neural processes and serves as the major locus of interneuronal synapses. Most cell bodies of sensory and interneurons are found in two ganglia anterior and posterior to the nerve ring. The two other neuropils of particular interest for epilepsy research with worms include the ventral and dorsal nerve cords; these function as motor neurons and, as such, form neuromuscular junctions (NMJs) with body wall muscles (BWMs) *(10)*. BWMs are located along four dorsolateral and ventrolateral quadrants and, much like individual *C. elegans* neurons, are well characterized *(11)*.

Recognizing the differences between the nerve ring and the nerve cords is of particular importance for appropriate use of *C. elegans* for research into mechanisms regulating neuronal excitability and, thus, seizures and epilepsy. Discussion of behavioral manifestations linked to aberrant synaptic transmission should clarify those differences and provide background on the identification of synaptic transmission mutants in worms.

3.2. Basic Characteristics of C. elegans Synapses

Appreciating the utility of *C. elegans* as a model for understanding epilepsy necessitates comprehension of more than the general organization of the *C. elegans* nervous system. Consideration of the kinds and composition of synapses at various regions along the *C. elegans* body allows one to effectively identify specific deficits in synaptic transmission, if present in mutants or otherwise normal worms exposed to pharmacological agents. Indeed, ultrastructural analysis performed to determine the gross organization of the *C. elegans* nervous system has also revealed presynaptic densities, filled with several hundred synaptic vesicles, as well as dense core vesicles, at worm NMJs *(10)*. The observation that *C. elegans* NMJs contain presynaptic and postsynaptic densities is further evidence of evolutionarily conserved neuronal functions.

The major hallmarks of mammalian synaptic transmission are conserved in *C. elegans (12)*. All 118 neuron types in the *C. elegans* nervous system may be categorized by expression of an array of neurotransmitters, including glutamate *(13)*, GABA *(14)*, acetylcholine *(15)*, dopamine *(16)*, and serotonin *(17)*. Additionally, some worm neurons also express neuropeptides *(18)*. The *C. elegans* nervous system may be small in comparison to the human nervous system, but it is certainly not “simple” at the molecular and cellular levels.

4. Strategies for Isolating Epilepsy Genes and Susceptibility Factors with a *C. elegans* Model

A more thorough discussion of how *C. elegans* may be used to specifically probe evolutionarily conserved biological phenomena pertinent to epilepsy will begin with Brenner's original efforts to identify *C. elegans* mutants with overt neuromuscular defects and continue through the diversity of approaches developed thereafter. It is important to note that the intent behind this review is not to provide a description of synaptic transmission in *C. elegans*; there are numerous review articles on this subject *(4)*. Instead, this discussion of synaptic transmission will be presented in the context of epilepsy and seizures.

4.1. Behavioral Screening

Brenner and colleagues began the identification of synaptic transmission genes in the 1960 s by searching for mutants with abnormal locomotory behaviors after ethyl methanesulfonate treatment. The specific phenotype sought was termed "uncoordination" and led to the systematic naming of 77 "*unc*" genes. At this level of analysis, functional and developmental defects, as well as neuronal and muscular defects, could not be dissociated. For example, muscle-specific genes, encoding either a myosin class II heavy chain or a troponin I isoform, were isolated as "*unc*" mutants and named *unc-54* and *unc-27*, respectively. Despite identifying muscle-specific loci, genes that would later be shown to play important, evolutionarily conserved roles in neuron-autonomous regulation of synaptic transmission, such as *unc-13*, were also isolated *(2)*. Along with identifying important neuronal genes through screens for clear uncoordinated phenotypes, additional approaches to identifying more subtle defects in synaptic transmission have been developed.

4.1.1. Foraging, Locomotion, and Shrinking Assays to Identify Inhibitory GABA Signaling Molecules

GABA is the principal inhibitory neurotransmitter in the human central nervous system and at *C. elegans* NMJs *(14)*. Alterations in GABA-mediated synaptic inhibition in humans *(19)*, as well as in rodent models *(20, 21)*, have been correlated with a reduced seizure threshold. However, despite our understanding that GABA signaling plays an important role in seizures and epilepsy, detailed understanding of precisely how GABA signaling molecules contribute to a seizure threshold remains an active area of investigation. Because overt *C. elegans* behavioral phenotypes, described below, are associated with inhibitory GABA, *C. elegans* may be an important model to further our understanding of GABA and epilepsy.

Antibody staining against GABA has revealed precisely 26 GABAergic neurons in the *C. elegans* nervous system *(14)*. Of these 26 GABAergic neurons, 4, called RMEs, reside in the nerve

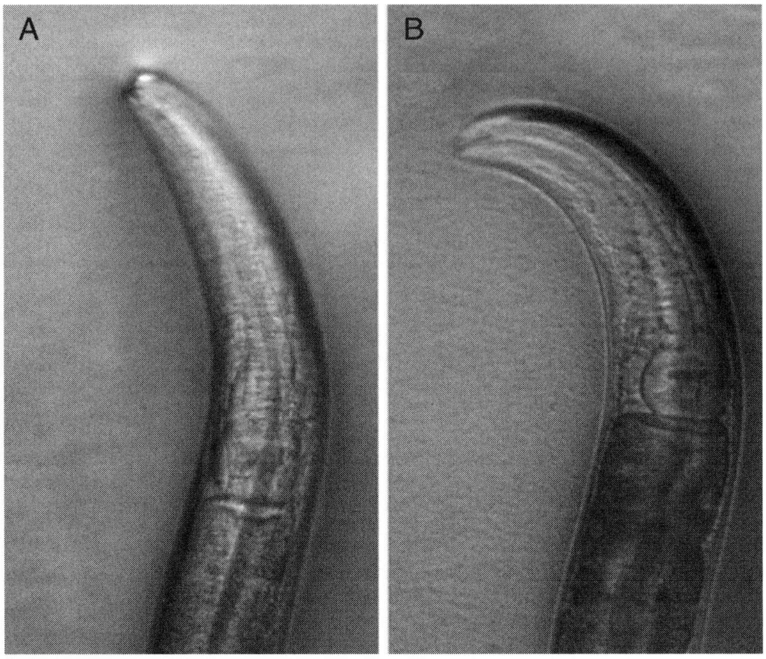

Fig. 1.1. **Inhibitory GABA transmission mediates a foraging behavior in *C. elegans*.** (**A**) WT worms narrowly sweep their heads from side to side during foraging. (**B**) Worms such as *unc-25* (lf) that lack inhibitory GABA transmission fail to limit head movements during foraging and exhibit exaggerated or "loopy" foraging.

ring and synapse onto head muscles *(10, 14)*. The anterior tip or "nose" of a WT worm moves sideways in a narrow arc during foraging (**Fig. 1.1A**). Laser ablation of RMEs disrupts normal foraging and, instead, causes "loopy" foraging, characterized by exaggeration of flexures at the worm nose. Mutants carrying loss-of-function (lf) alleles of genes essential for inhibitory GABA function, including *unc-25*, glutamic acid decarboxylase (GAD), or *unc-47*, a vesicular GABA transporter, also phenocopy the foraging defects of *C. elegans* with ablated RMEs (**Fig. 1.1B**). These results imply that RMEs function to limit head movements during foraging and implicate inhibitory GABA signaling in this behavior.

A second inhibitory GABA-mediated behavior in *C. elegans* involves 19 ventral GABAergic neurons. These 19 neurons are called D-type motor neurons and are further categorized by the side of the worm onto which they synapse and were found to work together with BWMs to coordinate locomotion in *C. elegans*. Locomotion is a result of the propagation of body bends from one end of a worm to the other. Body bends are generated when muscles on one side of a worm contract via acetylcholine (ACh), while muscles on the other side relax via GABA *(14)*. This coordination of body bending engenders a sinusoidal pattern of

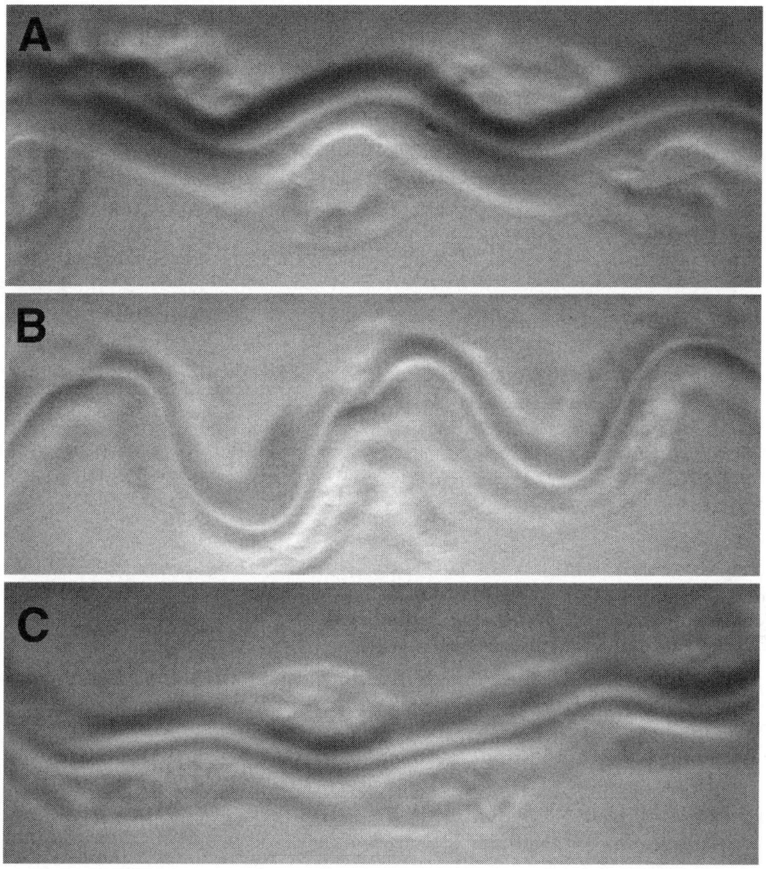

Fig. 1.2. **Imbalance between excitatory and inhibitory transmission at worm body wall neuromuscular junctions results in distinctive locomotory patterns.** (**A**) Body bends that characterize WT worm locomotion are governed by coordinated stimulation and relaxation of body wall muscles. This coordination is achieved through a balance of antagonistic excitatory ACh and inhibitory GABA signaling, respectively, at body wall NMJs. (**B**) Worms such as *goa-1*(lf) that are deficient in negative regulation of synaptic transmission have excessive ACh transmission at body wall NMJs and produce "tracks" of body bends with abnormally high amplitudes. (**C**) Worms such as *unc-25* (lf) that lack inhibitory GABA at body wall NMJs leave tracks with abnormally short amplitudes.

locomotion, which is easily noted by simply observing worm "tracks" on an agar plate (**Fig. 1.2A**). For example, loss of *goa-1*, which encodes G(o)alpha (GOA-1), a subunit of the heterotrimeric G-protein G(o), results in increased track amplitudes from an inability to negatively regulate neurotransmitter secretion at *C. elegans* NMJs (**Fig. 1.2B**) *(22)*. Although *goa-1*(lf) mutants are thought to have increased excitatory to inhibitory transmission ratios at worm NMJs *(23)*, these animals exhibit a different kind of track from worms completely lacking only inhibitory GABA transmission. Laser ablation of the D-type motor neurons or genetic loss of inhibitory GABA transmission allows only excitatory ACh input to BWMs and causes worms to bend their bodies with shorter amplitudes (**Fig. 1.2C**) *(14)*. Despite an ability to automate nematode locomotion assays, as performed with *goa-1*(lf) mutants *(22)*, perhaps a more economical technique for identifying inhibitory GABA mutants exists. This technique relies on simple observation of nematode "shrinking".

For example, the "shrinker" phenotype is independent of the previously described GABAergic RME-mediated foraging behavior and is typified by pulling in of the head and shortening of the body from hypercontraction of BWMs. Shrinking is most easily observed after touching a worm on its nose with a platinum wire. Along with deficits in GABA-mediated foraging and locomotion, *unc-25* and *unc-47*(lf) mutants, as well as worms with laser-ablated D-type motor neurons, also display shrinker phenotypes, suggesting that the shrinker phenotype could be used to identify other genes important for inhibitory GABA function *(14)*. Accordingly, a transcription factor, UNC-30, that has been shown to establish GABAergic neuron identity in D-type motor neurons, was discovered during a screen for mutant shrinker worms *(14, 24)*. Similarly, an auxiliary protein for vesicular GABA transport, called UNC-46, was also isolated following observation of an associated shrinker phenotype *(14, 25)*.

4.1.2. Defecation Assay to Identify Excitatory GABA Signaling Molecules

Although GABA is typically inhibitory, the effects of neurotransmitters are dependent upon target cell receptors and intracellular Cl^- levels. Accordingly, most GABA signaling in *C. elegans* involves an anion-selective GABA(A) receptor, UNC-49. Mutants carrying lf alleles of *unc-49* exhibit abnormal inhibitory GABA-mediated phenotypes. Yet, unlike *unc-25* or*unc-47*(lf) mutants, *unc-49*(lf) mutants fail to exhibit another conspicuous phenotype associated with generalized GABA signaling deficits – the expulsion defective phenotype *(14)*.

Expulsion is the terminal step in a series of rhythmic worm behaviors known as the defecation motor program (DMP) (**Fig. 1.3**, http://www.carpedb.ua.edu/defecation/N2.mov). Approximately every 45–50 s,*C. elegans* initiates the DMP with a posterior body contraction (pBoc). Unlike the other steps of the DMP, the pBoc occurs independently of synaptic transmission, as mutants defective in synaptic transmission are still capable of performing pBocs *(26)*. Despite its independence from the worm nervous system, the pBoc is followed by the second step in the DMP, a subtle and neuron-dependent anterior body contraction (aBoc). Specifically, the aBoc appears to be dependent upon signals from the AVL neuron, which expresses GABA. Like the pBoc, though, the aBoc still occurs normally in generalized GABA mutants, *unc-25* and *unc-47*, implying that the aBoc does not depend on GABA signaling and, instead, probably signals through another neurotransmitter (http://www.carpedb.ua.edu/defecation/unc-25.mov). More importantly for epilepsy research, expulsion, which

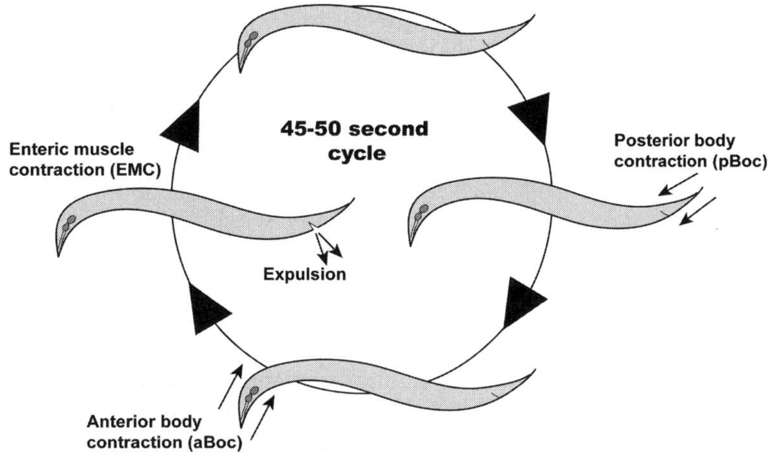

Fig. 1.3. **The *C. elegans* defecation motor program is a cycle of rhythmic behaviors with neuron-dependent and neuron-independent steps.** An average worm defecation cycle occurs every 45–50 s and consists of three muscle contractions. The first step of the defecation cycle is the posterior body contraction (pBoc), which is initiated by a neuron-independent intestinal calcium wave that is propagated by gap junctions. The second step of the defecation cycle is the anterior body contraction (aBoc), which appears to be mediated by an uncharacterized neurotransmitter. The last step of the defecation cycle is defined by enteric muscle contractions (EMC), which lead to expulsion of intestinal contents. Expulsion is dependent upon excitatory GABA transmission and has been used to identify and characterize an array of GABA-specific genes in *C. elegans*, including a cation-selective GABA receptor and others with essential roles in GABA transmission.

immediately follows the aBoc, occurs when enteric muscles, including intestinal, sphincter, and anal depressor muscles, contract. Enteric muscle contraction (EMC) squeezes the intestines and opens the anus to complete the defecation cycle *(14, 27)*. As suggested by laser ablation and genetic analysis, the EMCs are not only dependent upon the AVL but also on another GABAergic neuron, termed the DVB. The failure of EMCs in *unc-25* (lf) and *unc-47*(lf) mutants, but not in *unc-49*(lf) mutants, suggested the presence of an uncharacterized GABA receptor (not UNC-49) in the regulation of EMCs. In a forward genetic screen for factors mediating the DMP, Thomas isolated a mutant for *exp-1*, which, similar to *unc-25*(lf) and *unc-47*(lf) mutants, only lacked the expulsion step of the DMP *(27)*. Consistent with a failure to excite, not to inhibit, muscle contraction, *exp-1* was later shown to be a cation-selective GABA receptor *(28)*. Subsequent genetic analysis with the defecation assay led to the characterization of *tax-6*, a worm homolog of the phosphatase calcineurin (CaN), as a negative regulator of *exp-1 (29)*. Suggesting the existence of a conserved mechanism, CaN has been shown to similarly down-regulate GABA(A) receptors in a rat hypoxia-induced seizure model *(30)*.

These findings imply that genes with significant and perhaps evolutionarily conserved roles in mediating GABA transmission may be isolated through analysis of the DMP, in addition to locomotion, foraging, and shrinking. Furthermore, an understanding that excitatory GABA transmission occurs in a manner that is dependent upon a distinct pair of GABAergic neurons, the AVL and DVB neurons, instead of inhibitory RMEs and D-type motor neurons, is imperative. As will be discussed later, determining whether a given gene is particularly important for GABA function in *C. elegans* is sometimes, albeit to a lesser degree than with mammals, unclear. Demonstrating roles for such a gene in both inhibitory and excitatory GABA function places that gene in a class with *unc-25* and *unc-47* and reduces the probability that neuromuscular deficits are a result of signaling outside of the GABA system. It is true that general synaptic function mutants, like GABA mutants, also fail to expulse *(31)*. Yet, a mutant with reduced motility and failed expulsions, but not failed aBocs, would strongly implicate an associated gene in GABA function, not general synaptic function.

On a final note, appreciation of the cyclical properties of the DMP and recognition that its periodicity is not completely dependent upon synaptic transmission also has potential implications for epilepsy research. Data supporting regulation of the DMP by *C. elegans* intestinal calcium waves now exist. The importance of this finding is augmented by the discovery of a mutant, which exhibits increased defecation cycle lengths, as defined by the time between pBocs, as well as failed aBocs and expulsions. This mutant is neither defective for general synaptic transmission nor GABA transmission. Instead, this mutant lacks INX-16, a gap junction component, which was shown to be necessary for the propagation of intestinal calcium waves *(32)*. Ultrastructural evidence that gap junctions exist between cortical neurons in a mammalian system is now available *(33)*, and a role for gap junctions in seizure propagation is fairly well established *(34)*. Indeed, *connexin-36*, which encodes a gap junction component, has been associated with juvenile myoclonic epilepsy *(35)*. Considering these valuable links between gap junctions and epilepsy, the *C. elegans* defecation assay could plausibly be manipulated to further characterize factors with important roles in the regulation of gap junction activity.

4.2. Pharmacological Modulation of C. elegans *Behaviors*

Behavioral assays with *C. elegans* offer an excellent opportunity to isolate genes affecting neuronal excitability and, thus, have powerful implications for the identification of epilepsy genes and susceptibility factors. Yet, *C. elegans* behaviors tend to be subtle, leaving some concern about the subjectivity of data interpretation. To alleviate such concerns and accelerate discovery, various pharmacological assays have been developed for use with *C. elegans*. These assays utilize particular pharmacological agents, including seizure inducers, with known cognate targets in the *C. elegans* neuromuscular system. These drugs, readily absorbed through worm cuticles, now play key roles in refining our understanding of the specific deficits associated with genes of interest. Each drug is presented in its appropriate epilepsy-related context in the following sections.

4.2.1. Aldicarb Assay

The most successful approach to characterizing synaptic transmission genes in *C. elegans* has been the use of aldicarb, an acetylcholinesterase (AChE) inhibitor. When exposed to millimolar concentrations of aldicarb, wild-type (WT) worms accumulate ACh at NMJs. Such an increase in ACh results in paralysis, and ultimately death, over several hours of exposure. These phenotypes have served as the foundation of both forward and reverse genetic screens to isolate and characterize synaptic transmission genes. Two major approaches to the aldicarb assay are available and can provide distinct advantages for epilepsy researchers.

The original approach to the aldicarb assay was "chronic" exposure. Because aldicarb eventually kills WT animals, chronic exposure to aldicarb has been used as a selection for genes required for the positive regulation of ACh transmission. If a gene product normally functions to enhance ACh transmission, then disruption of that gene product should reduce ACh transmission. As a result, less ACh would be received at NMJs, leading to less stimulation of BWMs, when compared to muscle stimulation in WT worms. Therefore, "Ric" (resistant to inhibitors of cholinesterase) mutants, lacking normal ACh transmission, would be viable and capable of reproduction in the presence of aldicarb. Chronic exposure to aldicarb has led to the isolation of key synaptic transmission genes such as *snb-1* (synaptobrevin) *(36)*. Moreover, chronic aldicarb exposure has also revealed cholinergic and, likely, general functions for *unc-13* and other genes identified in Brenner's original *C. elegans* screen *(2, 37)*. Despite the utility of chronic aldicarb exposure in identifying ACh and, often, general synaptic function genes, this approach is limited in its usefulness for epilepsy research. That is, genes with functions in the negative regulation of synaptic transmission would be undetectable due to lethality from extreme degrees of paralysis.

As an alternative to chronic aldicarb exposure, an "acute" aldicarb exposure assay has also been developed. With acute aldicarb exposure, genetic backgrounds of interest are exposed to aldicarb and tested for locomotory impairment over the course of several hours. As a paragon of the usefulness of *C. elegans* in synaptic transmission and perhaps epilepsy research, a systematic reverse genetics approach combining RNAi with acute aldicarb exposure was employed *(38)*. From this screen, 185 Ric genes were identified, of which 71% (132) had not been previously implicated in synaptic transmission. Despite such efforts to identify

Ric genes, a complementary screen to identify "Hic" (hypersensitive to inhibitors of cholinesterase) mutants has not yet been performed. Hic mutants become completely paralyzed, as revealed by prodding of the head or tail, before WT or Ric worms. Notable Hic genes include *goa-1*(lf), *dgk-1*(lf), and *unc-43*(lf), all considered important negative regulators of synaptic transmission *(39)*. After initial characterization of these Hic genes with *C. elegans*, their mammalian homologs were found to be associated with seizures and epilepsy *(40, 41, 42)* Along with Hic genes, yet another class of lf mutants with implications for epilepsy, inhibitory GABA mutants, was also shown to confer hypersensitivity to acute aldicarb exposure *(43)*. These data are consistent with antagonistic signaling between ACh and GABA at body wall NMJs and provide evidence that, in addition to genes negatively regulating synaptic transmission, inhibitory GABA genes can also be isolated through application of the acute aldicarb exposure assay. Ultimately, complementary approaches to acute aldicarb exposure with other behavioral paradigms, including defecation, can be effectively employed to help distinguish GABA mutants from mutants deficient in regulating general synaptic transmission *(31)*.

4.2.2. PTZ Assay

We recently developed a complementary pharmacological assay, which introduces the possibility of anterior and/or posterior body contractions in addition to paralysis associated with the aldicarb assay *(44)*. This assay, which utilizes pentylenetetrazole (PTZ), a common seizure inducer and GABA receptor antagonist *(45)*, allows further characterization of Hic and Ric mutants. Likewise, the PTZ assay also allows separation of GABA mutants from Hic mutants. This distinction between GABA mutants and other Hic mutants is likely made possible by the inhibitory nature of the worm nerve ring, rich with GABA transmission from RMEs *(10, 44)*. Moreover, because PTZ is also used in mammalian seizure models, PTZ exposure with *C. elegans* may afford researchers a novel opportunity to isolate epilepsy genes and susceptibility factors.

Consistent with the notion that PTZ disrupts GABA transmission, mutant worms carrying lf alleles of GABA genes (*unc-25*, *unc-46*, *unc-47*, or *unc-49*) exhibit significant sensitivity to PTZ. Their sensitivity to PTZ is not simply manifested as paralysis. Instead, GABA mutants on PTZ not only are paralyzed, but also display strong anterior convulsions. These convulsions are best described as repetitive "head-bobs" and are quite obvious to an observer (**Fig. 1.4**, http://www.carpedb.ua.edu/PTZ/unc-25.mov). Despite their substantial sensitivity to PTZ, GABA mutants are not the only mutants that respond to PTZ with anterior convulsions. Similar to GABA mutants, worms carrying lf alleles of Ric genes, *snb-1*, encoding the v-SNARE, synaptobrevin, or *unc-104*, encoding a kinesin-like protein used for

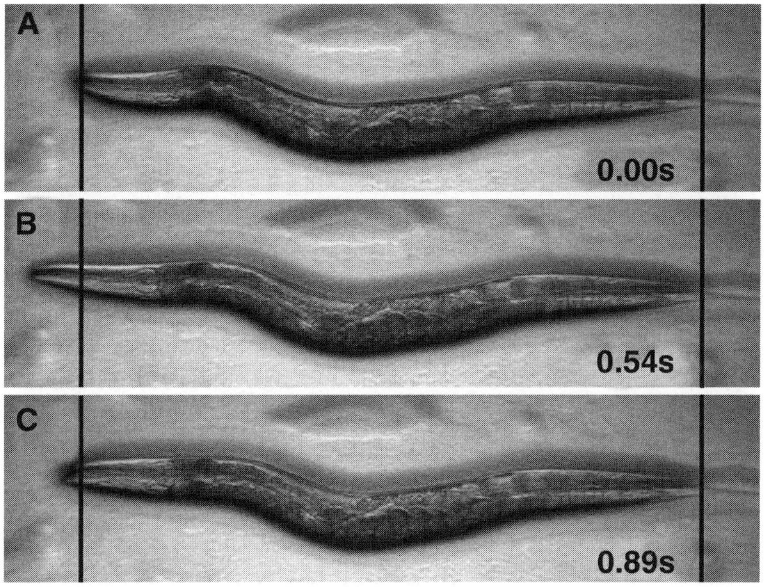

Fig. 1.4. ***C. elegans* mutants without inhibitory GABA transmission exhibit strong anterior convulsions in response to a common seizure inducer, PTZ.** Still frame images demonstrating *C. elegans* undergoing convulsions following exposure to 10 mg/mL PTZ. The still frame images were selected from videos (25 frames/s) available at http://www.carpedb.ua.edu/PTZ/unc-25.mov. The **black lines** represent stationary reference points for visualization of anterior and/or posterior movements in relation to time (indicated in s). Ventral is down, and anterior is to the left in all images, where lines are placed perpendicular to the original position of the nose of a worm. *C. elegans* mutants lacking only inhibitory GABA transmission display anterior convulsions on PTZ. (**A**) Both the head and tail of an *unc-25* (lf) mutant, representative of other inhibitory GABA mutants, are positioned behind the stationary reference points. (**B**) Exposure to PTZ causes only the head of an immobilized inhibitory GABA mutant to extend across a reference point. (**C**) Following PTZ-induced head extension, the head of the *unc-25* (lf) mutant retracts to its original position behind the line.

anterograde transport of synaptic vesicles, also exhibit anterior convulsions *(44)*. Because both *snb-1*(lf) and *unc-104*(lf) mutants are likely deficient in positive regulation of synaptic function in all neurons *(37)*, their responses to PTZ implicate all positive regulators of general synaptic function in PTZ-induced anterior convulsions. However, some Ric mutants, including *unc-4*(lf) mutants, which are exclusively lacking in cholinergic motor neuron identity, do not exhibit anterior convulsions in the presence of PTZ. These findings suggest that, like GABA mutants, generalized Ric mutants, but not WT or specific ACh mutants with Ric phenotypes, have a propensity for PTZ-induced head-bobs *(44)*. Hence, PTZ may be used not only to isolate inhibitory GABA mutants, but also to further categorize Ric genes according to general or specific neuronal abnormalities. Additionally, Hic mutants may also be distinguished from GABA mutants via PTZ exposure.

Hic mutants, defective for negative regulation of general synaptic transmission, respond distinctively to PTZ. Nevertheless, there are some uncertainties surrounding the particular response of this class of mutants to PTZ. Our group has found that null *goa-1* mutants, known to lack negative regulation of synaptic transmission *(23, 46)*, respond to PTZ (and aldicarb) with significant paralysis. Yet, this paralysis of *goa-1*(lf) mutants is not accompanied by repetitive convulsions (Locke *et al.*, unpublished data). Similar to *goa-1*(lf) mutants, *unc-43*(lf) mutants, also thought to lack negative regulation of synaptic transmission due to aberrant signaling through G(o)alpha, are sensitive to PTZ. In contrast to other mutants in this class, though, *unc-43*(lf) mutants have spontaneous full-body convulsions *(39)*. Furthermore, the frequency and severity of these spontaneous full-body convulsions, associated with BWMs, is enhanced by PTZ exposure (**Fig. 1.5**, http://www.carpedb.ua.edu/PTZ/unc-43.mov) *(44)*. The full-body convulsions of *unc-43*(lf) mutants are distinct from convulsions exhibited by other PTZ-exposed worm synaptic transmission mutants currently represented in the literature. Full-body convulsions of *unc-43*(lf) mutants may suggest that UNC-43 functions independently of GOA-1 in a discrete biological process, which contributes to a worm "seizure" threshold. Some pertinent evidence for GOA-1-independent functions of UNC-43 in synaptic transmission does exist, as *unc-43*(lf) mutants, like GABA mutants, have shrinker phenotypes *(14)*, while *goa-1*(lf) mutants do not (Locke *et al.*, unpublished data). Furthermore, *unc-43* (lf) mutant males have been shown to exhibit spontaneous muscle cell-autonomous male spicule protractions *(47)*. UNC-43 modulation of neurotransmitter release through a BK channel, SLO-1, has also been shown to occur independently of DGK-1 and EAT-16, G(o)alpha signaling regulators *(48)*. Moreover, UNC-43 has been found to regulate AMPA-type ionotropic glutamate receptor density at worm NMJs through UNC-2, which encodes the alpha subunit of a calcium channel *(49)*. Interestingly, *CACNA1A*, an *unc-2* orthologue, has been linked to idiopathic generalized epilepsy *(50)*. Likewise, BK channels are associated with temporal lobe epilepsy *(51, see* **Chapter** 7, this book), while AMPA receptor-mediated synaptic transmission has been implicated in autosomal-dominant partial epilepsy with auditory features *(52)*. Despite ambiguities regarding *C. elegans* mutant responses to PTZ exposure, variations in these responses likely allow for the establishment of novel mutant categories.

Considering the significant sensitivity of general synaptic function mutants to PTZ, one might wonder if PTZ is truly useful for studying epilepsy with *C. elegans*. However, substantial evidence for the utility of PTZ exposure with a *C. elegans* model of epilepsy now exists. Work from our laboratory demonstrates that *lis-1*, the

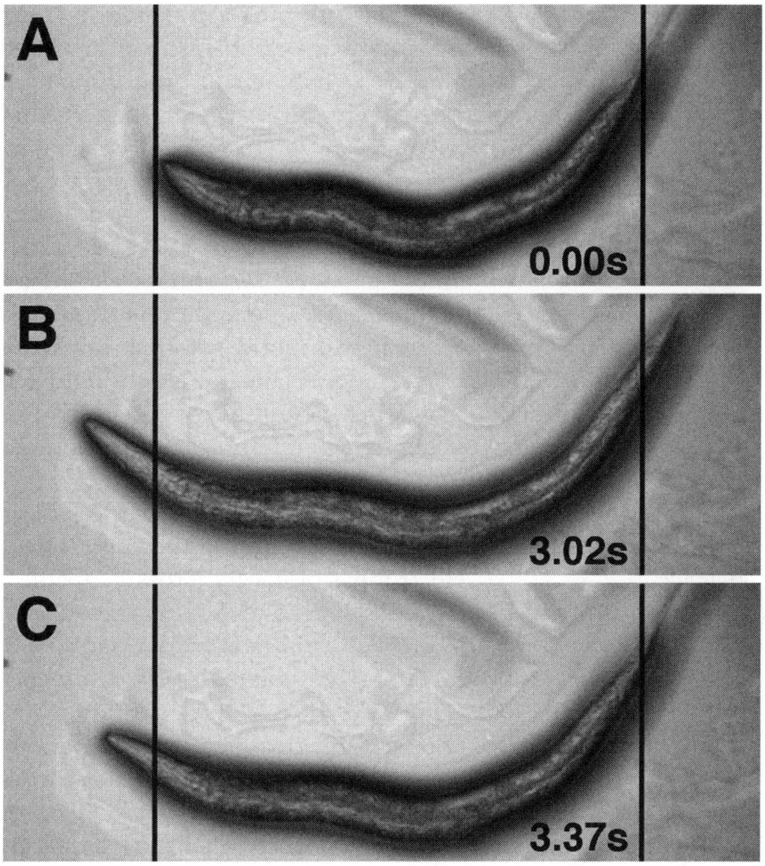

Fig. 1.5. ***C. elegans* CaMKII(lf) mutants exhibit full-body convulsions in response to PTZ.** Still frame images demonstrating *C. elegans* undergoing convulsions following exposure to 10 mg/mL PTZ as described in the legend for Fig. 5; corresponding video is available at http://www.carpedb.ua.edu/PTZ/unc-43.mov. *C. elegans* calcium/calmodulin-dependent kinase II (UNC-43) mutants display full-body convulsions on PTZ. (**A**) Both the head and tail of an *unc-43* (lf) mutant are positioned behind the stationary reference points. (**B**) Exposure to PTZ causes both the head and tail of an immobilized *unc-43*(lf) mutant to extend across reference points. (**C**) Following PTZ-induced muscle contractions, the head and tail of the *unc-43*(lf) mutant retract to their original positions behind the lines.

C. elegans homolog of *LIS1*, also exhibits anterior convulsions in response to PTZ. This discovery is potentially important for epilepsy research, as deletion of *LIS1* is the major cause of Miller–Dieker lissencephaly, a neuronal migration disorder accompanied by intractable epilepsy. Along with *lis-1*(lf) mutants exhibiting anterior convulsions on PTZ, we also observed misaccumulation of a synaptic vesicle marker, SNB-1::GFP, specifically in GABAergic D-type motor neurons. Similar observations were also made following RNAi against *dhc-1*, a dynein heavy chain. Such data implicate *C. elegans lis-1* and the dynein motor complex in GABAergic vesicle transport, a functional consequence that was found to be independent of developmental anomalies *(44)*. Although no obvious impairment of vesicle transport resulted from *LIS1* haploinsufficiency in a mouse model of lissencephaly, electrophysiological data suggest that *LIS1* heterozygous mice have altered GABA function in inhibitory circuits *(53)*. Indeed, the *C. elegans* epilepsy model may play a considerable role in further characterizing *LIS1* modifier genes. Additional work from our laboratory has revealed genetic interactions among *lis-1* and other *C. elegans* homologs of neuronal migration genes, such as *nud-1* (*NUDC*) and *cdka-1* (*p35*), through the PTZ assay *(54)*. As an animal model with considerable resilience to neurodevelopmental abnormalities and sensitivity to seizure inducers, *C. elegans* may offer a unique opportunity to isolate epilepsy-associated genes, which might otherwise go unnoticed with rodent models.

4.3. Electrophysiological Methods

By definition, an epilepsy model must demonstrate abnormal electrical discharges in the central nervous system *(55)*. However, electrophysiological methods capable of revealing abnormal electrical discharges in *C. elegans* are lacking in comparison to those available for other animal models of epilepsy. The various advantages of *C. elegans* in the laboratory are accompanied by an adverse trade-off. The tiny size of this nematode, and particularly its nervous system, greatly impedes electrophysiology. Likewise, a tough proteinaceous cuticle serves as a formidable barrier to electrodes. Consequently, worms must be dissected and glued to a surface to allow precise entry of electrodes, precluding behavioral analysis in conjunction with electrophysiology. These caveats imply that *C. elegans* is recalcitrant to epilepsy research. Despite this assertion, a limited, but growing, number of *C. elegans* researchers have been successful with in vivo electrophysiological methods. Moreover, in vitro electrophysiological methods and optical imaging techniques, which do allow behavioral and physiological correlation, have also been developed for use with *C. elegans.*

4.3.1. In Vivo Electrophysiological Methods

Behavioral and pharmacological assays with *C. elegans* and other model organisms have illuminated neurobiological phenomena with implications for epilepsy. Yet, it is insufficient to merely identify cells and/or macromolecules implicated in epilepsy, as this disorder is defined by electrophysiological characteristics. Although sufficient electrophysiological data to define *C. elegans* as an epilepsy model are unavailable, a few examples demonstrating the use of in vivo electrophysiological methods in the nematode may demonstrate how such data could, and should, eventually be obtained.

Data within a collection of *C. elegans* mutants indicate that the inhibitory GABAergic RMEs in the *C. elegans* nerve ring may be responsible for anterior convulsions following acute exposure to PTZ *(44)*. These neurons are proximal to a pair of sensory neurons, called ASEs, also found in the *C. elegans* nerve ring. Goodman and colleagues developed a preparation in which they exposed ASEs in live animals and recorded from them with whole-cell patch-clamp electrodes. Moreover, these researchers also performed electrophysiological recordings with 42 other neurons, proximal to the ASEs, and determined that *C. elegans* neurons exhibit voltage-dependent membrane currents *(56)*. With this preparation, it should be possible to characterize the electrophysiological properties of RMEs in mutants with demonstrated susceptibilities to seizure inducers perhaps before, during, or after worm "seizure" induction.

As previously discussed, frequencies of paralysis following exposure to PTZ or aldicarb suggests an E/I imbalance at body wall NMJs *(38, 54)*. To address E/I transmission at *C. elegans* body wall NMJs, researchers developed an alternative electrophysiological preparation. Their work showed that WT worm BWMs express one GABA receptor, UNC-49, and two ACh receptors (AChRs), only one of which is sensitive to levamisole, an AChR agonist *(57)*. Because *C. elegans* BWMs are considered to be more accessible to electrodes than the *C. elegans* nerve ring, this preparation may be most useful for rapid analysis of electrophysiological properties associated with genes or neurons of interest.

4.3.2. In Vitro Electrophysiological Methods

Despite the simplicity of the *C. elegans* nervous system, the inherent interconnectivity of this intact animal limits use of *in vivo* electrophysiological methods for addressing cell autonomy of specific activity-related effector proteins and pathways. To alleviate this quandary, researchers established a primary culture method for *C. elegans* neurons and muscle cells. This work demonstrated, through whole-cell patch clamping of *C. elegans* chemosensory neurons and body wall muscle cells, that intrinsic properties of these cultured cells are similar to those *in vivo*. In fact, muscle cells and motor neurons appeared to form synapses in culture with this method *(58)*. Because *C. elegans* is transparent, one might imagine using flow cytometry to isolate and culture cells with specific identities to determine spatial and temporal gene expression patterns that underlie their electrical activity. Accordingly, a GABAergic neuron-specific expression profiling study with *C. elegans*, in which specific neurons were extracted and cultured via expression of GFP, driven by an *unc-25* promoter for GABAergic neurons, has been reported *(59)*. *C. elegans* researchers have also profiled cholinergic neuron-specific gene expression by isolating UNC-4::GFP-expressing cells *(60)* and muscle-specific gene expression by isolating MYO-3::GFP-expressing cells *(61)*.

4.3.3. Optical Imaging

Although successful application of electrophysiological methods may help define *C. elegans* as a genuine epilepsy model, another technique, particularly useful for *C. elegans*, could be further exploited to enhance our understanding of modifiers of neuronal synchrony. Because *C. elegans* is transparent, it is amenable to imaging via fluorescent optical indicators. Genetically encoded calcium probes, including cameleon *(62)* and G-CaMP *(63)*, have been developed for use with *C. elegans*. As one of several examples of the utility of calcium probes in the *C. elegans* model, cameleon was used to confirm a role for *goa-1*, a negative regulator of synaptic transmission and possible susceptibility factor for epilepsy, in serotonergic silencing of motor neurons essential to egg laying *(46)*. Cameleon also confirmed the existence of a gap junction-dependent calcium wave in the worm defecation cycle, as previously discussed *(32)*, while G-CaMP has been used to explain worm foraging behaviors at a neuronal circuit level *(64)*. In addition to calcium imaging, worm researchers have also successfully employed a pH-sensitive synaptobrevin–GFP fusion protein, called synaptopHluorin, to identify and characterize a plasma membrane reservoir of synaptobrevin *(65)* and to correlate thermotaxis with synaptic activity of a specific pair of ciliated neurons, called the AFDs *(66)*. Unlike electrophysiological recordings, optical imaging can be performed without dissection in *C. elegans*. Hence, neural activity may not only be recorded in live animals, but also be correlated with behavioral states.

4.4. Bioinformatics

Conventional behavioral, pharmacological, and electrophysiological methods with *C. elegans* have become important for the isolation and characterization of epilepsy genes and susceptibility factors. Yet, the development of tools to predict which of the approximately 20,000 *C. elegans* genes to assay for aberrant neuronal excitability will more rapidly advance this goal. To complement conventional approaches, numerous bioinformatics tools have been developed specifically for use with *C. elegans* (**Fig. 1.6**).

The nucleus for worm bioinformatics is WormBase, which encompasses information on *C. elegans* genes and associated phenotypes, gene ontologies, protein structures, RNAi experiments, phylogenetic trees, physical mapping data, and published abstracts. WormBase is also being populated with similar information for other emerging nematode model organisms such as *C. briggsae* and *C. remanei* *(67)*. Moreover, WormBase now includes *in vivo* expression patterns for many genes. These genes consist of at least 900 genes, whose expression patterns or “chronograms” were determined via flow cytometry with transgenic *C. elegans*, expressing GFP under predicted *cis* regulatory sequences *(68)*. Perhaps most interesting for the discovery of novel epilepsy genes and susceptibility factors, WormBase features gene expression profiling data from over 400 microarray experiments *(67)*. These microarray experiments include, but are not limited to, those performed to determine gene expression profiles from cholinergic motor

***Caenorhabditis elegans* WWW Server (http://elegans.swmed.edu)**
Collection of hyperlinks to sites relevant to the study of *C. elegans*

WormBook (http://www.wormbook.org)
Open-access, online collection of original peer-reviewed chapters on *C. elegans* biology

WormBase (http://www.wormbase.org)
Comprehensive gene summary pages with structural, functional, and phylogenetic information
Bibliography section with published articles and peer-reviewed meeting abstracts

WormAtlas (http://www.wormatlas.org)
Database of behavioral and structural anatomy of *C. elegans*, including neuron wiring information

WorfDB (http://worfdb.dfci.harvard.edu)
Over 12,000 sequences for predicted protein-encoding open reading frames in *C. elegans*

***C. elegans* Interactome Project (http://vidal.dfci.harvard.edu/interactomedb)**
Approximately 5,500 confirmed protein-protein interactions with *C. elegans* proteins

***C. elegans* Localizome Project (http://vidal.dfci.harvard.edu/localizome)**
Spatiotemporal expression patterns or "chronograms" for roughly 2,000 *C. elegans* genes

Expression Patterns for *C. elegans* Promoter::GFP Fusion Database (http://gfpweb.aecom.yu.edu)
Additional spatiotemporal expression patterns for approximately 2,000 *C. elegans* genes

***C. elegans* Promoterome Project (http://vidal.dfci.harvard.edu/promoteromedb)**
Predicted sequences for promoters of over 6,000 *C. elegans* genes

Genetic Interaction Predictor (http://tenaya.caltech.edu:8000/predict)
Predictions of *C. elegans* genetics interactions by integration of data from yeast, flies, and worms

CarpeDB (http://www.carpedb.ua.edu)
Comprehensive database with gene summary pages for all known epilepsy genes and susceptibility loci
Bibliography section with published articles pertinent to the study of epilepsy in animal models
Downloadable videos of *C. elegans* synaptic transmission mutants with seizure-like behaviors

Fig. 1.6. **Bioinformatic databases available for use in *C. elegans* research.**

neurons *(60)*, GABAergic motor neurons *(59)*, and scores of other cell types under various conditions *(69)*. Along with spatial and temporal gene expression data, WormBase also includes protein–protein interaction data, much of which were generated from high-throughput yeast two-hybrid screens *(70)*. After computationally integrating disparate information from *C. elegans*, *Saccharomyces cerevisiae*, and *Drosophila melanogaster*, worm researchers have also developed a genetic interactions predictor to facilitate understanding of genes and their shared functions. These predictions for genetic interactions, especially dependent upon data from WormBase, may be accessed through a variety of easily accessible online web sites *(71, 72*, **Fig.** 1.6). These bioinformatic resources typify the systems approach toward integrated discovery of complex biological phenomena. Coordination of neuronal activity and bioinformatic data should aid in the identification and characterization of genetic modifiers of neuronal activity, thus accelerating treatments and cures for epilepsy.

5. Conclusion

As a result of a US executive branch-initiated conference in March 2000, various goals for epilepsy research were established. These goals point to a need for new animal models with the capability of more rapidly providing insights into developmental, genetic, and neurophysiological characteristics of epilepsy. Moreover, a suggested requirement for elucidating basic mechanisms of epilepsy was to establish national collaborations for identifying epilepsy genes and susceptibility factors *(55)*. Although it is unrealistic for worms to mimic overt clinical phenotypes, we contend that *C. elegans*, a model system that is renowned for its worth in genetic, genomic, and chemogenomic screening, is capable of contributing to these goals and, thus, has emerged as a valuable system for advancing epilepsy research.

Acknowledgments

We would like to thank all members of the Caldwell lab, especially Bwarenaba Kautu, Kyle Lee, and Kalen Berry, for experimental contributions and insightful discussions and KB for his assistance with generating figures and videos. A Basil O'Connor Scholar Award from the March of Dimes and a CAREER Award from the National Science Foundation to GAC have funded epilepsy research in the Caldwell lab.

References

1. Scott RA, Lhatoo SD, Sander JW. The treatment of epilepsy in developing countries: Where do we go from here? *Bull World Health Organ* 2001;79(4):344–51.
2. Brenner S. The genetics of *Caenorhabditis elegans. Genetics* 1974;77(1):71–94.
3. *C. elegans* Sequencing Consortium. Genome sequence of the nematode *C. elegans*: A platform for investigating biology. *Science* 1998;282(5396):2012–8.
4. Girard LR, Fiedler TJ, Harris TW, Carvalho F, Antoshechkin I, Han M, Sternberg PW, Stein LD, Chalfie M. *WormBook*: The online review of *Caenorhabditis elegans* biology. *Nucleic Acids Res* 2007;35(Database Issue):D472–5.
5. Ogg S, Paradis S, Gottlieb S, Patterson GI, Lee L, Tissenbaum HA, Ruvkun G. The Fork head transcription factor DAF-16 transduces insulin-like metabolic and longevity signals in *C. elegans. Nature* 1997;389(6654):994–9.
6. Nakae J, Park BC, Accili D. Insulin stimulates phosphorylation of the forkhead transcription factor FKHR on serine 253 through a Wortmannin-sensitive pathway. *J Biol Chem* 1999;274(23):15982–5.
7. Barr MM and Sternberg PW. A polycystic kidney-disease gene homologue required for male mating behavior in *C. elegans. Nature* 1999;401(6751):386–9.
8. Nauli SM, Alenghat FJ, Luo Y, Williams E, Vassilev P, Li X, Elia AE, Lu W, Brown EM, Quinn SJ, Ingber DE, Zhou J. Polycystins 1 and 2 mediate mechanosensation in the primary cilium of kidney cells. *Nat Genet* 2003;33(2):129–37.

9. Sulston JE and Horvitz HR. Post-embryonic cell lineages of the nematode, *Caenorhabditis elegans*. *Dev Biol* 1977;56(1):110–56.
10. White JG, Southgate E, Thomson JN, Brenner S. The structure of the nervous system of *Caenorhabditis elegans*. *Phil Trans R Soc Lond [Biol]* 1976;275:327–48.
11. Lee RY, Lobel L, Hengartner M, Horvitz HR, Avery L. Mutations in the alpha1 subunit of an L-type voltage-activated Ca2+ channel cause myotonia in *Caenorhabditis elegans*. *EMBO J* 1997;16(20):6066–76.
12. Bargmann CI. Neurobiology of the *Caenorhabditis elegans* genome. *Science* 1998; 282(5396):2028–33.
13. Li H, Avery L, Denk W, Hess GP. Identification of chemical synapses in the pharynx of *Caenorhabditis elegans*. *Proc Natl Acad Sci USA* 1997;94(11):5912–6.
14. McIntire SL, Jorgensen E, Horvitz HR. Genes required for GABA function in *Caenorhabditis elegans*. *Nature* 1993;364(6435):334–7.
15. Alfonso A, Grundahl K, McManus JR, Rand JB. Cloning and characterization of the choline acetyltransferase structural gene (*cha-1*) from *C. elegans*. *J Neurosci* 1994;14(4):2290–300.
16. Sulston JE, Dew M, Brenner S. Dopaminergic neurons in the nematode *Caenorhabditis elegans*. *J Comp Neurol* 1975;163(2):215–26.
17. Horvitz HR, Chalfie M, Trent C, Sulston JE, Evans PD. Serotonin and octopamine in the nematode *Caenorhabditis elegans*. *Science* 1982;216(4549):1012–4.
18. Schinkmann K and Li C. Localization of FMRFamide-like peptides in *Caenorhabditis elegans*. *J Comp Neurol* 1992;316(2):251–60.
19. Cossette P, Liu L, Brisebois K, Dong H, Lortie A, Vanasse M, Saint-Hilaire JM, Carmant L, Verner A, Lu WY, Wang YT, Rouleau GA. Mutation of *GABRA1* in an autosomal dominant form of juvenile myoclonic epilepsy. *Nat Genet* 2002;31(2):184–9.
20. DeLorey TM, Handforth A, Anagnostaras SG, Homanics GE, Minassian BA, Asatourian A, Fanselow MS, Delgado-Escueta A, Ellison GD, Olsen RW. Mice lacking the beta3 subunit of the GABAA receptor have the epilepsy phenotype and many of the behavioral characteristics of Angelman syndrome. *J Neurosci* 1998;18(20):8505–14.
21. Kobayashi M and Buckmaster PS. Reduced inhibition of dentate granule cells in a model of temporal lobe epilepsy. *J Neurosci* 2003;23(6):2440–52.
22. Cronin CJ, Mendel JE, Mukhtar S, Kim YM, Stirbl RC, Bruck J, Sternberg PW. An automated system for measuring parameters of nematode sinusoidal movement. *BMC Genet* 2005;6(1):5.
23. Miller KG, Emerson MD, Rand JB. G(o)alpha and diacylglycerol kinase negatively regulate the Gqalpha pathway in *C. elegans*. *Neuron* 1999;24(2):323–33.
24. Eastman C, Horvitz HR, Jin Y. Coordinated transcriptional regulation of the *unc-25* glutamic acid decarboxylase and the *unc-47* GABA vesicular transporter by the *Caenorhabditis elegans* UNC-30 homeodomain protein. *J Neurosci* 1999;19(15):6225–34.
25. Schuske K, Palfreyman MT, Watanabe S, Jorgensen EM. UNC-46 is required for trafficking of the vesicular GABA transporter. *Nat Neurosci* 2007;10(7):846–53.
26. Liu JW and Thomas JH. Regulation of a periodic motor program in *C. elegans*. *J Neurosci* 1994;14(4):1953–62.
27. Thomas JH. Genetic analysis of defecation in *Caenorhabditis elegans*. *Genetics* 1990; 124(4):855–72.
28. Beg AA and Jorgensen EM. EXP-1 is an excitatory GABA-gated cation channel. *Nat Neurosci* 2003;6(11):1145–52.
29. Lee J, Song HO, Jee C, Vanoaica L, Ahnn J. Calcineurin regulates enteric muscle contraction through EXP-1, excitatory GABA-gated channel, in *C. elegans*. *J Mol Biol* 2005;352(2):313–8.
30. Sanchez RM, Dai W, Levada RE, Lippman JJ, Jensen FE. AMPA/kainate receptor-mediated downregulation of GABAergic synaptic transmission by calcineurin after seizures in the developing rat brain. *J Neurosci* 2005;25(13):3442–51.
31. Miller KG, Alfonso A, Nguyen M, Crowell JA, Johnson CD, Rand JB. A genetic selection for *Caenorhabditis elegans* synaptic transmission mutants. *Proc Natl Acad Sci U S A* 1996;93(22):12593–8.
32. Peters MA, Teramoto T, White JQ, Iwasaki K, Jorgensen EM. A calcium wave mediated by gap junctions coordinates a rhythmic behavior in *C. elegans*. *Curr Biol* 2007;17(18):1601–8.
33. Hamzei-Sichani F, Kamasawa N, Janssen WG, Yasumura T, Davidson KG, Hof PR, Wearne SL, Stewart MG, Young SR, Whittington MA, Rash JE, Traub RD. Gap junctions on hippocampal mossy fiber axons demonstrated by thin-section electron microscopy and freeze fracture replica immunological labeling. *Proc Natl Acad Sci U S A* 2007;104(30):12548–53.
34. Perez-Velazquez JL, Valiante TA, Carlen PL. Modulation of gap junctional

mechanisms during calcium-free induced field burst activity: A possible role for electrotonic coupling in epileptogenesis. *J Neurosci* 1994;14(7):4308–17.

35. Hempelmann A, Heils A, Sander T. Confirmatory evidence for an association of the *connexin-36* gene with juvenile myoclonic epilepsy. *Epilepsy Res* 2006;71(2–3):223–8.
36. Nonet ML, Saifee O, Zhao H, Rand JB, Wei L. Synaptic transmission deficits in *Caenorhabditis elegans* synaptobrevin mutants. *J Neurosci* 1998;18(1):70–80.
37. Nguyen M, Alfonso A, Johnson CD, Rand JB. *Caenorhabditis elegans* mutants resistant to inhibitors of acetylcholinesterase. *Genetics* 1995;140(2):527–35.
38. Sieburth D, Ch'nq Q, Dybbs M, Tavazoie M, Kennedy S, Wang D, Dupuy D, Rual JF, Hill DE, Vidal M, Ruvkun G, Kaplan JM. Systematic analysis of genes required for synapse structure and function. *Nature* 2005;436(7050):510–7.
39. Robatzek M and Thomas JH. Calcium/calmodulin-dependent protein kinase II regulates *Caenorhabditis elegans* locomotion in concert with a G(o)/G(q) signaling network. *Genetics* 2000;156(3):1069–82.
40. Iwasa H, Kikuchi S, Mine S, Sugita K, Miyagishima H, Hasegawa S. Functional significance of stimulatory GTP-binding protein in hippocampus is associated with kindling-elicited epileptogenesis. *Psychiatry Clin Neurosci* 2000;54(2):191–7.
41. Leach NT, Sun Y, Michaud S, Zheng Y, Ligon KL, Ligon AH, Sander T, Korf BR, Lu W, Harris DJ, Gusella JF, Maas RL, Quade BJ, Cole AJ, Kelz MB, Morton CC. Disruption of diacylglycerol kinase delta associated with seizures in humans and mice. *Am J Hum Genet* 2007;80(4):792–9.
42. Singleton MW, Holbert WH 2nd, Lee AT, Bracey JM, Churn SB. Modulation of CaM kinase II activity is coincident with induction of status epilepticus in the rat pilocarpine model. *Epilepsia* 2005;46(9):1389–400.
43. Jiang G, Zhuang L, Miyauchi S, Miyake K, Fei YJ, Ganapathy V. A Na^{+}/Cl^{-}-coupled GABA transporter, GAT-1, from *Caenorhabditis elegans*: Structural and functional features, specific expression in GABA-ergic neurons, and involvement in muscle function. *J Biol Chem* 2005;280(3):2065–77.
44. Williams SN, Locke CJ, Braden AL, Caldwell KA, Caldwell GA. Epileptic-like convulsions associated with LIS-1 in the cytoskeletal control of neurotransmitter signaling in *Caenorhabditis elegans*. *Hum Mol Genet* 2004;13(18):2043–59.
45. Huang RQ, Bell-Horner CL, Dibas MI, Covey DF, Drewe JA, Dillon GH. Pentylenetetrazole-induced inhibition of recombinant gamma-aminobutyric acid type A (GABA(A)) receptors: Mechanism and site of action. *J Pharmacol Exp Ther* 2001;298(3):986–95.
46. Shyn SI, Kerr R, Schafer WR. Serotonin and Go modulate functional states of neurons and muscles controlling *C. elegans* egg-laying behavior. *Curr Biol* 2003;13(21):1910–5.
47. LeBoeuf B, Gruninger TR, Garcia LR. Food deprivation attenuates seizures through CaMKII and EAG K+ channels. *PLoS Genet* 2007;3(9):1622–32.
48. Liu Q, Chen B, Ge Q, Wang ZW. Presynaptic Ca2+/calmodulin-dependent protein kinase II modulates neurotransmitter release by activating BK channels at *Caenorhabditis elegans* neuromuscular junction. *J Neurosci* 2007;27(39):10404–13.
49. Umemura T, Rapp P, Rongo C. The role of regulatory domain interactions in UNC-43 CaMKII localization and trafficking. *J Cell Sci* 2005;118(Pt 5):3327–38.
50. Chioza B, Wilkie H, Nashef L, Blower J, McCormick D, Sham P, Asherson P, Makoff AJ. Association between the alpha(1a) calcium channel gene *CACNA1A* and idiopathic generalized epilepsy. *Neurology* 2001;56(9):1245–6.
51. Brenner R, Chen QH, Vilaythong A, Toney GM, Noebels JL, Aldrich RW. BK channel beta4 subunit reduces dentate gyrus excitability and protects against temporal lobe seizures. *Nat Neurosci* 2005;8(12):1752–9.
52. Fukata Y, Adesnik H, Iwanaga T, Bredt DS, Nicoll RA, Fukata M. Epilepsy-related ligand/receptor complex LGI1 and ADAM22 regulate synaptic transmission. *Science* 2006;313(5794):1792–5.
53. Jones DL and Baraban SC. Characterization of inhibitory circuits in the malformed hippocampus of *Lis1* mutant mice. *J Neurophysiol* 2007;98(5):2737–46.
54. Locke CJ, Williams SN, Schwarz EM, Caldwell GA, Caldwell KA. Genetic interactions among cortical malformation genes that influence susceptibility to convulsions in *C. elegans*. *Brain Res* 2006;1120(1):23–34.
55. Jacobs MP, Fischbach GD, Davis MR, Dichter MA, Dingledine R, Lowenstein DH, Morrell MJ, Noebels JL, Rogawski MA, Spencer SS, Theodore WH. Future

directions for epilepsy research. *Neurology* 2001;57(9):1536–42.

56. Goodman MB, Hall DH, Avery L, Lockery SR. Active currents regulate sensitivity and dynamic range in *C. elegans* neurons. *Neuron* 1998;20(4):763–72.
57. Richmond JE and Jorgensen EM. One GABA and two acetylcholine receptors function at the *C. elegans* neuromuscular junction. *Nat Neurosci* 1999;2(9):791–7.
58. Christensen M, Estevez A, Yin X, Fox R, Morrison R, McDonnell M, Gleason C, Miller DM 3rd, Strange K. A primary culture system for functional analysis of *C. elegans* neurons and muscle cells. *Neuron* 2002;33(4):503–14.
59. Cinar H, Keles S, Jin Y. Expression profiling of GABAergic motor neurons in *Caenorhabditis elegans. Curr Biol* 2005;15(4):340–6.
60. Fox RM, Von Stetina SE, Barlow SJ, Shaffer C, Olszewski KL, Moore JH, Dupuy D, Vidal M, Miller DM 3rd. A gene expression fingerprint of *C. elegans* embryonic motor neurons. *BMC Genomics* 2005;6(1):42.
61. Fox RM, Watson JD, Von Stetina SE, McDermott J, Brodigan TM, Fukushige T, Kraus M, Miller DM 3rd. The embryonic muscle transcriptome of *C. elegans. Genome Biol* 2007;8(9): R188.
62. Miyawaki A, Griesbeck O, Heim R, Tsien RY. Dynamic and quantitative Ca2+ measurements using improved cameleons. *Proc Natl Acad Sci U S A* 1999;96(5):2135–40.
63. Nakai J, Ohkura M, Imoto K. A high signal-to-noise Ca(2+) probe composed of a single fluorescent protein. *Nat Biotechnol* 2001;19(2):137–41.
64. Chalasani SH, Chronis N, Tsunozaki M, vGray JM, Ramot D, Goodman MB, Bargmann CI. Dissecting a circuit for olfactory behaviour in *Caenorhabditis elegans. Nature* 2007;450(7166):63–70.
65. Dittman JS and Kaplan JM. Factors regulating the abundance and localization of synaptobrevin in the plasma membrane. *Proc Natl Acad Sci U S A* 2006;103(30):11399–404.
66. Samuel AD, Silva RA, Murthy VN. Synaptic activity of the AFD neuron in *Caenorhabditis elegans* correlates with thermotactic memory. *J Neurosci* 2003;23(2):373–6.
67. Bieri T, Blasiar D, Ozersky P, Antoshechkin I, Bastiani C, Canaran P, Chan J, Chen N, Chen WJ, Davis P, Fiedler TJ, Girard L, Han M, Harris TW, Kishore R, Lee R, McKay S, Muller HM, Nakamura C, Petcherski A, Rangarajan A, Rogers A, Schindelman G, Schwarz EM, Spooner W, Tuli MA, Van Auken K, Wang D, Wang X, Williams G, Durbin R, Stein LD, Sternberg PW, Spieth J. WormBase: New content and better access. *Nucleic Acids Res* 2007;35(Database Issue): D506–10.
68. Dupuy D, Bertin N, Hidalgo CA, Venkatesan K, Tu D, Lee D, Rosenberg J, Svrzikapa N, Blanc A, Carnec A, Carvunis AR, Pulak R, Shingles J, Reece-Hoyes J, Hunt-Newbury R, Viveiros R, Mohler WA, Tasan M, Roth FP, Le Peuch C, Hope IA, Johnsen R, Moerman DG, Barabasi AL, Baillie D, Vidal M. Genome-scale analysis of *in vivo* spatiotemporal promoter activity in *Caenorhabditis elegans. Nat Biotechnol* 2007;25(6): 663–8.
69. Kim SK, Lund J, Kiraly M, Duke K, Jiang M, Stuart JM, Eizinger A, Wylie BN, Davidson GS. A gene expression map for *Caenorhabditis elegans. Science* 2001;293(5537): 2087–92.
70. Li S, Armstrong CM, Bertin N, Ge H, Milstein S, Boxem M, Vidalain PO, Han JD, Chesneau A, Hao T, Goldberg DS, Li N, Martinez M, Rual JF, Lamesch P, Xu L, Tewari M, Wong SL, Zhang LV, Berriz GF, Jacotot L, Vaglio P, Reboul J, Hirozane-Kishikawa T, Li Q, Gabel HW, Elewa A, Baumgartner B, Rose DJ, Yu H, Bosak S, Sequerra R, Fraser A, Mango SE, Saxton WM, Strome S, Van Den Heuvel S, Piano F, Vandenhaute J, Sardet C, Gerstein M, Doucette-Stamm L, Gunsalus KC, Harper JW, Cusick ME, Roth FP, Hill DE, Vidal M. A map of the interactome network of the metazoan *C. elegans. Science* 2004; 303(5657): 540–3.
71. Zhong W and Sternberg PW. Genome-wide prediction of *C. elegans* genetic interactions. *Science* 2006;311(5766): 1481–4.
72. Galperin MY. The molecular biology database collection:2007 update. *Nucleic Acids Res* 2007;35(Database Issue): D3–4.

Chapter 2

The Genetics and Molecular Biology of Seizure Susceptibility in Drosophila

Juan Song and Mark A. Tanouye

Abstract

There has been an increased interest recently in various Drosophila models of human disease. The goals are to uncover the fundamental biological principles underlying causes and cures of human pathology utilizing the power of Drosophila genetics. A particularly exciting prospect is that new therapeutics could be forthcoming through identification of disease-causing genes followed by targeted drug development or by the development of platforms for high-throughput drug screening. This chapter reviews a model of human epilepsy based on a set of seizure-sensitive Drosophila mutants that exhibit features resembling some human epilepsies. Especially interesting are descriptions of a novel class of mutations that are second-site seizure-suppressor mutations. These mutations revert epilepsy phenotypes back to the wild-type range of seizure susceptibility. The genes responsible for seizure suppression are cloned with the goal of identifying targets for lead compounds that may be developed into new anti-epileptic drugs (AEDs).

Key words: epilepsy, seizure, bang sensitive, *Drosophila*, seizure suppressor, AED.

1. Introduction

Genetic and molecular analyses of *Drosophila melanogaster* (fruit fly) mutants have provided models for examining difficult problems in biology, particularly in genetics, developmental biology, and neurobiology *(1)*. An important lesson has been that fundamental processes, and many of the essential gene products, are conserved across species. Thus, findings are generally applicable to other biological systems such as mouse and human *(2)*. An implication from cross-species conservation is that Drosophila has the potential to be an attractive system for examining human pathologies *(3, 4)*. The goals are to uncover fundamental biological principles underlying

Scott C. Baraban (ed.), *Animal Models of Epilepsy: Methods and Innovations, vol. 40*

DOI 10.1007/978-1-60327-263-6_2 Springerprotocols.com

causes and cures of human pathology. Of particular interest is also the prospect that new therapeutics might be forthcoming through (i) identification of target genes that could be utilized for new drug development or (ii) the development of platforms for high-throughput drug screening.

This chapter examines the genetics and molecular biology of seizure susceptibility in Drosophila as a model for human epilepsy. Three major topics are discussed: (1) Drosophila seizure-sensitive mutants, seizure-like electrical activity, and seizure-like behaviors; (2) seizure-suppressor mutations; and (3) treatment with anti-epileptic drugs (AEDs).

These topics are fit into a background context, i.e., that the power of Drosophila genetics can be brought to bear on the development of new epilepsy therapeutics. In particular, a central thesis is to explore a general approach that leads from

$$[\text{epilepsy mutant}] \rightarrow [\text{seizure} - \text{suppressor gane}] \rightarrow [\text{drug target}] \rightarrow [\text{anti} - \text{epileptic drug}]$$

At first glance, it seems this would be a roundabout way for drug discovery; however, initial indications are that it could be a surprisingly productive approach. More importantly, it appears to be capable of identifying unexpected classes of compounds leading toward development of novel AEDs: possibly drugs with reduced neurological side effects or drugs with efficacy against intractable epilepsies when developed for human therapeutics.

2. Using *Drosophila* for Seizure Studies

2.1. Drosophila Seizure-Sensitive Mutants, Seizure-Like Electrical Activity, and Seizure-Like Behaviors

The idea that fruit flies can have seizures is an odd notion; and even odder is the notion that these might be able to model human epilepsy. Nonetheless, there are similarities at the cellular and subcellular levels between fly and mammalian nervous systems, particularly in excitable membranes, e.g., voltage-gated and ligand-gated signaling molecules such as sodium (Na), potassium (K), and calcium (Ca) channels and acetylcholine (ACh), glutamate, and γ-aminobutyric acid (GABA) receptors. The differences in the nervous systems of flies and mammals are also salient. For example, differences exist in general central nervous system (CNS) organization. Most prominently, a ganglionic organization in the fly brain with synaptic neuropilar regions rather than organization into synaptic layers was found, for example, throughout mammalian cortex. Nevertheless, electrical shock of sufficient intensity delivered to the brain of adult Drosophila elicits CNS spiking activity that is seizure like in appearance (**Fig. 2.1**, *5–9*). This activity is similar to

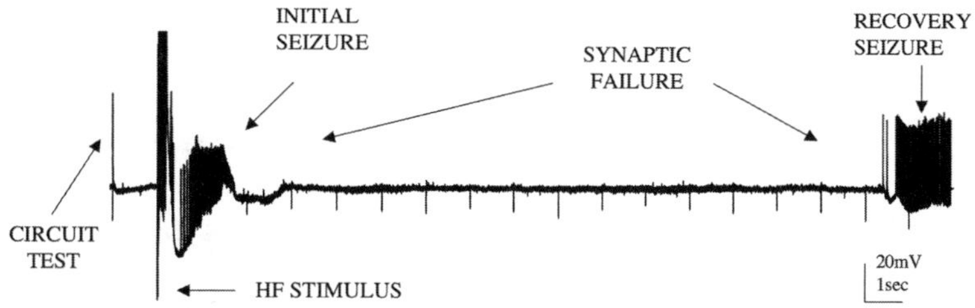

Fig. 2.1. **Seizure-like activity in a *sda* fly.** Seizure-like activity is elicited by a high-frequency stimulus (HF stimulus = 8 V) and displayed at a low sweep speed. Stimulus is a short wave train (0.5 ms pulses at 200 Hz for 300 ms) of electrical stimuli delivered to the brain. Recording is from a dorsal longitudinal flight muscle (DLM) fiber and reflects the activity of the single DLM motoneuron that innervates it. An initial seizure is observed immediately following delivery of the stimulus. In this recording, the HF stimulus and initial seizure are followed by a quiescent period that is characterized by synaptic failure within the giant fiber (GF) neurocircuit. During this period, there are stimulus artifacts (downward-going) from continuous single-pulse stimulation of the GF (0.5 Hz) but no evoked DLM potentials. A recovery seizure appears as additional seizure-like activity occurring just after the synaptic failure period and just prior to recovery (recovery not evident in this trace). Figure adapted from Kuebler and Tanouye *(6)*.

other animals with complex nervous systems, including humans. Seizure-like activity in Drosophila is uncontrolled with abnormal neuronal firing that approaches 100 Hz for 3 s. Seizure-like activity is extensive with all neurons thus far examined, about 30 different neurons participating in the seizure *(6)*.

The discovery of Drosophila seizure-sensitive mutants began with the initial characterization of a class of behavioral mutants called bang-sensitive (BS) paralytics *(10–12)*. BS mutants display a prominent abnormal behavioral phenotype indicative of neurological dysfunction. In response to a mechanical stress stimulus, BS mutants initially show seizure-like behavior lasting 2–3 s. This behavioral seizure-like activity is characterized by intense, uncoordinated motor activity including wing flapping, leg shaking, and abdominal muscle contractions *(10, 13)*. Electrophysiological analysis shows that seizure-like behavior corresponds to intense seizure-like neuronal firing throughout the CNS *(5, 6)*. Seizure-like behavior is followed by a brief behavioral paralytic period termed a "coma" by Seymour Benzer *(10)*. At the beginning of the paralytic period, flies of all genotypes are completely quiescent. Electrophysiological analysis shows that paralysis corresponds to synaptic failure at many CNS central synapses *(5, 6)*. The duration of the paralytic period varies by genotype, age, rearing temperature, and possibly other factors. The shortest duration is for the *sda* mutant, about 30 s for 3-day-old flies at 24°C. In contrast, the duration for *bss* mutants at 3 days is about 2 min, with initial complete quiescence followed by multiple rounds of seizure and quiescence resembling tonic–clonic activity. The paralytic period is followed by a short bout of seizure-like activity called a "recovery seizure" and shortly thereafter the BS fly awakens and resumes normal behavioral activity. Interestingly, immediately following the paralytic period, the BS fly temporarily loses seizure sensitivity called a "refractory period." That is, flies cannot be reparalyzed by mechanical stress stimulation *(5, 6)*. Refractory period duration varies according to genotype, with *sda* mutants having a refractory period of about 5–7 min and *bss* mutants about 10–15 min. Electrophysiological analysis shows that refractory period corresponds to a transient increase in BS seizure threshold *(6)*.

BS mutants are especially sensitive to seizures evoked by electrical stimulation (**Fig. 2.2**; **Table 2.1**). For example, in normal flies, seizure-like activity is evoked by 30 V stimulation, while

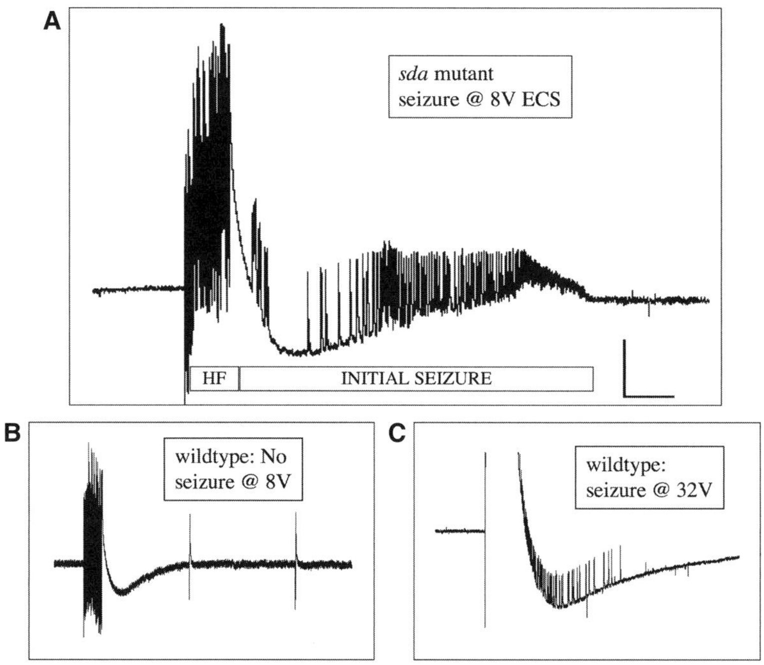

Fig. 2.2. **The bang-sensitive (BS) mutant *sda* fly is more susceptible to seizures than a normal wild-type fly and therefore has a lower seizure threshold.** (**A**) Seizure-like activity (initial seizure) is elicited in a *sda* fly by a high-frequency stimulus of low strength (8 V) and displayed at a high sweep speed. (**B**) A low-voltage HF stimulus of 8 V fails to elicit a seizure in a wild-type CS fly because the stimulus is below the seizure threshold. Following the HF stimulus artifact, there is no seizure activity observed in this recording displayed at a high sweep speed. Note also that there is no period of synaptic failure, and single-pulse stimulation of the GF (0.5 Hz) continues to evoke DLM potentials. Two such effective single-pulse stimuli are depicted in this trace; each was effective in evoking a DLM potential. (**C**) Seizure-like activity is elicited in a wild-type CS fly by a high-voltage HF stimulus (32 V), which is above the threshold for seizure. The seizure in this recording begins within the large stimulus artifact and is displayed at a high sweep speed. Vertical calibration bar is 20 mV. Horizontal calibration bar is 300 ms. Adapted from Kuebler et al. *(7)*.

Table 2.1
Seizure-sensitive mutants and their affected gene products. Seizure threshold is voltage of high-frequency stimulation. For comparison the seizure threshold is 30.1 V for normal Canton-Special wild-type flies *(7)*

Seizure-sensitive mutant	Threshold	Gene product	Reference
bang senseless (*bss1, bss2*)	3.2, 3.7	Unknown	*6, 9*
easily shocked (*eas*)	3.4	Ethanolamine kinase	*13*
slamdance (*sda*)	6.7	Aminopeptidase N	*14*
bang sensitive (*bas1, bas2*)	7.6, 3.8	Unknown	*9*
technical knockout (*tko*)	9.9	Ribosomal protein S12	*9, 15*
jitterbug (*jbug*)	10.5	Unknown	9
couch potato (*cpo*)	11.1	RNA-binding protein	*16*
kazachoc (*kcc*)	17.0	K^+ Cl^- cotransporter	*17*
knockdown (*kdn*)	20.2	Citrate synthase	*18*
stress-sensitive (*sesB*)	Unknown	Adenine nucleotide translocase	*19*
Na,K ATPase	Unknown	Na, K ATPase	*20*

a 10-fold lower stimulation (3 V) is effective in *bss*^{1} mutants *(6)*. A collection of 13 BS mutants representing 11 genes encoding a variety of products forms the basis for a Drosophila seizure model (**Table 2.1**). As one example, the *sda* BS mutation is described here in more detail. Flies carrying *sda* have a defect in aminopeptidase N *(14)*. Aminopeptidase N is a transmembrane ectoenzyme that catalyzes the removal of neutral and basic amino acids from the N termini of several small peptide substrates, such as angiotensin III and enkephalin. The catalytic domain of aminopeptidase N faces the exterior of the plasma membrane and is responsible for specific peptide cleavages. The enzyme is anchored by a transmembrane domain. The original mutation *sda*$^{iso7.8}$ arose spontaneously and was identified based on its BS paralytic phenotype. Molecular analysis showed that it was due to a small DNA insertion located in the open reading frame of exon 3 *(14)*. The BS behavioral phenotype of *sda* flies is completely penetrant; 100% of flies are paralyzed by mechanical stress stimulation. The seizure threshold of *sda* is low compared to the wild type (6.7 V, about 5-fold lower than wild type), making *sda* a desirable genetic background to test for gene interactions with other seizure-suppressor and seizure-enhancer genes and for drug treatment studies.

Seizures in flies and humans have similarities providing support for the utility of the investigation. For seizure-like activity in flies (i) all individuals have a seizure threshold; (ii) genetic mutations can modulate seizure susceptibility; (iii) electroconvulsive shock treatment (ECT) raises the threshold for subsequent seizure-like activity; (iv) seizure-like activity spreads through the CNS along particular pathways that depend on functional synaptic connections and recent electrical activity; (v) seizure-like activity in flies can be spatially segregated into particular regions of the CNS; (vi) Drosophila seizure phenotypes can be ameliorated by the human AEDs valproate, phenytoin, gabapentin, and potassium bromide (KBr); and (vii) Na channel mutations are excellent seizure suppressors, consistent with the notion that many AEDs are targeting Na channels *(6, 7, 9, 21–26)*.

2.2. Seizure-Suppressor Mutations

The usual starting material for genetic analysis is the single gene mutation that gives rise to a mutant phenotype such as seizure sensitivity. In some cases, the presence of two mutations in an individual causes a phenotype that is different from that caused by either mutation alone; that is, the two mutations interact genetically. Two important types of genetic interaction are enhancement and suppression *(11, 27–29)*. *Enhancers* are modifying mutations that make the mutant phenotype more extreme. *Suppressors* are modifying mutations that yield individuals phenotypically more like the wild type (normal), e.g., the mutant phenotype is "suppressed." In some cases the gross wild-type phenotype is completely restored; in other cases, restoration is only partial. Thus, suppressors are mutations that elicit a revertant or partially revertant phenotype. But suppressor mutations can be genetically separated, by recombination, from the mutation that they suppress. This section examines a novel class of genes called "seizure suppressors." Mutations defining these suppressors revert the seizure-sensitivity phenotype of BS mutants. Among the several questions that arise are the following: (i) Are there seizure-suppressor genes, and how might they lead to new therapeutics? (ii) What is the entire range of potential gene products that can act as seizure suppressors? (iii) Is this range limited to nervous system-specific gene products, such as signaling molecules, or does it include non-nervous system gene products as well? The answers to some of these questions would help to evaluate better the potential of this approach.

Recently considerable attention has been focused on genetic animal model systems of epilepsy *(30, 31)*. The different models have their respective strengths and weaknesses, but for the particular approach of seizure suppression genetics, investigations in Drosophila or zebrafish model systems appear to be better choices than two other excellent general genetically tractable models, the nematode *Caenorhabditis elegans* and mouse.

Phenotypic characterization of seizures is a problem in *C. elegans* because electrophysiology recording is difficult. Therefore, it is difficult to exploit the superior mutant isolation and genetic capabilities available there. Mouse phenotypes are more relevant to human seizures and adult neurophysiology is excellent; however, the large number of genes, potential genetic complexities, and positional cloning requirements do not make them an ideal choice. In Drosophila or zebrafish, there is the possibility for advanced electrophysiology, mutant isolation, genetic analysis, molecular biology, and genomics. Thus, although this investigation remains a challenging one for Drosophila or zebrafish, it appears to be more promising than for some other possible model systems.

In Drosophila, an unexpected finding from initial study of seizure-suppressor mutations is their apparent abundance and ease of identification. A collection of 20 suppressor mutations in 11 genes has already been identified (**Table 2.2**) *(7, 9, 24, 25, 32, 33)*. Continued identification of seizure-suppressor mutations through genetic screens promises to provide a rich source of genetic aberrations that can serve to dissect apart seizure susceptibility and act as a basis for discovering new AED targets. The following is a more detailed description of selected seizure-suppressor mutations.

Table 2.2
Seizure-suppressor mutants and their affected gene products

Seizure-suppressor mutant	Gene product	References
paralyzed ($para^{ST76}$, $para^{JS1}$)	Na channel	*7, 24*
male lethal (mle^{napts})	Na channel regulator	*7*
shakingB ($shakB^{2}$)	Gap junction channel	*7, 9*
Shaker (Sh^{KS133})	K channel	*7*
escargot (esg^{EP684} + 4 alleles)	Zn-finger transcription factor	*32*
meiosis-P26 ($mei\text{-}P26^{EG16}$, $mei\text{-}P26^{1}$)	Ring finger B-box-NHL protein	*33*
topoisomerase I ($top1^{JS}$ + 3 alleles)	DNA topoisomerase type I	*25*
kazal-domain protein-1 (kdp1)	Serine protease inhibitor	*32*
kazal-domain protein-2 (kdp2)	Serine protease inhibitor	*32*
suppressor of eas7 (su(eas7))	Unknown	*33*
suppressor of eas13 (su(eas13))	Unknown	*33*

2.3. Sodium Channel Mutations Are Potent Seizure Suppressors

Search for seizure-suppressor genes initially utilized a so-called reverse genetics approach *(7)*. That is, extant Drosophila neurological mutations which might act as seizure suppressors were selected and tested in double mutant combinations to determine if they could suppress BS phenotypes. This approach identified several seizure suppressors including *Sh* (K channel) and $shakB^2$ (gap junction connexin). In addition, Na channel mutations were tested, inspired by observations that sodium currents are major targets for several first-line AEDs, including phenytoin, carbamazepine, lamotrignine, and valproate *(34, 35)*. Two Drosophila Na channel mutants are $para^{ST76}$ and mle^{napts} that affect the channel structural gene and regulate channel expression, respectively *(36–38)*. Both mutants are seizure-resistant mutants showing electrophysiological seizure thresholds that are about two–three times higher than for normal flies *(7)*. Double mutant combinations between each of the Na channel mutants and various BS mutants show a reversion of seizure susceptibility to wild-type levels *(7)*. The $para^{ST76}$ and mle^{napts} mutations both cause a reduction of functional Na channels *(37, 38)*, suggesting that they may suppress seizures in much the same way as the AEDs, phenytoin and carbamazepine *(39, 40)*. During repetitive firing, these drugs are thought to stabilize Na channels in the inactive state, thereby reducing the number of channels that can be activated. This leads to reduced capacity for high-frequency (HF) firing *(41, 42)*. Indeed, reduced capacity for HF firing has been observed in $para^{ST76}$ and mle^{napts} mutants *(7, 43)*.

2.4. Gap Junction Mutations Suppress Seizure-Like Activity in Some, but Not All, BS Mutants

The selection of gap junction mutations to test by a reverse genetics approach was inspired by the idea that electrical synaptic transmission is an important mechanism for synchronizing signaling in the brain contributing to the generation and maintenance of seizures *(44)*. Experimental observations have reinforced this notion. Pharmacology that reduces electrical transmission diminishes seizures, and enhanced electrical transmission increases the frequency and severity of seizures *(44–46)*. The *shakB* gene encodes a gap junction protein, and mutations in this gene have impaired electrical transmission *(47, 48)*. The $shakB^2$ mutant was found to be a seizure-resistant Drosophila mutant displaying a seizure threshold about three times higher than for normal flies *(7, 9)*. Double mutant combinations between $shakB^2$ and various BS mutants show that it is an effective seizure-suppressor mutation, e.g., it completely suppresses seizure-like activity caused by *sda*, *kdn*, and *jbug* mutations. Seizure-like activity caused by *eas* and *tko* mutations are partially suppressed by $shakB^2$. Interestingly, $shakB^2$ cannot suppress seizure-like activity caused by bas^2, bss^1, and bss^2 mutations *(7, 9)*. Synapses in the $shakB^2$ mutant have been shown to be impaired in electrical transmission *(48, 49)*. Thus, $shakB^2$ is proposed to suppress seizure-like activity by a mechanism similar to that suggested for drugs such as carbenoxolone that block gap junction activity *(50)*. Impaired transmission at electrical synapses is thought to interfere with synchronous activation of neuronal populations, leading to decreased seizure susceptibility. Observations on *shakB*^{2} are generally consistent with this mechanism.

2.5. A Mutation of DNA Topoisomerase I Is an Unexpected Seizure Suppressor

Reverse genetics has been well suited for validating the phenomenon of seizure suppression and the testing of suspected candidates. For example, the finding that sodium channel mutations act as seizure suppressors is a strong validation of the suppressor genetics approach. Usefulness for therapeutics may be limited, however, since Na channels are already known to be a target for AEDs. For therapeutics, much more significant are unexpected classes of seizure-suppressor genes that do not compromise CNS function, and therefore might be developed into drugs with minimal neurological side effects. The most powerful approach for discovering unexpected suppressor genes is to use an unbiased "forward genetics" approach in which random mutagenesis is used, and new mutations that suppress seizures are selected directly. This approach could identify, without preconceptions, suppressors that are completely novel, thereby revealing unexpected suppressor mechanisms and defining the entire range of gene products that can sub-serve seizure suppression. An expectation is that there would be two general types of genes identified in forward genetics screens: (i) some mutations may identify non-nervous system-specific gene products. These might be anticipated, for example, because most of the seizure-sensitive BS genes are, themselves, not nervous system specific. These types of gene products could be particularly important for developing new AEDs with minimal neurological side effects or (ii) some mutations may identify nervous system-specific gene products that affect fundamentally important features of CNS structure and function. These may give additional insight into the types of CNS dysfunction responsible for seizure disorder and allow molecular access into these processes.

The general strategy employed in screens is straightforward and standard for Drosophila *(9, 24, 25, 32, 33)*. The analysis begins with flies carrying a BS genetic background (*eas*, *sda*, or *bss*). Flies are mutagenized, and selection is for exceptional animals (new mutants) that are no longer seizure sensitive. Primary selection is behavioral, a loss of BS paralysis; then a secondary electrophysiological screen used for seizure-threshold determination. The molecular identification of the suppressor gene is then determined by standard cloning methods.

An interesting example of forward genetic screening discovered a novel seizure suppressor *top1*^{JS} that is a mutation affecting Drosophila DNA topoisomerase I (called topo I in mammals) *(25)*.

The *top1*^{JS} mutation was isolated in a fly with an *eas* genetic background. For a mutagenic agent, the screen used a mobilized DNA transposable element called a P-element. Seizure suppression of the *eas* background was caused by transposition of the DNA element into the 5′ untranslated region of the *top1* gene, 257 bp upstream of the translation start site. Seizure suppression is due to reduced transcription of the *top1* gene; mutant transcription is about 12.5-fold less than normal; the *top1* mRNA is otherwise normal. The *top1* gene is an essential gene: of several known *top1* mutations, *top1*^{JS} is the only viable mutation, while all others are homozygous lethal *(51, 52)*.

The *top1*^{JS} mutation is a general seizure suppressor ameliorating phenotypes of *sda*, *eas*, and *bss (25)*. For example, *top1*^{JS} suppresses *sda* seizure-like behaviors and paralysis by about 73%. The threshold for evoking seizures is raised about 2.5-fold by the suppressor mutation. For *eas*, behavioral phenotypes are suppressed in 63% of animals and seizure threshold is raised about 3.5-fold. The *bss* seizure and paralytic behavior phenotypes are not suppressed by *top1*^{JS} significantly. That is, most *top1*^{JS} *bss* double mutants showed bang sensitivity. However, there is some indication that *top1*^{JS} acts to reduce the severity of *bss* seizure-like behavior, mainly about a 2-fold decrease in tonic–clonic-like activity *(25)*. Taken together, these results are consistent with *top1*^{JS} being a general seizure suppressor, with *sda* and *eas* mutants being more easily suppressed and *bss* mutants being more resistant to suppression. Interestingly, *sda, eas*, and *bss* encode very different products. As described, *eas* encodes ethanolamine kinase which is involved in the synthesis of phosphatidyl ethanolamine in neuronal membrane, and *sda* encodes aminopeptidase N *(13, 14)*. The gene product of *bss* is still unknown but is thought to encode another very different product (Parker and Tanouye, unpublished results). Therefore, these three BS mutations are likely increasing seizure sensitivity by different mechanisms. How can it be that *top1*^{JS} can act on multiple different seizure mechanisms (albeit, not very impressively on *bss* suppression)?

The identification of *top1*^{JS} as a seizure suppressor is surprising, since DNA topoisomerases have not been previously associated with seizure, seizure control, or any other electrical excitability functions. Type I DNA topoisomerase is thought to resolve the torsional tension associated with DNA replication and transcription by binding to the DNA and relaxing the helix *(53)*. One possible explanation is that transcription in active neurons generates supercoiled DNA that must be continuously relaxed to sustain high levels of RNA synthesis. The binding of *top1* enzyme to the DNA forms a cleavable complex leading to relaxation of the DNA supercoil, thereby allowing sustained high levels of RNA synthesis *(53)*. Reduced levels of *top1* activity due to mutation might lead to partial inhibition of transcription in active neurons.

There is no obvious evidence that this could be the case, but even if it were, through what excitability mechanism might this become manifest? The *top1*^{JS} mutation that acts to suppress seizures causes no other phenotypes whether in a wild-type or a BS mutant genetic background *(25)*. Thus, *top1*^{JS} flies do not display any obvious general nervous system excitability defects, e.g., they are not temperature-sensitive paralytics (possible evidence for fly hypoexcitability dysfunction), and they do not exhibit leg shaking under ether anesthesia (possible evidence for fly hyperexcitability dysfunction). The *top1*^{JS} flies show no notable defects in specific behaviors, e.g., they eat, jump, fly, groom, court, and mate normally *(25)*. This is not true of *top1* mutations generally. All *top1* mutations besides *top1*^{JS} are lethal mutations, indicating an essential role for the *top1* gene in Drosophila development *(51, 52)*. The *top1*^{JS} mutation is the first and only homozygous viable *top1* mutation thus far identified *(25)*. The only known phenotype for *top1*^{JS} is seizure suppression which occurs without any demonstrable effect on electrical excitability. To date, explanations for seizure suppression remain unknown.

3. *Drosophila* Can Provide Insights into Potential Anti-seizure Treatments

3.1. Drug Treatments for Seizure-Sensitive Mutants: Treatment with Anti-epileptic Drugs

Flies have been treated with AEDs in three main ways: (i) injection of drug directly into the brain, (ii) acute drug feeding, and (iii) chronic drug treatment *(9, 21–26)*. An apparently effective, albeit technically difficult, method of drug delivery in flies is by micropuncturing the fly head and infusing valproate solution directly into the head capsule *(21)*. At a concentration of 5 mM valproate, 75% of *sda* flies and 15% of *bss* flies show wild-type seizure thresholds. At 10 mM valproate, 100% of *sda* flies and 58% of *bss* flies show wild-type seizure thresholds. At 25 mM sodium valproate and higher, 100% of *sda* and *bss* flies show wild-type seizure thresholds. These valproate injection experiments provide validation for the Drosophila model showing that AEDs suppress seizure sensitivity in BS mutants.

Acute drug treatment of AEDs involves feeding of drug for one or a few days, usually as part of a sucrose solution applied to filters. Feeding of drugs is less effective than head injection experiments. For example, despite its effectiveness in injection studies, valproate feeding shows little or no effect in ameliorating BS mutant phenotypes at feeding concentrations as high as 10 mM *(26)*. Part of this problem may be due to low concentrations of the drug actually reaching the brain following feeding. Flies fed the anticonvulsant KBr at 0.25% were shown by ion chromatography to accumulate only 0.016 mg of bromide per fly head *(23)*.

Chronic treatment feeding methods are also used in several studies. In this procedure, BS mutant larvae are raised on drug-laced food throughout development and tested as adults. As argued by Reynolds et al. *(22)*, many AEDs only exert their therapeutic effects in humans over weeks of treatment and so the chronic treatment paradigm could more closely approximate this circumstance. Chronic drug treatment could affect nervous system development, in addition to its normal operation. Drug feeding experiments have shown general agreement between acute and chronic treatments, with the latter generally being more effective. BS phenotypes are ameliorated by phenytoin, gabapentin, and KBr *(22, 23)*. The gap junction blocker carbenoxolone has also been found to be effective *(9)*. Not effective when fed to BS mutants are the AEDs carbamazepine, ethosuximide, and vigabactrin *(22)*. In general, the phenotype most readily suppressed by AED feeding is tonic–clonic-like activity in *bss* mutants, usually indicated by a prominent decrease in the paralytic recovery time. In some cases, for *eas* and *sda* mutants, there are small percentages of flies (usually less than 10%) that are resistant to BS paralysis.

3.2. Treatment with top1 Inhibitors Inspired from top1^{JS} Seizure Suppressor

Seizure suppression by *top1*^{JS} is through reduced transcription of the top1 gene suggesting that other treatments lowering *top1* enzymatic activity, such as top1 inhibitor drugs, may also be effective at ameliorating seizure phenotypes. Camptothecin (CPT) is a quinoline-based plant phytochemical that acts as a potent inhibitor of top1 activity *(54, 55)*. CPT, similar phytochemicals, and their derivatives are thought to work by interfering with the reassociation of DNA after cleavage by top1, trapping the enzyme in a covalent DNA linkage *(56)*. This religation interference can lead to apoptosis, generating considerable interest for top1 inhibitors in the cancer clinic *(57)*.

Interestingly, top1 inhibitor effects resemble those of AEDs in reducing the paralytic recovery time of BS mutants in drug-rearing experiments and in short-term drug feeding experiments *(21–23, 25, 26)*. For example, *bss* flies fed CPT in short-term drug feeding recover from paralysis about two-thirds faster than control flies *(25)*. In addition, tonic–clonic-like activity is nearly completely suppressed. Similar reductions of *bss* recovery time are observed by feeding two other phytochemicals that are potent top1 inhibitors, apigenin and kaempferol *(26)*. Electrophysiological recordings corroborate these behavioral results. CPT treatment causes a modest increase in seizure threshold. Synaptic failure time is greatly decreased *(25, 26)*. Taken together, the findings using top1 inhibitors show general consistency among the findings *(25, 26)*. All three inhibitors, CPT, apigenin, and kaempferol, have similar results. All three of the drugs are effective on *bss*, *eas*, and *sda* mutants. The effectiveness of all three drugs is similar; paralytic recovery time is greatly decreased in *bss* mutants, and paralytic behavior is reverted to a similar extent, i.e., in a few percent of flies in *sda* and *eas* mutants. In each case, drug is more effective for *sda* than for *eas*. In each case, CPT is the most effective of the top1 inhibitors, apigenin is nearly as potent, and kaempferol is slightly less effective. Thus far, genetic suppression has always been more effective than drug feeding. This may be because of inducible detoxification mechanisms present in Drosophila and/or possible blood–brain barrier difficulties. Nevertheless, the consistencies among all these pharmacological and genetic findings suggest strongly that it is the inhibition of top1 function that is responsible for the amelioration of mutant phenotypes.

However, a comparison of drug feeding experiments in Drosophila suggests that, in general, top1 inhibitors may be a less effective anti-epileptic agent compared to phenytoin, but may be better than valproate, potassium bromide, and carbenoxolone *(9, 21–23)*. These features suggest the intriguing and attractive possibility that CPT derivatives may be especially valuable for co-treatment of certain brain tumors that also present with epilepsy *(58)*.

4. Concluding Remarks

Despite the frequency of seizure disorders in the human population, the genetic and physiological basis for these defects has been difficult to resolve. Although many genetic defects that cause seizure susceptibility have been identified, these involve disparate biological processes, many of which are not neural specific. The large number and heterogeneous nature of the genes involved make it difficult to understand the complex factors underlying the etiology of seizure disorders. Examining the kinds of effects that known genetic mutations can have on seizure susceptibility is one approach that may prove fruitful. This approach can be helpful both in understanding how different physiological processes affect seizure susceptibility and in identifying novel therapeutic treatments. In this chapter, the goal has been to present one view of how a genetically tractable system can be used to examine factors that influence seizure susceptibility. This particular presentation has been to show how one might (i) validate a model through a sizable collection of Drosophila seizure-sensitive mutants with phenotypes that resemble those seen in human epilepsy and are treatable with human AEDs; (ii) identify and characterize molecularly a sizable collection of seizure-suppressor mutations; (iii) validate the suppressor genetics approach with mutations that correspond to AEDs, that is, especially Na channel, GABA system, and gap junction seizure suppressors; (iv) choose from among the suppressor collection candidates that might define novel AED targets such as *top1*; and (v) investigate new compounds for AED efficacy such as CPT and other top1 inhibitors. The *top1* gene is an interesting candidate, presenting a distinct prospect for AED development via top1 inhibitor compounds. However, at present, there is no reason to believe that *top1* is the best candidate, merely the first one discovered through the suppressor genetics approach.

Currently available AEDs exert their effects through the targeting of Na channels, Ca channels, or elements of GABA inhibition. Although they are associated with some neurological side effects and are not effective in cases of intractable epilepsy, they have generally served the epilepsy patient population very well. The overall effectiveness of these compounds tends to leave us within a paradigm that focuses on these target molecules or a few others that sub-serve similar signaling functions. The enormous value of the Drosophila suppressor genetics approach is that it provides a logic that allows us to rationally branch out to consider other classes of compounds not ordinarily considered to be involved in neuronal excitability. The approach presented here shows how the ability of mutations to suppress seizures may be quantified and provides a baseline for exhaustive future studies involving both forward and reverse genetic approaches to identify suppressor mutations. An understanding of how these various genetic factors suppress seizure susceptibility may be a gateway to dissecting the tremendously complex and heterogeneous problem of seizure disorders. These examples demonstrate the utility of Drosophila as a model system for studying human seizure disorders and provide insights into the possible mechanisms by which seizure susceptibility is modified.

Acknowledgments

We thank Dr. Daria Hekmat-Scafe and Dr. Louise Parker for discussion. This work was supported by an NIH research grant and an Epilepsy Foundation grant to M.T.

References

1. Rubin GM, Lewis EB. A brief history of *Drosophila*'s contributions to genome research. *Science* 2000;287:2216–2218.
2. Veraksa A, Del Campo M, McGinnis W. Developmental patterning genes, and their conserved functions: From model organisms to humans. *Mol Genet Metab* 2000;69:85–100.
3. Tickoo S, Russell S. *Drosophila melanogaster* as a model system for drug discovery, and pathway screening. *Curr Opin Pharmacol* 2002;2:555–560.

4. Bier E. *Drosophila*, the golden bug, emerges as a tool for human genetics. *Nat Rev Genet* 2005;6:9–23.
5. Pavlidis P, Tanouye MA. Seizures and failures in the giant fiber pathway of *Drosophila* bang-sensitive paralytic mutants. *J Neurosci* 1995;15:5810–5819.
6. Kuebler D, Tanouye MA. Modifications of seizure susceptibility in *Drosophila*. *J Neurophysiol* 2000;83:998–1009.
7. Kuebler D, Zhang HG, Ren X, Tanouye MA. Genetic suppression of seizure susceptibility in *Drosophila*. *J Neurophysiol* 2001;86: 1211–1225.
8. Lee J, Wu C-F. Electroconvulsive seizure behavior in *Drosophila*: Analysis of the physiological repertoire underlying a stereotyped action pattern in bang-sensitive mutants. *J Neurosci* 2002;22: 11065–11079.
9. Song J, Tanouye MA. Seizure suppression by *shakB*^{2}, a gap junction mutation in Drosophila. *J Neurophysiol* 2006;95: 627–635.
10. Benzer S. From the gene to behavior. *J Am Med Assoc* 1971;218:1015–1022.
11. Ganetzky B, Wu C-F. Indirect suppression involving behavioral mutants with altered nerve excitability in *Drosophila melanogaster*. *Genetics* 1982;100:597–614.
12. Engel JE, Wu C-F. Altered mechanoreceptor response in *Drosophila* bang-sensitive mutants. *J Comp Physiol A* 1994;175: 267–278.
13. Pavlidis P, Ramaswami M, Tanouye MA. The *Drosophila easily shocked* gene: A mutation in a phospholipid synthetic pathway causes seizure, neuronal failure, and paralysis. *Cell* 1994;79:23–33.
14. Zhang HG, Tan J, Reynolds E, Kuebler D, Faulhaber S, Tanouye MA. The *Drosophila slamdance* gene: A mutation in an aminopeptidase can cause seizure, paralysis and neuronal failure. *Genetics* 2002;162:1283–1299.
15. Royden CS, Pirrotta V, Jan LY. The *tko* locus, site of a behavioral mutation in *D. melanogaster*, codes for a protein homologous to prokaryotic ribosomal protein S12. *Cell* 1987;51:165–173.
16. Glasscock E, Tanouye MA. Drosophila *couch potato* mutants exhibit complex neurological abnormalities including epilepsy phenotypes. *Genetics* 2005;169: 2137–2149.
17. Hekmat-Scafe DS, Lundy MY, Ranga R, Tanouye MA. Mutations in the K+/Cl- cotransporter gene *kazachoc (kcc)* increase seizure susceptibility in *Drosophila*. *J Neurosci* 2006;26:8943–8954.
18. Fergestad T, Bostwick B, Ganetzky B. Metabolic disruption in *Drosophila* bang-sensitive seizure mutants. *Genetics* 2006;173: 1357–1364.
19. Zhang YQ, Roote J, Brogna S, Davis AW, Barbash DA, Nash D, Ashburner M.Stress sensitive B encodes an adenine nucleotide translocase in *Drosophila melanogaster*. *Genetics* 1999;153:891–903.
20. Schubiger M, Feng Y, Fambrough DM, Palka J. A mutation of the Drosophila sodium pump α subunit results in bang-sensitive paralysis. *Neuron* 1994;12: 373–381.
21. Kuebler D, Tanouye MA. Anticonvulsant valproate reduces seizure-susceptibility in mutant *Drosophila*. *Brain Res* 2002;958: 36–42.
22. Reynolds ER, Stauffer EA, Feeney L, Rojahn E, Jacobs B, McKeever C. Treatment with the antiepileptic drugs phenytoin and gabapentin ameliorates seizure and paralysis of *Drosophila* bang-sensitive mutants. *J Neurobiol* 2003;58:503–513.
23. Tan JS, Lin F, Tanouye MA. Potassium bromide, an anticonvulsant, is effective at alleviating seizures in the *Drosophila* bang-sensitive mutant *bang senseless*. *Brain Res* 2004;1020:45–52.
24. Song J, Tanouye M. A role for *para* sodium channel gene 3' UTR in the modification of *Drosophila* seizure susceptibility. *Dev Neurobiol* 2007;67:1944–1956.
25. Song J, Hu J, Tanouye M. Seizure suppression by *top1* mutations in *Drosophila*. *J Neurosci* 2007;27:2927–2937.
26. Song J, Parker L, Hormozi L, Tanouye MA. DNA topoisomerase I inhibitors ameliorate seizure-like behaviors and paralysis in a Drosophila model of epilepsy. *Neuroscience* 2008;156:722–728.
27. Hartman PE, Roth JR. Mechanisms of suppression. *Adv Genet* 1973;17:1–105.
28. Jarvik J, Botstein D. Conditional-lethal mutations that suppress genetic defects in morphogenesis by altering structural proteins. *Proc Natl Acad Sci U S A* 1975;72: 2738–2742.
29. Guarente L. Synthetic enhancement in gene interaction: A genetic tool come of age. *TIG* 1993;9:362–366.
30. Baraban SC. Emerging epilepsy models: Insights from mice, flies, worms, and fish. *Curr Opin Neurol* 2007;20: 164–168.

31. Song J, Tanouye MA. From bench to drug: Human seizure modeling using *Drosophila*. *Prog Neurobiol* 2008;84: 182–191.
32. Hekmat-Scafe DS, Dang KN, Tanouye MA. Seizure suppression by gain-of-function *escargot* mutations. *Genetics* 2005;169: 1477–1493.
33. Glasscock E, Singhania A, Tanouye MA. The *mei-P26* gene encodes a RING finger B-box coiled-coil-NHL protein that regulates seizure susceptibility in *Drosophila*. *Genetics* 2005;170:1677–1689.
34. Ragsdale DS, Avoli M. Sodium channels as molecular targets for antiepileptic drugs. *Brain Res Brain Res Rev* 1998; 26:16–28.
35. Catterall WA. Molecular properties of brain sodium channels: An important target for anticonvulsant drugs. *Adv Neurol* 1999; 79:441–456.
36. Ramaswami M, Tanouye MA. Two sodium channel genes in *Drosophila*: Implications for channel diversity. *Proc Natl Acad Sci U S A* 1989;86:2079–2082.
37. Loughney K, Kreber R, Ganetzky B. Molecular analysis of the *para* locus, a sodium channel gene in *Drosophila*. *Cell* 1989;58: 1143–1154.
38. Kuroda MI, Kernan MJ, Kreber R, Ganetzky B, Baker BS. The maleless protein associates with the X chromosome to regulate dosage compensation in *Drosophila*. *Cell* 1991;66:935–947.
39. Kuo CC. A common anticonvulsant binding site for phenytoin, carbamazepine, and lamotrigine in neuronal sodium channels. *Mol Pharmacol* 1998;54: 712–721.
40. McNamara JO. Emerging insights into the genesis of epilepsy. *Nature* 1999;399 (Suppl):A15–A22.
41. McLean MJ, MacDonald RL. Multiple actions of phenytoin on mouse spinal cord neurons in cell culture. *J Pharmacol Exp Ther* 1983;227:779–789.
42. McLean MJ, Macdonald RL. Sodium valproate, but not ethosuximide, produces use- and voltage-dependent limitation of high frequency repetitive firing of action potentials of mouse central neurons in cell culture. *J Pharmacol Exp Ther* 1986;237: 1001–1011.
43. Nelson JC, Wyman RJ. Examination of paralysis in *Drosophila* temperature-sensitive paralytic mutations affecting sodium channels; a proposed mechanism of paralysis. *J Neurobiol* 1990;21: 453–469.
44. Carlen PL, Frances S, Zhang L., Naus C, Kushnir M, Velazquez JL. The role of gap junctions in seizures. *Brain Res Brain Res Rev* 2000;32:235–241.
45. Jahromi SS, Wentlandt K, Piran S, Carlen P. Anticonvulsant actions of gap junctional blockers in an *in vitro* seizure model. *J Neurophysiol* 2002;88: 1893–1902.
46. Gajda Z, Gyengesi E, Hermesz E, Ali KS, Szente M. Involvement of gap junctions in the manifestation and control of the duration of seizures in rats *in vivo*. *Epilepsia* 2003;44:1596–1600.
47. Krishnan SN, Frei E, Swain G, Wyman RJ. *Passover*, a gene required for synaptic connectivity in the giant fiber system of *Drosophila*. *Cell* 1993;73:967–977.
48. Phelan P, Nakagawa M, Wilkin MB, Moffat KG, O'Kane CJ, Davies JA, Bacon JP. Mutations in *shaking-B* prevent electrical synapse formation in the *Drosophila* giant fiber system. *J Neurosci* 1996;16: 1101–1113.
49. Thomas JB, Wyman RJ. Mutations altering synaptic connectivity between identified neurons in *Drosophila melanogaster*. *J Neurosci* 1984;4:530–538.
50. Szente M, Gajda Z, Said Ali K, Hermesz E. Involvement of electrical coupling in the *in vivo* ictal epileptiform activity induced by 4-aminopyridine in the neocortex. *Neuroscience* 2002; 115:1067–1078.
51. Lee MP, Brown SD, Chen A, Hsieh T-S. DNA topoisomerase I is essential in *Drosophila melanogaster*. *Proc Natl Acad Sci U S A* 1993;90:6656–6660.
52. Zhang CX, Chen AD, Gettel NJ, Hsieh T-S. Essential functions of DNA topoisomerase I in *Drosophila melanogaster*. *Dev Biol* 2000;222:27–40.
53. Champoux JJ. DNA topoisomerases: Structure, function and mechanism. *Annu Rev Biochem* 2001;70:369–413.
54. Boege F, Straub T, Kehr A, Boesenberg C, Christiansen K, Anderson A, Jacob F, Kohrle J. Selected novel flavones inhibit the DNA binding or the DNA religation step of eukaryotic topoisomerase I. *J Biol Chem* 1996;271: 2262–2270.
55. Pommier Y, Pourquier P, Fan Y, Strumberg D. Mechanism of action of eukaryotic DNA topoisomerase I and drugs targeted to the

enzyme. *Biochim Biophys Acta* 1998;1400: 83–105.

56. Li T, Liu L. Tumor cell death induced by topoisomerase-targeting drugs. *Annu Rev Pharmacol Toxicol* 2000;41:53–77.
57. Leppard JB, Champoux JJ. Human DNA topoisomerase I: Relaxation, roles, and damage control. *Chromosoma* 2005;114: 75–85.
58. Moots PL, Maciunas RJ, Eisert DR, Parker RA, Abou-Khalil B. The course of seizure disorders in patients with malignant gliomas. *Arch Neurol* 1995;52: 717–724.

Chapter 3

The Albino *Xenopus laevis* Tadpole as a Novel Model of Developmental Seizures

D. Sesath Hewapathirane and Kurt Haas

Abstract

Here we describe a novel model system based on the transparent albino *Xenopus laevis* tadpole which is particularly well suited for the study of seizures and their sequelae within the intact developing brain. This system allows in vivo imaging of neuronal circuit activity with single-cell resolution, as well as acute and long-term imaging of neuronal growth and synapse formation, within the intact unanesthetized brain. Mounting evidence supports a strong role for neuronal transmission in regulating major aspects of brain circuit formation, including synaptogenesis, synapse strengthening and elimination, as well as axonal and dendritic arbor growth. Given the high incidence of seizures during periods of early brain development in humans, such model systems are necessary to better understand how paroxysmal seizure activity may alter activity-dependent processes occurring during development and whether early-life seizures induce persistent aberrant alterations in neural circuitry.

Key words: epilepsy, neonatal seizures, *Xenopus laevis* tadpole, pentylenetetrazol, calcium imaging, two-photon microscopy, in vivo imaging, development.

1. Introduction

Seizures are extremely common during the first few years of life *(1)*. In the majority of individuals who develop epilepsy, the onset of spontaneous seizures often occurs during childhood *(2, 3)*. Prenatal fetuses, neonates, and infants are at a high risk of experiencing seizures provoked by pro-convulsant stimuli such as fever, hypoxia, drugs, or trauma *(4)*. While the high incidence of provoked early-life seizures is partly due to the increased likelihood of experiencing seizure-promoting brain insults, substantial experimental evidence demonstrates that the immature brain exhibits a reduced seizure threshold compared to the mature brain *(1, 4)*.

Scott C. Baraban (ed.), *Animal Models of Epilepsy: Methods and Innovations, vol. 40*

DOI 10.1007/978-1-60327-263-6_3 Springerprotocols.com

Unfortunately, the duration of heightened seizure susceptibility and seizure expression coincides with critical and sensitive periods of brain maturation. Early brain development involves a myriad of intricately coordinated, spatially and temporally defined programs culminating in the formation of complex neural circuits. Many of these fundamental processes underlying brain circuit formation, including axonal and dendritic arbor growth, synapse formation, and refinement, are regulated by neuronal transmission *(5–7)*. Whether common early-life seizures, characterized by abnormal and excessive neuronal activity, interfere with these normal developmental programs has been the subject of particular interest. The importance of investigating potentially deleterious effects of seizures on developmental processes in the immature brain is further underscored by mounting evidence for the developmental origins for numerous common neurological and psychiatric disorders.

To date, clinical studies examining whether early-life seizures are associated with long-term neurological dysfunction have yielded an incomplete picture. While it is recognized that children with epilepsy are at a heightened risk for developing neurological comorbidities later in life, such as behavior problems, cognitive deficits, and psychiatric disorders *(8–11)*, and that particular types of provoked seizures during development are associated with an increased risk for subsequent development of epilepsy *(4, 12)*, whether these pathological effects are *caused* by seizures has been difficult to determine *(13, 14)*. Associating direct effects of seizures on later outcomes is hampered by the inherent variability of patient populations, including differences in genetic backgrounds, types and frequency of seizures, age of seizure onset, medications, and environmental factors *(15)*. Indeed, an alternative explanation of these observations is that both seizures and neurological comorbidities share a common pathological origin. Such issues are difficult to resolve solely through human studies and are amenable for study using appropriate experimental animal models. Animal models offer an increased degree of control over variables confounding clinical studies, facilitating longitudinal prospective investigations, and providing an ever-growing arsenal of sophisticated and sensitive measures of seizure effects on neuronal anatomy and function.

To more directly address the question of whether early-life seizures lead to long-term neurological abnormalities or an increased risk for the development of spontaneous seizures, a large number of studies have been conducted, mostly utilizing juvenile rodents. Seizures induced in immature rodents are associated with (i) permanent changes in behavior, (ii) deficits in learning and memory, and (iii) an increased susceptibility to seizures in adulthood *(16, 17)*. These findings suggest that early-life seizures can permanently alter neural function resulting in neurological deficits. Several important questions arise from these observations. How do seizures induce these alterations in brain function? How are developing neural circuits altered by seizure activity? Are neurons at specific maturational states more prone to seizure-induced effects? Can protective strategies be developed to counter seizure-induced neural dysfunction? Although these questions have been the subject of intensive study, definitive answers have remained elusive. While numerous studies demonstrate a variety of abnormal structural and functional changes induced by seizures in immature rodents *(18, 19)*, it has been difficult to establish a *causal* link between seizures and morphological or functional alterations. Furthermore, the precise molecular mechanisms regulating the observed seizure-induced alterations remain unclear. These difficulties are, in part, a reflection of the limitations of existing model systems most commonly used in epilepsy research.

An attractive alternative approach is the development of non-mammalian in vivo model systems to examine the effects of seizures on brain development and function. Such model organisms enjoy widespread use in other areas of neurodevelopmental research due to the many advantages they confer and the large number of specialized experimental techniques designed for use in these animal species. Only recently have these advantages been exploited for the field of epilepsy research with the emergence of several new experimental models based on organisms such as zebrafish (*Danio rerio*), the fruit fly (*Drosophila melanogaster*), and the nematode *Caenorhabditis elegans* (discussed elsewhere in this volume and reviewed in *20*). We recently developed a novel model of developmental seizures based on the albino *Xenopus laevis* tadpole, an organism widely used in studies of early brain development *(21)*. This model organism is ideally suited for studies of seizure propagation, the acute and long-term effects of early-life seizures on neuronal growth and circuit formation, as well as the molecular mechanisms involved in these processes *(22)*.

1.1. The Albino X. laevis Tadpole – An Experimental Model of Early Brain Development

The African clawed frog, *X. laevis*, has served as an important model system for studies of vertebrate ontogeny for over a century. Indeed, the basis for much of our understanding of the early stages of central nervous system development comes from studies conducted in *X. laevis*, including seminal work conducted in the larval frog examining retinal ganglion cell axonal pathfinding, which provided critical evidence for the chemoaffinity model of axonal guidance *(23)*. The continued use of this organism as a convenient model system of brain development is due to evidence demonstrating that neuronal physiology and function, neuropharmacology, intra- and intercellular molecular signaling pathways, cellular growth and differentiation, and neural circuit development are all remarkably consistent with what is known of mammalian counterparts. *Xenopus* tadpoles have proven ideal particularly for the study of early development of vertebrate central neural circuits *(24–26)*. These studies have mainly focused on the development of the retinotectal system, in which the retinal ganglion cells of the eye project to optic tectum to create a functional circuit capable of processing visual stimuli. The optic tectum is a useful structure for study given its relatively large size, ordered cell body layers, proliferative zone which continuously adds new neurons to the tectum, a completely isolated afferent projection from the eye, and accessible dorsal position directly under the skin. Furthermore, tadpoles of naturally occurring albino *X. laevis* are completely transparent, providing a clear window allowing direct in vivo imaging of the intact developing brain. Thus, the albino tadpole offers a remarkably accessible preparation for the study of early development of brain neural circuits, which are currently not available in mammalian systems.

The use of *X. laevis* tadpoles also confers numerous practical advantages over mammalian systems, in terms of rearing and husbandry as well as cost-effectiveness. Adult frogs can be obtained from a number of commercial sources and can be induced to produce fertilized eggs throughout the year. Typically, a single mating pair will yield several hundred tadpoles, and adult frogs remain reproductively active for many years. After fertilization, embryos rapidly progress through early developmental stages without parental care and develop into freely swimming tadpoles in less than 1 week *(27)*.

1.2. Characterizing the Albino X. laevis *Tadpole as a New Model of Developmental Seizures*

Previous studies demonstrated the ability to induce seizures in the adult frog using electrical or chemical stimuli *(28–30)*. However, seizure induction in tadpoles had not previously been reported. Our initial experiments were designed to characterize chemoconvulsant-induced seizures in tadpoles to determine appropriateness as a new model of developmental seizures.

2. Techniques to Study Seizure Activity in Tadpoles

2.1. Behavioral Seizures in X. laevis *Tadpoles*

One advantage of the *X. laevis* tadpole is the relatively high bioavailability of a variety of centrally acting drugs following application to the normal bathing medium. As such, potential chemoconvulsant drugs can be applied to the bath of freely swimming tadpoles that can then be visually monitored for evidence of convulsive behavior. Developing tadpoles first demonstrate continuous free-swimming behavior approximately 3 days post-fertilization. The majority of our seizure studies are carried out on stage 47 tadpoles *(27)*, approximately 5 days post-fertilization, since they demonstrate reproducible, easy to classify motor behaviors, while possessing brain circuits undergoing dynamic neuronal growth. Six chemoconvulsants with a range of mechanisms of action were tested for ability to induce convulsive behavior, including pentylenetetrazol (PTZ), picrotoxin, bicuculline, pilocarpine, 4-aminopyiridine (4-AP), and kainic acid. Within specific concentration ranges, each chemoconvulsant reliably induced similar patterns of abnormal behavior, which were classified into five progressive stages corresponding to increasing severity (**Fig. 3.1A**). In control conditions, tadpole behavior is characterized by slow swimming interspersed with brief pauses, with tadpoles continuously remaining in an upright orientation. At low chemoconvulsant doses, abnormal alterations in normal motor patterns were observed, including rapid "darting" swimming, excessive immobility, and loss of postural control with tadpoles

a)

Seizure stage	Seizure-associated behavior
class I	intermittent rapid swimming, swimming in tight circles, sporadic rapid darting swimming behavior
class II	behavioral arrest/immobility
class III	loss of postural control — on side or upside down
class IV	repetitive side-to-side lateral movements of the head without propulsion
class V	C-shaped contraction — rapid, bilaterally alternating repetitive contractions of the axial musculature

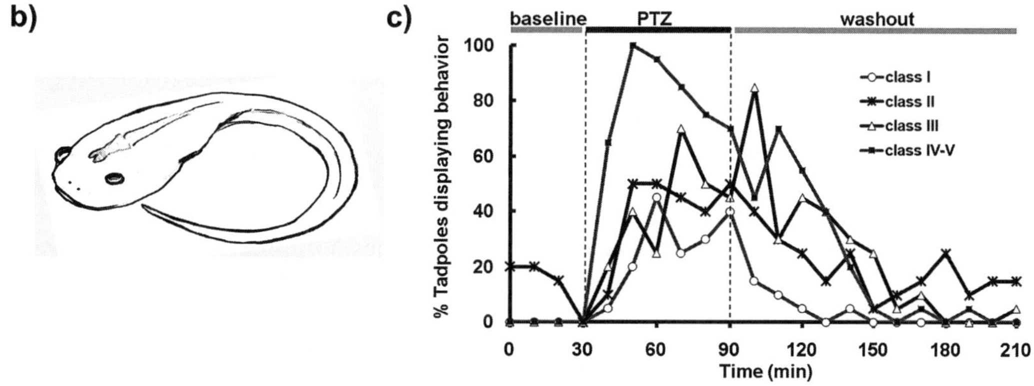

Fig. 3.1. **Behavioral seizures in *X. leavis* tadpoles. (A)** Chemoconvulsant-induced behavioral seizures categorized into five progressive classes. **(B)** C-shaped contraction – class V seizure. **(C)** Time course of seizure behavior in response to PTZ, 15 mM.

turning on their side or floating upside down. Higher doses elicited more severe convulsive behaviors with tonic and clonic components. These higher class behaviors involved rapid, repetitive, bilaterally alternating movements including head bobbing and pronounced unilateral contractions of the axial musculature resulting in "C-shaped" postures (**Fig. 3.1B**). C-shaped contractions (Class V behavioral seizures) were the maximal seizure-associated convulsive behaviors observed. The observed abnormal behaviors demonstrated clear dose-dependence, both in terms of the severity and latency to onset of the specific behaviors. Once initiated, chemoconvulsant-induced behaviors were observed for the entire duration of exposure to the drug. Abnormal behaviors gradually dissipated upon drug washout, accomplished by placing tadpoles in fresh rearing medium, with complete recovery observed within 1 hour. High drug doses were also marked by significant levels of lethality. Based on our results, chemoconvulsant doses consistently eliciting Class V behaviors within 20 minutes with no, or minimal, lethality are PTZ, 10–50 mM; bicuculline, 1–5 mM; picrotoxin, 1 mM; kainic acid, 0.25–1 mM; pilocarpine, 75 mM; and 4-AP, 0.5–2.5 mM. The observation that similar patterns of abnormal convulsive behavior were elicited by chemoconvulsant drugs with different mechanisms of action strongly suggests that these behaviors are associated with seizures, rather than direct effects of the drugs on motor patterns. In the tadpole, PTZ exhibited the widest range of doses which consistently elicited seizures in the absence of lethality or overt toxicity. PTZ, a non-competitive γ-aminobutyric acid (GABA) receptor antagonist, is a widely used convulsant in many animal seizure preparations *(31, 32)*. A concentration of 15 mM PTZ reliably elicits Class V seizure behaviors with a seizure-onset latency of approximately 10 minutes, demonstrating no lethality even after extended exposures to drug (>8 hours). A detailed analysis of the time course of specific seizure-related behaviors upon application of PTZ demonstrates the rapid onset and consistent expression of Class I–V behaviors, which gradually dissipate upon drug washout (**Fig. 3.1C**). PTZ-induced behavioral seizures could be suppressed by a variety of anticonvulsant drugs, including diazepam, phenytoin, and valproic acid.

2.2. Electrographic Recordings of Seizure Activity

Electrographic seizure activity can be recorded from tadpole brains using extracellular field electrodes. One advantage of this preparation is the ability to immobilize tadpoles without the use of anesthetic agents. Tadpoles can be paralyzed by bath application of a reversible paralytic agent (pancuronium dibromide, 10 μM) and embedded in agarose (1%, prepared in normal rearing medium). Immobilization in this manner eliminates the confounding effects of anesthetic agents on neural transmission and seizure expression *(33, 34)*. The immobilized tadpoles are subsequently immersed in oxygenated rearing medium and are able to survive for over 24 hours. Agarose is porous, allowing ready diffusion drugs applied to the bath, and following experiments tadpoles can be released into the normal bathing medium where normal behavior resumes over the course of 30–60 minutes. Extracellular field recordings from the tadpole brain following PTZ exposure demonstrate characteristic seizure-like electrographic discharges (**Fig. 3.2A**). Electrographic seizure activity characterized by robust high-amplitude spikes gradually appears 20–30 minutes after application of PTZ. Following drug washout, seizure activity gradually dissipates, reaching baseline levels over the course of 30–60 minutes. Application of valproate (5 mM) during PTZ-induced seizures readily suppresses electrographic discharges (**Fig. 3.2B**). Similar results were induced by the chemoconvulsants kainic acid, pilocarpine, or 4-AP *(21)*.

2.3. In Vivo Imaging of the Circuit Properties of Seizure Activity Using Calcium-Sensitive Dyes

Cell-permeable calcium-sensitive fluorescent indicators offer a noninvasive method to monitor seizure activity throughout the intact brain with single-cell resolution. Following loading of neurons with calcium-sensitive dyes, fluorescence microscopy is used to detect changes in dye fluorescence which correspond to changes in cytoplasmic calcium concentration. Measures of somatic calcium spikes have been shown to correlate to neuronal discharge, with spike amplitudes directly correlating to the number of action potentials fired *(35–37)*. Here we demonstrate detection of seizure activity in the unanesthetized tadpole brain using Oregon Green BAPTA-1 AM and in vivo two-photon time-lapse microscopy (**Fig.3.3**). As with electrophysiological recording, unanesthetized tadpoles are paralyzed and embedded in a thin layer of agarose within a perfused imaging chamber. Imaging captures an X–Y optical section through the tadpole brain, allowing simultaneous

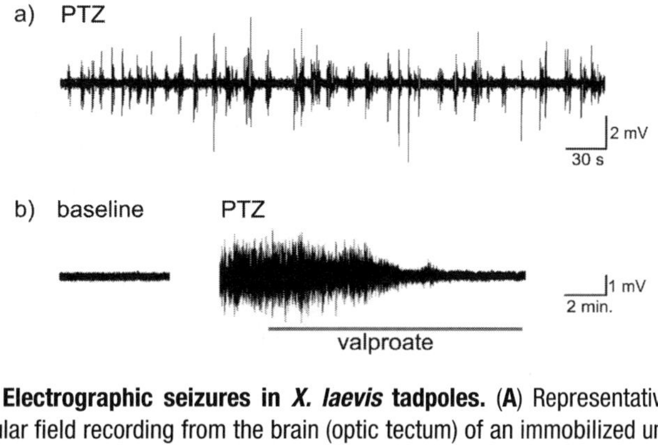

Fig. 3.2. **Electrographic seizures in *X. laevis* tadpoles.** (**A**) Representative in vivo extracellular field recording from the brain (optic tectum) of an immobilized unanesthetized tadpole showing epileptiform discharges induced by application of PTZ, 15 mM. (**B**) Bath application of the anti-epileptic drug valproate (5 mM) suppresses PTZ-induced seizures.

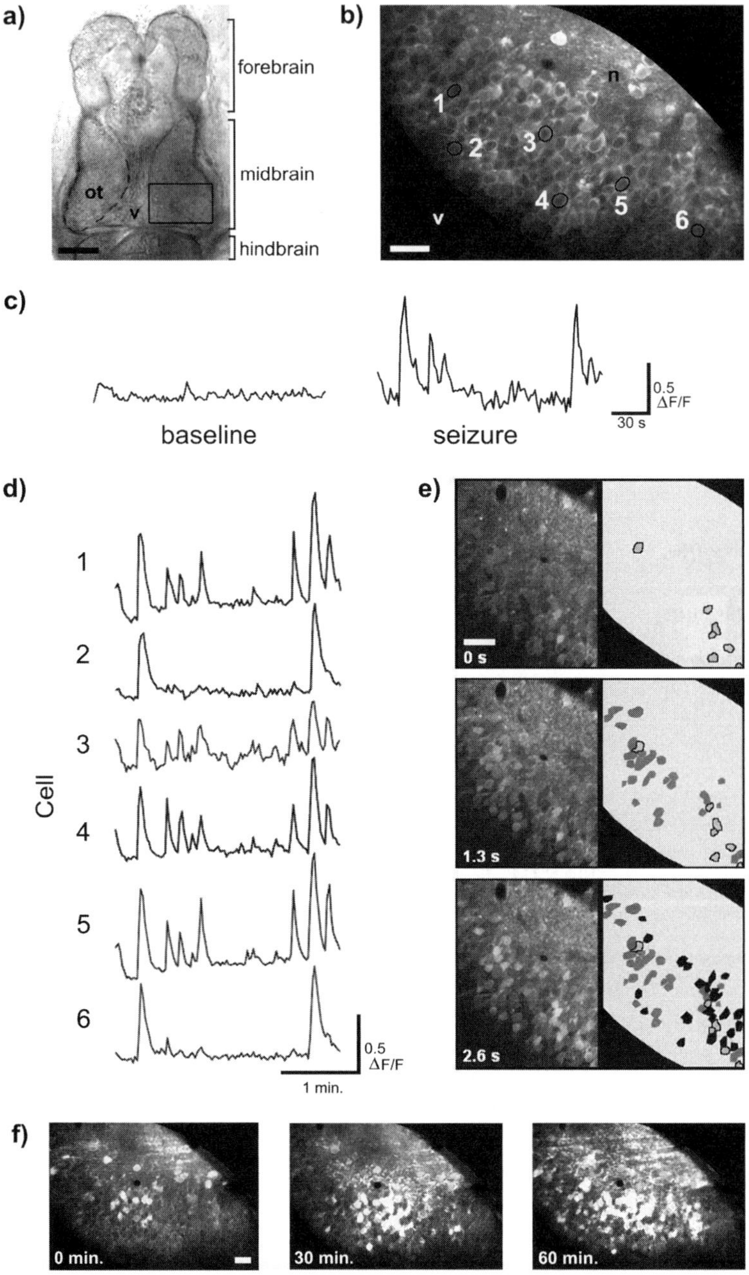

Fig. 3.3. **In vivo imaging of seizure activity in *X. laevis* tadpoles.** (**A**) Dorsal view of the intact albino tadpole brain using bright field microscopy (rostral end toward top of image). *ot* – optic tectum; *v* – ventricle. Rectangle delineates region of optic tectum imaged for changes in intracellular calcium levels. Scale bar = 200 μm. (**B**) In vivo two-photon fluorescence microscopy image of the tadpole optic tectum loaded with the calcium-sensitive fluorescent indicator Oregon Green BAPTA. *n* – neuropil; *v* – ventricle. (**C**) Representative example of quantified change in fluorescence ($\Delta F/F$) of calcium-sensitive indicator within a single cell during baseline period and 60 minutes after application of PTZ. (**D**) Calcium responses of six cells identified in (**B**) demonstrating synchronized high-amplitude spikes during seizure. (**E**) Time-lapse images demonstrating wave-like progression of calcium activity progressing from the rostral-to-caudal tectum. Cells first spiking at $t = 0$ s outlined in black, at $t = 1.3$ s shaded gray, and at $t = 2.6$ s shaded black. (**F**) Number of simultaneously spiking cells increases with duration of PTZ exposure, suggesting progressive recruitment into synchronized waves of activity during seizures. Scale bar in (**B**), (**E**), and (**F**) = 20 μm.

monitoring of neuronal activity in hundreds of neurons with single-cell resolution (**Fig. 3.3B**). During baseline recordings, intermittent small-amplitude fluorescence fluctuations were observed within individual cells, without discernible pattern. Upon application of PTZ, however, large somatic fluctuations in calcium dye fluorescence were observed from individual cells throughout the tectum (**Fig. 3.3C**). Analysis of selected cells distributed across the entire extent of the tectum revealed that these high-amplitude spikes are synchronous (**Fig. 3.3D**) and propagate from the rostral-to-caudal tectum in large repetitive waves of calcium activity (**Fig. 3.3E**). As the seizure progressed over time, the number of neurons recruited into these large propagating waves of calcium activity increased (**Fig.3.3F**). Thus, the powerful combination of calcium-sensitive dyes and in vivo time-lapse two-photon microscopy can be effectively employed for direct imaging of seizure propagation through the developing brain and serves as a useful tool to probe alterations in neural circuitry following seizures.

3. In Vivo Imaging of Seizure-Induced Effects on Neuronal Development

As discussed earlier, functionally significant effects of early-life seizures and the mechanisms mediating these aberrant alterations of normal brain development remain to be determined. We have developed techniques based on the transparent tadpole to specifically address these issues. One of these important experimental tools is single-cell electroporation (SCE) , a method first tested and described in the tadpole brain for restricted delivery of dyes, DNA, peptides, and other macromolecules to individual cells within intact tissues *(38, 39)*. SCE can be used to fluorescently label individual neurons within the tadpole brain with dyes or, when transfecting cells with plasmid DNA encoding protein fluorophores, fluorescent proteins (**Fig. 3.4A–C** and Color Plate 1, middle of book). Once fluorescently labeled, individual neurons can be imaged using in vivo two-photon microscopy to achieve high-resolution three-dimensional images for accurate morphometric measurements of neuronal growth *(5, 40)*. Repeated time-lapse imaging of the same neuron at interval ranging from seconds to days allows capture and measurement of both dynamic and protracted periods of neuronal growth (**Fig. 3.4D, E**). A particularly useful feature of the *Xenopus* tadpole model is that throughout tadpole development, new neurons are added to the brain at a well-defined proliferative zone in the caudal–medial tectum. Thus, SCE can be used to target this region to label immature neurons for subsequent in vivo imaging of their growth (**Fig. 3.4B**). These methods have proven effective for imaging neuronal morphology in the developing brain to characterize growth behavior and identify molecular mechanisms of developmental plasticity.

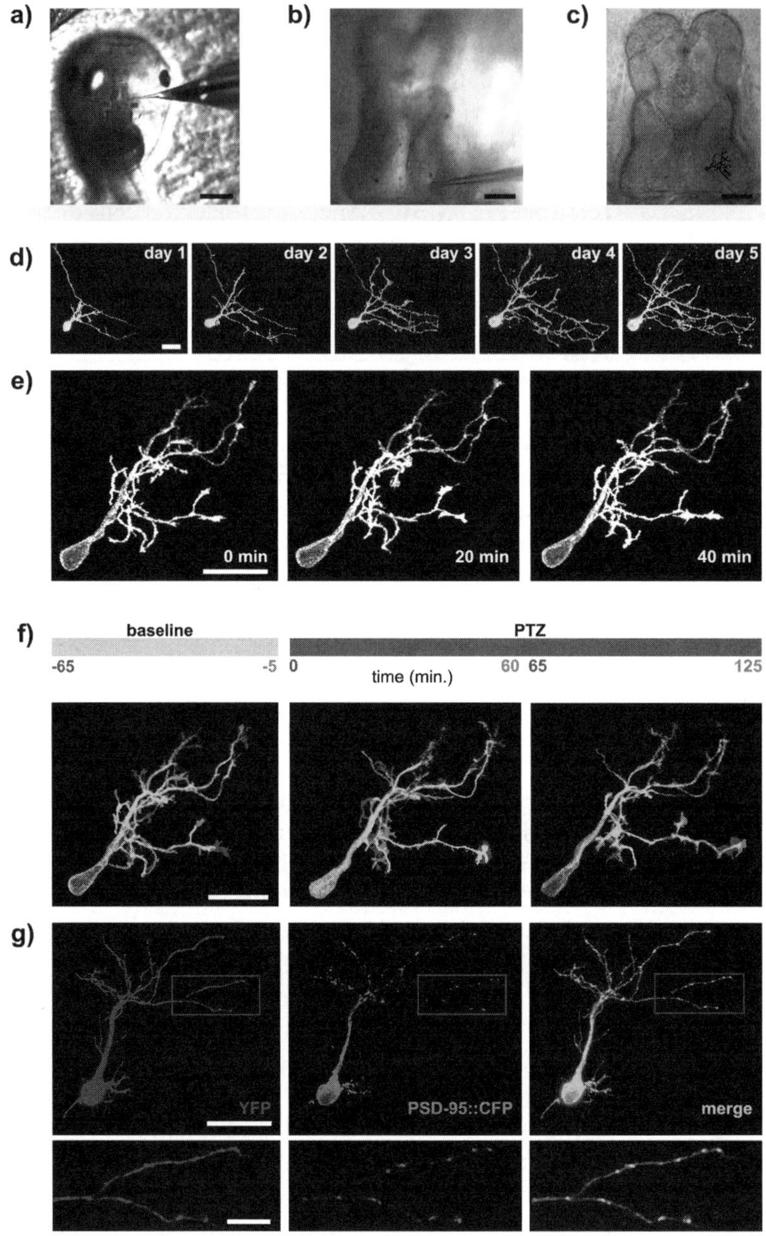

Fig. 3.4. **In vivo imaging of dendritic arbor growth and synapse formation in *X. laevis* tadpoles.** (**A**) Single-cell electroporation allowing targeted neuronal labeling and transfection in vivo. Scale bar = 2 mm. (**B**) Higher magnification image of (**A**) showing insertion of single-cell electroporation micropipette into tadpole brain. (**C**) Schematic representation of a single neuron within the tadpole brain labeled with AlexaFluor 488 dextran 3000. Scale bar in (**B**) and (**C**) = 200 μm. (**D**) Time lapse in vivo two-photon microscopy imaging of neuronal growth using 24 hour intervals. (**E**) Rapid time-lapse imaging in immobilized unanesthetized tadpoles allows examination of dynamic dendritic growth processes characterized by dramatic structural rearrangements over short time scales. (**F**) Rapid time-lapse imaging in vivo used to examine PTZ seizure-induced effects on dynamic dendritic growth processes within the intact unanesthetized brain. Images are color coded to highlight dendritic segments that are added (green) and retracted (red) during each period. (**G**) Co-electroporation of two plasmid vectors encoding different proteins to concurrently visualize excitatory synapses (PSD-95-CFP) and cellular morphology (YFP) of an individual neuron in vivo. Boxed region showed at higher magnification in lower panels. Scale bars (**D**–**G**) = 20 μm; lower panel of (**G**) = 5 μm. (*see* Color Plate 1, middle of book)

This system can be applied to the study of developmental seizures by directly imaging neuronal growth during seizures using in vivo two-photon time-lapse imaging in the immobilized unanesthetized tadpole (**Fig. 3.4F**). Furthermore, the long-term consequences of seizures can be measured by imaging the same neurons over hours or days following controlled seizure induction. Another important application of SCE is the co-expression of multiple proteins by transfecting with multiple independent plasmids. Here we demonstrate the expression of a space-filling protein fluorophore along with a fluorescently tagged synaptic protein (**Fig. 3.4G**). Imaging the formation, maintenance, and elimination of synapses during seizures will provide needed insight as to how early-life seizures alter the morphology and synapse formation of growing neurons. Additionally, the ability to conduct single-cell genetic manipulations using SCE is a powerful tool that will facilitate the examination of molecular mechanisms underlying observed seizure-induced effects.

4. Conclusion

The high degree of similarity between the fundamental neurobiology of *Xenopus* with higher vertebrates together with distinct experimental and practical advantages have established the albino *X. laevis* tadpole as an important in vivo model of early brain development. This organism has been widely used in studies of activity-dependent neuronal growth, synaptogenesis, neuronal plasticity, and circuit formation – issues of particular importance with respect to epilepsy research. Use of this model in epilepsy research holds much promise to address long-standing issues related to the effects of seizures on brain development and, by complementing and contributing to work conducted in mammalian model systems, may prove useful in the development of novel therapeutic strategies to suppress seizures and their adverse effects on brain function.

References

1. Ben-Ari Y, Holmes GL. Effects of seizures on developmental processes in the immature brain. *Lancet Neurol* 2006;5:1055–63.
2. Cowan LD. The epidemiology of the epilepsies in children. *Ment Retard Dev Disabil Res Rev* 2002;8:171–81.
3. Hauser WA. Epidemiology of epilepsy in children. *Neurosurg Clin N Am* 1995;6: 419–29.
4. Jensen FE, Baram TZ. Developmental seizures induced by common early-life insults: Short- and long-term effects on seizure susceptibility. *Ment Retard Dev Disabil Res Rev* 2000;6:253–7.
5. Sin WC, Haas K, Ruthazer ES, Cline HT. Dendrite growth increased by visual activity requires NMDA receptor and Rho GTPases. *Nature* 2002;419:475–80.

6. Ruthazer ES, Akerman CJ, Cline HT. Control of axon branch dynamics by correlated activity in vivo. *Science* 2003;301:66–70.
7. Cohen-Cory S. The developing synapse: Construction and modulation of synaptic structures and circuits. *Science* 2002;298: 770–6.
8. Cornaggia CM, Beghi M, Provenzi M, Beghi E. Correlation between cognition and behavior in epilepsy. *Epilepsia* 2006;47(Suppl) 2:34–9.
9. Gaitatzis A, Trimble MR, Sander JW. The psychiatric comorbidity of epilepsy. *Acta Neurol Scand* 2004;110:207–20.
10. Vestergaard M, Pedersen CB, Christensen J, Madsen KM, Olsen J, Mortensen PB. Febrile seizures and risk of schizophrenia. *Schizophr Res* 2005;73:343–9.
11. Vestergaard M, Pedersen CB, Sidenius P, Olsen J, Christensen J. The long-term risk of epilepsy after febrile seizures in susceptible subgroups. *Am J Epidemiol* 2007;165: 911–18.
12. Annegers JF, Hauser WA, Shirts SB, Kurland LT. Factors prognostic of unprovoked seizures after febrile convulsions. *N Engl J Med*1987;316:493–8.
13. Shinnar S, Hauser WA. Do occasional brief seizures cause detectable clinical consequences? *Prog Brain Res* 2002;135:221–35.
14. Swinkels WA, Kuyk J, van Dyck R, Spinhoven P. Psychiatric comorbidity in epilepsy. *Epilepsy Behav* 2005;7:37–50.
15. Tsopelas ND, Saintfort R, Fricchione GL. The relationship of psychiatric illnesses and seizures. *Curr Psychiatry Rep* 2001;3: 235–42.
16. Holmes GL, Ben-Ari Y. The neurobiology and consequences of epilepsy in the developing brain. *Pediatr Res* 2001;49:320–5.
17. Lynch M, Sayin U, Bownds J, Janumpalli S, Sutula T. Long-term consequences of early postnatal seizures on hippocampal learning and plasticity. *Eur J Neurosci* 2000; 12:2252–64.
18. Holmes GL. Effects of seizures on brain development: Lessons from the laboratory. *Pediatr Neurol* 2005;33:1–11.
19. Holopainen IE. Seizures in the developing brain: Cellular and molecular mechanisms of neuronal damage, neurogenesis and cellular reorganization. *Neurochem Int* 2008;52:935–47.
20. Baraban SC. Emerging epilepsy models: Insights from mice, flies, worms and fish. *Curr Opin Neurol* 2007;20:164–8.
21. Hewapathirane DS, Dunfield D, Yen W, Chen S, Haas K. In vivo imaging of seizure activity in a novel developmental seizure model. *Exp Neurol* 2008;211(2):480–8.
22. Holmes GL. Seizure-induced brain damage: From tadpoles to children. *Exp Neurol* 2008;213(1):7–9.
23. Meyer RL. Roger Sperry and his chemoaffinity hypothesis. *Neuropsychologia* 1998;36: 957–80.
24. Aizenman CD, Cline HT. Enhanced visual activity in vivo forms nascent synapses in the developing retinotectal projection. *J Neurophysiol* 2007;97:2949–57.
25. Tao HW, Zhang LI, Engert F, Poo M. Emergence of input specificity of ltp during development of retinotectal connections in vivo. *Neuron* 2001;31:569–80.
26. Vislay-Meltzer RL, Kampff AR, Engert F. Spatiotemporal specificity of neuronal activity directs the modification of receptive fields in the developing retinotectal system. *Neuron* 2006;50:101–14.
27. Nieuwkoop PD, Faber J. *Normal table of Xenopus laevis.* Amsterdam: Elsevier North Holland Publishing Company; 1994.
28. Bennett JM. Modification of strychnine-induced convulsions by anticonvulsants in the frog. *Neuropharmacology* 1972;11:297–9.
29. Servi Z, Strejckova A. Epileptogenic focus in the frog telencephalon. Seizure irradiation from the focus. *Physiol Bohemoslov* 1967;16:522–30.
30. Morrell F, Tsuru N. Kindling in the frog: Development of spontaneous epileptiform activity. *Electroencephalogr Clin Neurophysiol* 1976;40:1–11.
31. Ahmad A, Dhawan BN. Metrazol test for rapid screening of anticonvulsants. *Jpn J Pharmacol* 1969;19:472–4.
32. Velisek L, Veliskova J, Moshe SL. Developmental seizure models. *Ital J Neurol Sci* 1995;16:127–33.
33. Gallagher TJ, Galindo A, Richey ET. Inhibition of seizure activity during enflurance anesthesia. *Anesth Analg* 1978;57:130–2.
34. Ishizawa Y. Mechanisms of anesthetic actions and the brain. *J Anesth* 2007;21:187–99.
35. Smetters D, Majewska A, Yuste R. Detecting action potentials in neuronal populations with calcium imaging. *Methods* 1999;18: 215–21.
36. Ramdya P, Reiter B, Engert F. Reverse correlation of rapid calcium signals in the zebrafish optic tectum in vivo. *J Neurosci Methods* 2006;157:230–7.

37. Badea T, Goldberg J, Mao B, Yuste R. Calcium imaging of epileptiform events with single-cell resolution. *J Neurobiol* 2001;48:215–27.

38. Haas K, Jensen K, Sin WC, Foa L, Cline HT. Targeted electroporation in Xenopus tadpoles in vivo–from single cells to the entire brain. *Differentiation* 2002;70:148–54.

39. Haas K, Sin WC, Javaherian A, Li Z, Cline HT. Single-cell electroporation for gene transfer in vivo. *Neuron* 2001;29:583–91.

40. Haas K, Li J, Cline HT. AMPA receptors regulate experience-dependent dendritic arbor growth in vivo. *Proc Natl Acad Sci U S A* 2006;103:12127–31.

Chapter 4

Zebrafish as a Simple Vertebrate Organism for Epilepsy Research

Scott C. Baraban

Abstract

For many years, scientists have developed animal models of epilepsy to study specific aspects of the human condition. Rodents are the species of choice in the vast majority of these studies. In a departure from these rodent-centric models, we here describe zebrafish (*Danio rerio*), a genetically tractable vertebrate that is particularly well suited to epilepsy research. Zebrafish do not possess the complex central nervous system we have come to expect in other animal models and will never be mistaken for miniature versions of the complex mammalian brain. Nonetheless, the nervous system of this "simple" vertebrate is comprised of individual elements (glutamatergic excitatory neurons and GABAergic inhibitory neurons, for example) that we routinely study as critical to the generation of abnormal electrical discharge in all higher species. In electrophysiological and behavioral studies, we have shown that seizures elicited by acute convulsant treatment (i.e., pentylenetetrazole, PTZ) or knockdown of zebrafish homologs for human genetic epilepsy disorders (i.e., tuberous sclerosis complex, TSC) reliably evoke seizures in developing zebrafish. Using tectal or forebrain recordings, we also show that this epileptiform activity is sensitive to well-established antiepileptic drugs (many initially identified in rodents). Overall, our studies highlight two important aspects of this model: (i) electrical features reminiscent of mammalian seizure syndromes can be reproduced in zebrafish and (ii) zebrafish may be useful in an antiepileptic drug-screening strategy. Given the untapped potential for high-throughput drug discovery, analysis of genetic forms of epilepsy, or large-scale mutagenesis screens to identify novel seizure-modifying genes, zebrafish could provide an essential tool for understanding (and treating) epilepsy.

Key words: *Danio rerio*, pentylenetetrazole, immature, developing, electrophysiology, morpholino, TSC.

1. Introduction

Understanding seizure phenomena in the immature brain requires careful consideration of neurodevelopment. Traditionally, this analysis has focused on the biophysical

Scott C. Baraban (ed.), *Animal Models of Epilepsy: Methods and Innovations, vol. 40*

DOI 10.1007/978-1-60327-263-6_4 Springerprotocols.com

properties of immature neurons, the function and distribution of their synapses, and the unique developmentally regulated expression patterns of ion channel and receptor subunits. From these observations it is now well established that basic nervous system components necessary to generate seizure activity are present relatively early in development. For example, abnormal repetitive electrical discharge patterns have been observed in neonates and infants *(1–3)*. Although seizures in the newborn human exhibit a number of characteristics that are unlike those seen in adults, their presence suggests that a fully mature and complex nervous system is not strictly required for seizure genesis. Taken one step further, electrophysiological studies of immature mammalian cerebral tissue explants in culture *(4, 5)* demonstrate a clear capacity to generate long-duration, rhythmic, and highly synchronized electrical activity that mimics clinical electrographic seizure patterns.

If neurodevelopment is not yet complete, and immature forms of excitatory and inhibitory neurotransmission prevail, how is it possible to generate abnormal electrical activity in the immature brain? For the most part, ontogenetic studies relevant to this question rely exclusively on immature rodents. In the rat brain, expression of the NMDA glutamate receptor subunit peaks late in the first postnatal week and AMPA subunit density somewhere around postnatal day 10 *(6)*. Metabotropic glutamate receptor activity is more robust in the neonatal hippocampus *(7)*, and δ-aminobutyric acid (GABA), the primary inhibitory neurotransmitter in the adult nervous system, is depolarizing up to the second postnatal week. Each of these features favors a hyperexcitable state and creates a foundation for the generation of abnormal synchronized electrical discharge in the immature rodent. Indeed, seizure thresholds to chemoconvulsant agents or electrical kindling stimuli are much lower in young mice or rats *(8–10)*. While much remains to be explored about seizure genesis in developing rodents, the present review strays from this traditional approach and explores underlying mechanisms of seizure generation in developing zebrafish. By studying the function and organization of a simple immature nervous system capable of generating abnormal electrical discharge, we seek to provide insight into the more complex human condition. This is not to infer that additional immature rodent studies are of little value. On the contrary, applying what we know from the vast rodent literature, appropriately designed experiments in a genetically tractable organism such as zebrafish could reveal unifying concepts of how seizures are generated (and may be stopped) in the immature nervous system.

2. Zebrafish as a Model Organism

Zebrafish (*Danio rerio*), a simple vertebrate species, have emerged as a popular organism in developmental biology research. From a logistical viewpoint, this popularity is linked to the ease of maintaining large colonies of fish, relatively short generation times, and that a single spawning can yield hundreds of experimentally useful offspring *(11)*. Considerable genomic resources already exist for zebrafish, including high-density genetic maps and a cadre of mutants generated by exposure to ethylnitrosourea (ENU) or retroviral insertion. A further appeal is that zebrafish eggs are fertilized externally allowing early embryonic developmental stages to be accessible for study and observation; optical transparency of zebrafish larvae at these early time points is an additional technical advantage. Owing to this transparency, early embryos can easily be injected with DNA, RNA, protein, or morpholino oligonucleotides (MO; **Fig. 4.2B**). The latter have recently emerged as a powerful antisense tool to selectively knockdown gene function in the developing zebrafish *(12)*.

2.1. Organization of the Developing Zebrafish Brain

Primary neurogenic (*Notch*, *delta*) and basic helix-loop-helix (*neurogenin, achaete scute, Dlx*) genes necessary to generate a highly organized central nervous system (CNS) in mammals have also been shown to guide brain development in zebrafish. Early larval stages (0–24 hr post-fertilization) of forebrain and hindbrain development, especially development of retinal projections to the diencephalon and optic tecta (a typically dominant structure in teleosts), incorporate mechanisms similar to those found in higher vertebrates, are well studied, and reviewed elsewhere *(13–15)*. By 5 days post-fertilization (dpf), all the major components of the zebrafish brain, e.g., telencephalon, diencephalon, mesencephalon, and rhombencephalon, are established. Of particular relevance to seizure studies described here, the zebrafish telencephalon (forebrain) can be histologically divided into (i) an area dorsalis telencephali where most cells lie remote from the ventricular surface and occasionally form distinct laminae and (ii) a basal area ventralis telencephali primarily organized into nuclei that lie close to the ventricle *(16)*. Not unlike their counterparts in mammals, these forebrain structures are populated by neuronal subtypes e.g., glutamatergic, GABAergic, glycinergic, catecholaminergic, and serotonergic *(15, 17–19)*. Associated receptors, transporters, and synaptic vesicle proteins can also be found in the developing zebrafish brain *(20–22)*. Adding to the basic anatomical data provided by in situ hybridization and antibody staining studies, zebrafish mutants affecting the formation of specific neuronal subtypes (i.e., m865 for dopaminergic cells) *(23)* offer novel insights into how the zebrafish CNS is organized. From a structural perspective, transgenic zebrafish expressing green or yellow fluorescent proteins in subpopulations of neurons *(24, 25)* and advanced in vivo confocal imaging techniques offer a tremendous opportunity to further define brain structure in the developing zebrafish. As these finer details emerge, it is likely that the overall organization and neurochemical content of the zebrafish nervous system will continue to correspond with like structures accepted as important for neural function (and epilepsy) in higher vertebrates.

3. Seizures in Developing Zebrafish

This next section briefly summarizes an acute seizure model in immature zebrafish *(26)*. Prior to the establishment of this model, relatively little effort had been made to model epilepsy in simple organisms *(27–29)* and whether seizures could be observed in zebrafish was unknown. In all experiments described, seizure activity was elicited using a common convulsant agent, pentylenetetrazole (PTZ), and wild-type zebrafish from the Tübingen line; subsequent studies have determined that other common convulsants (e.g., pilocarpine, bicuculline, or 4-aminopyridine) work equally well. Behavioral studies were performed on freely swimming larvae between 6 and 7 dpf; electrophysiology experiments were conducted on agar-immobilized fish.

First, zebrafish larvae were placed individually in one well of a 96-well Falcon plate. In normal bathing medium, fish were observed to move infrequently and in small dart-like motions. Normal bathing medium was then removed and replaced with medium containing 15 mM PTZ. Pharmacokinetic data on drug distribution in immature zebrafish are not available and concentrations were chosen empirically; fish do not develop scales until several weeks of age and compounds are thought to be readily absorbed across the skin or gastrointestinal tract *(30, 31)*. In stepwise progression, PTZ-exposed zebrafish undergo a period of increased swim activity (Stage I), rapid whirlpool-like swimming (Stage II), and finally, brief violent clonus-like convulsions accompanied by a short post-ictal loss of posture (Stage III). Locomotion tracking software (Noldus Technology Inc.) was used to reliably quantify these seizure behaviors. PTZ-induced seizures are attenuated or abolished by co-application of *(26)* or overnight pre-incubation with *(31)* antiepileptic drugs, e.g., valproic acid, diazepam, lamotrigine, or topiramate.

How is it possible to generate seizures in immature zebrafish? The simple answer is that the zebrafish brain (despite notable differences in size and laminar pattern) and the mammalian brain share many basic structures and functional capacities. Excitatory neurotransmission in the zebrafish brain is mediated by three ionotropic glutamate receptor subtypes *(20)*, and GABAergic interneurons exert inhibitory actions via activation of postsynaptic, bicuculline-sensitive, $GABA_A$ receptors *(32, 33)*. These basic excitatory and synaptic components appear to underlie "second-order motion perception" *(34)*, long-term potentiation *(35)*, and odor-evoked neural activity *(36, 37)*, suggesting that higher-level CNS processing capacities are present in zebrafish.

Next, electrophysiology experiments were performed to demonstrate that PTZ exposure elicits abnormal electrical discharge in the central nervous system of immature zebrafish. Field recording electrodes placed in the forebrain, midbrain (at the level of the optic tectum), or hindbrain of agar-immobilized zebrafish larvae (3–8 dpf) were used to detect electrical signals originating in these structures. Fish were exposed to PTZ at concentrations between 5 and 15 mM. Early in development (3 dpf), PTZ elicited events that were extremely large in amplitude and duration. Events are essentially "all-or-none" and consisted of prolonged, clustered bursts of complex population spikes, ranging from 4 to 10 seconds. This burst-suppression pattern resembles electrographic patterns typically observed in pediatric epilepsy syndromes *(38, 39)*. Later in development (5–8 dpf) these long-duration "ictal-like" events were accompanied by brief monophasic deflections occurring at rates up to 30 per minute. Under these conditions, high-frequency "interictal-like" events culminated in a large multiphasic "ictal-like" discharge followed by a short period of quiescence. Such electrical events are never seen under control recording conditions and were designated as "epileptiform". Representative recordings of PTZ-evoked discharge in wild-type fish can be found elsewhere *(26, 40)* and are remarkably similar to those reported for PTZ-exposed rodents *(41–43)*. In simultaneous field recordings, epileptiform discharge in the forebrain precedes that observed in the mid- or hindbrain. Epileptiform events were abolished by co-application of kynurenate (2.5 mM) or CNQX/APV (20 μM, 50 μM), consistent with a requirement for postsynaptic glutamate receptors.

3.1. Available Methods for Monitoring Seizures in Zebrafish

Although the millimeter scale size of immature zebrafish (5–7 mm in length) prohibits simultaneous video and electroencephalographic monitoring, a number of sophisticated electrophysiological and imaging approaches are feasible. Like their counterparts in rodents, these techniques offer a clear strategy to analyze the single-cell and network activities underlying seizure initiation and propagation in the zebrafish CNS.

Extracellular field recordings, discussed above, can be used to detect the summed potential changes of multiple neurons in the vicinity of a recording electrode. These are commonly used to study network activity underlying the generation of burst discharge in vitro *(44–46)*. As expected, these field potential measurements have relatively low spatial resolution and are best used to detect population bursts or the effect of pharmacological manipulations on these bursts. In some instances the electrode tip can be placed in close proximity to a neuron and effectively record single-unit action potentials at the soma. However, precise cellular measurement of membrane currents underlying burst generation requires intracellular recording. Again, similar to rodent in vitro and in vivo preparations, these recordings can be obtained in developing zebrafish using sharp intracellular or whole-cell patch electrode strategies. In many cases, single-cell recordings have been limited to the large mechanosensory Rohon–Beard neurons identified along the spinal cord in semi-intact zebrafish larvae preparations *(47–49)*. These seminal studies, spanning current-clamp evaluation of neuronal firing activity and voltage-clamp analysis of potassium currents, suggest that application of single-cell recording strategies to neurons located in zebrafish forebrain networks could offer the same level of experimental investigation commonly employed in rodent epilepsy research. If one considers combining single-cell recording techniques with genetic manipulation (transgenic fish expressing a human epilepsy gene mutation, for example) a wide range of in vivo experiments designed to evaluate genes (and signaling pathways) involved in acute or chronic seizure generation emerge.

Neural imaging techniques, primarily those that rely on optical detection of changes in the fluorescence of calcium-sensitive dyes, offer an opportunity to examine synchronous firing patterns in epileptic zebrafish. As discussed elsewhere in this volume (see Trevelyan and Yuste, Chapter 12), optical imaging has an important spatial advantage for detecting populations of neurons that are active throughout an entire neural network during seizures. The advent of two-photon microscopy is a particular advancement in that neurons of interest that are located deep (>100 μm) within CNS tissue can be evaluated. Importantly, calcium indicators can be loaded into zebrafish using a variety of different techniques and the "health" of dye-loaded cells in the developing (and intact) zebrafish brain is superior to similar slice-loading methodologies. An easy method of loading large numbers of CNS neurons is to inject a dextran-coupled calcium indicator into early blastomeres using procedures similar to injection of DNA or RNA *(50, 51)*. During subsequent cell divisions the dye is passively distributed to cells and the high molecular weight dextran prevents dye extrusion, thus trapping the indicator. In practice, this approach may result in uneven distribution and dye concentration over time and is best confined to the first 3 or 4 dpf. A further refinement, taking full advantage of the genetic tractability of this species, is to generate transgenic fish expressing a calcium-sensitive derivative of green fluorescent protein (cameleon) in neurons *(52)*. Cameleon is a hybrid protein in which cyan and yellow fluorescent proteins are linked by calmodulin and an M13 calmodulin-binding domain. Through a conformational change this protein can serve as a ratiometric calcium indicator for in vivo imaging studies *(53)*. Using these types of approaches, functional neural networks activated in response to odorant or light stimulation have been mapped in the immature zebrafish brain *(54, 55)*. The application of this technology to seizure initiation and propagation in intact zebrafish could yield significant insights.

3.2. Evaluation of Antiepileptic Drugs in Developing Zebrafish

Preclinical screening of compounds with anti-convulsant or anti-epileptogenic properties has primarily been accomplished using rodents in which seizures are induced by chemical or electrical methods. The two most common, maximal electroshock (MES) or pentylenetetrazole (Metrazol), tests *(56, 57)* have successfully identified several classes of antiepileptic drugs. Both models screen adult and otherwise normal rodents and are designed to mimic seizures observed clinically. Though recent developments in mouse genetics now provide a variety of single-gene mutants homologous to gene defects found in rare forms of human epilepsy (e.g., *KCNA1*, *TSC*, *Nav1*) these mice are not commonly used in antiepileptic drug (AED) screening protocols. Moreover, an additional limitation of rodent models is that high-throughput testing is prohibitively expensive and time consuming. A pilocarpine-induced seizure model in Drosophila larvae was recently proposed as an alternative *(58)*, but as a non-vertebrate species may not identify compounds translating into clinical use in humans. In contrast, zebrafish, easily scaled up to generate thousands of off-spring each year, offer a simple vertebrate species with tremendous potential for preclinical AED screening. Compounds can be added to the bathing medium in which zebrafish larvae develop allowing for rapid assessment of toxicity. Moreover, with clear behavioral and electrophysiological manifestations of seizure activity, anti-convulsant (in locomotion tracking assays) and antiepileptic (in extracellular field recording assays) approaches can be rapidly adapted for large-scale compound screening. Based strictly on behavioral locomotion assays, Berghmans and coworkers recently screened 14 standard antiepileptic drugs *(31)*. These studies, though limited in scope, hint at the possibilities of using zebrafish for AED discovery. As a "first pass" this behavioral strategy may be effective in identifying compounds with the ability to suppress seizure-like behavior. However, as sedative effects of some compounds may also arrest or preclude PTZ behaviors sole reliance on this approach is open to false-positive compound identification. For example, ethosuximide showed a significant decrease in locomotor activity in a PTZ behavioral assay *(31)*, but failed to inhibit PTZ-induced seizure discharge in tectal recordings *(26)*.

Nonetheless, suppression of PTZ-evoked seizure activity in immature zebrafish could serve as a rapid method to screen large numbers of candidate compounds. This type of program would not necessarily compete with established rodent AED testing, but could serve as a first step in narrowing thousands of potential compounds (rapidly tested in immature zebrafish) to a more manageable hundred or less compounds (to be thoroughly tested in rodents), with ultimate identification of one or two drug candidates that could move directly to clinical trials (**Fig. 4.1**). To demonstrate this approach we screened 20 potential compounds in a series of behavioral and electrophysiology assays. Compounds were chosen to include AEDs and commercially available channel openers such as pinacidil *(59)* and cromakalim *(60)*. In behavioral testing, wild-type zebrafish (7 dpf) were placed in a single well and bathed in 15 mM PTZ. Locomotion tracking software was used to quantify seizure behavior and after a stable level of activity was obtained (~15 min) bathing medium was replaced with one containing PTZ plus a test compound. If tracking indicated a reduction in Stage II or III activity the compound was designated as a potential "positive" and moved on to the electrophysiology assay. Here PTZ was bath applied and brain electrical activity was monitored in agar-immobilized zebrafish (7 dpf). Again, once a stable level of

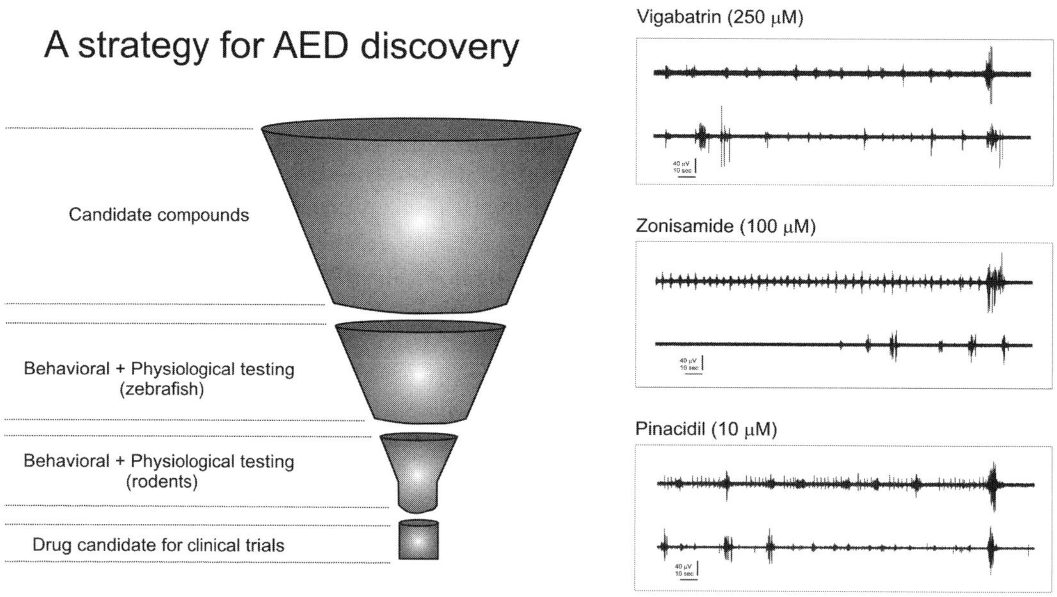

Fig. 4.1. **AED screening strategy in zebrafish.** Schematic showing the steps that could be taken to incorporate zebrafish as a high-throughput "first-pass" filter, in a larger effort to identify new antiepileptic drugs. At each step in this process the number of potential compounds will be reduced, ultimately resulting in a relatively small number of compounds available for clinical trials. At right are sample electrophysiological recordings from PTZ-exposed zebrafish (top traces in each example) showing the development of a stable level of electrographic seizure activity and the effect of putative AEDs on this activity. Note that zonisamide reduced burst discharge activity, but vigabatrin and pinacidil had little effect in this model.

"ictal- and "interictal-like" bursting was established in wild-type zebrafish, the test compound was added. Near-complete suppression of electrical activity was seen with common AEDs such as valproate and diazepam *(26)*; for some drugs reduced activity, primarily small-amplitude bursts, was noted but long-duration low-frequency events persisted (**Fig. 4.1**, *zonisamide*) and the remaining compounds had little effect on PTZ-evoked epileptiform activity (**Fig. 4.1**, *vigabatrin* and *pinacidil*). These experiments in zebrafish are fairly simple and with several recording setups could easily be used to test a range of drug concentrations in a single day. Whether this information could directly translate to human drug trials or, more likely, serve as a pre-filter for compounds entering into further rodent trials remains to be fully examined.

4. A Zebrafish Model of Tuberous Sclerosis Complex

Although valuable as acute seizure models or as system for rapid drug screening, these applications fail to exploit the full potential of zebrafish in epilepsy research. Namely, can zebrafish be used to model (and study) specific clinical epileptic conditions? To address this issue we began with investigation of a human genetic mutation known to cause seizures in nearly 90% of patients, e.g., tuberous sclerosis complex (TSC). TSC is an autosomal dominant multi-organ disorder caused by mutations in *TSC1* or *TSC2* *(61, 62)*. Epileptic seizures are the most common neurological problem in these patients and tend to be both progressive and medically intractable in children. *TSC1* and *TSC2* gene products, hamartin and tuberin respectively, form a heteroduplex that regulates many cellular functions including cell growth *(63)*, proliferation *(64)*, and size *(65)*. Neuropathological features of TSC include cortical tubers, subependymal nodules, and subependymal giant cell astrocytomas *(66)*. Although it is commonly believed that these CNS abnormalities underlie cognitive deficit and seizures associated with TSC *(67, 68)*, the mechanism(s) necessary for generation of abnormal electrical discharge remain unknown. Rodent models are the most common approach to address this question and are reviewed elsewhere in this volume (see Ess Chapter 5). Taking a reductionist approach, we recently initiated a project to study the relationship between seizures, brain architecture, and loss of TSC function in immature zebrafish. This strategy allowed us to examine a fundamental question in the TSC field – namely, are cortical tubers necessary for seizure generation. The identification of *TSC1* and *TSC2* homologs in *Drosophila melanogaster* (fruit flies) resulted in an exciting series of experiments elucidating a role for the insulin signaling pathway in control of cell growth *(69)*. Using

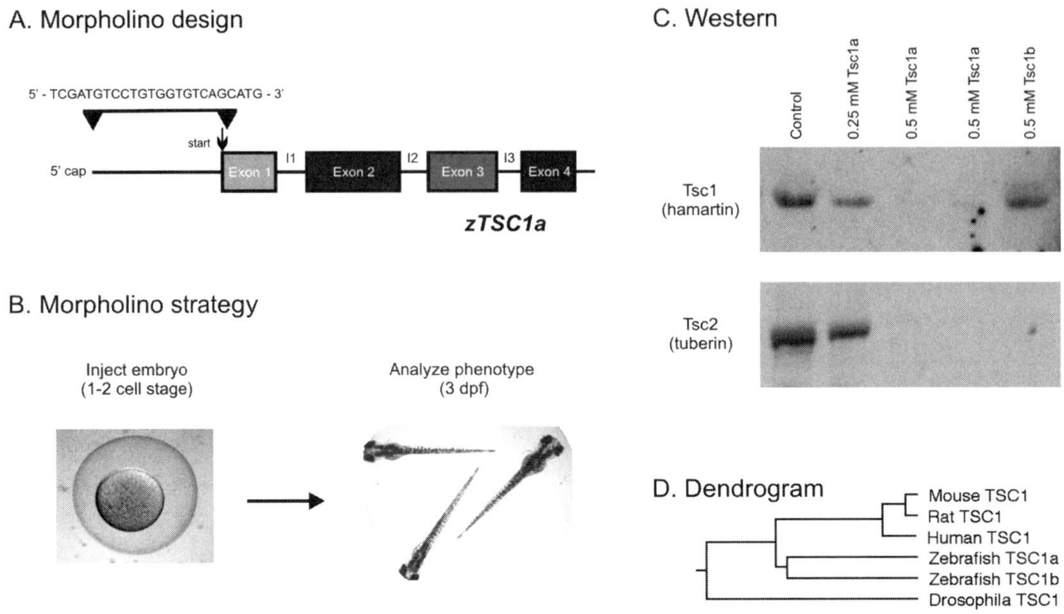

Fig. 4.2. **Morpholino antisense strategy. (A)** Cartoon showing the zebrafish TSC1a gene (*zTSC1a*) and sequence design for a "start site" morpholino. **(B)** Zebrafish embryo at the 1 cell stage ready for MO injection; zebrafish larvae at 3 days post-fertilization. **(C)** Western blot analysis of TSC morphants. Control (0.5 mM), *tsc1a* (0.25 mM and 0.5 mM), and *tsc2* (0.5 mM) morpholinos were injected into embryos at the 1–2 cell stage and raised to 3 dpf ($n = 30$ for each group). The equivalent of one larvae per lane was probed with antibodies against hamartin (TSC1) and tuberin (TSC2). Note the effectiveness of *tsc1a* morpholinos at knocking down both hamartin and tuberin levels, whereas *tsc2* knocks down only tuberin. **(D)** Dendrogram showing homology between human, rat, mouse, zebrafish, and drosophila TSC1 sequences.

a similar database mining strategy, we identified several *TSC* genes in zebrafish designated *tsc1a*, *tsc1b*, and *tsc2* (**Fig. 4.2D**). RT-PCR analysis of whole zebrafish larvae indicate that mRNAs for these genes are inherited maternally and expressed at 3 dpf, e.g., a developmental time point at which we can observe seizure discharge in zebrafish *(40)*. Next, antisense morpholino oligonucleotides were designed to disrupt *TSC* gene function in zebrafish. Morpholinos are short oligonucleotides consisting of a nucleic acid base, a morpholine ring, and a non-ionic phosphorodiamidate inter-subunit linkage that block protein translation by targeting the 5′ untranslated region of messenger RNA *(12)* (**Fig. 4.2A**). MO strategies have been successfully used to generate zebrafish models of juvenile cystic kidney disease *(70)*, Alzheimer's disease *(71)*, and Huntington's disease *(72)*.

Morpholinos directed against *tsc1a*, *tsc1b*, and *tsc2* were injected into zebrafish embryos at the 1–2 cell stage and resultant larvae (or morphants) were raised to 3 dpf (**Fig. 4.2B**). Western blot analysis confirmed that 0.25 mM *tsc1a* MO reduced hamartin protein expression by roughly one-half, and 0.5 mM *tsc1a* MO abolished hamartin expression in morphants. Zebrafish *TSC1* morpholinos also knocked down tuberin levels (**Fig. 4.2C**), suggesting that in the absence of hamartin, tuberin is unstable and may be targeted for degradation. Injection of 0.5 mM *tsc2* MO knocked down tuberin levels, but had no effect on hamartin (**Fig. 4.2C**). In contrast, *tsc1b* and scrambled MO control injections had no effects in any morphant tested. We also examined the phosphorylation state of ribosomal protein S6 (not shown), as previous studies suggest that loss of *TSC1* or *TSC2* leads to increased S6 kinase activity *(73, 74)*, but did not detect elevated levels in *tsc1a* or *tsc1b* morphants. Taken together, these results suggest that morpholino injections offer an efficient system to knockdown TSC protein expression in developing zebrafish.

In TSC, clinical studies often suggest that cortical tubers are the primary source of epileptogenesis. When the offending cortical tuber can be identified and surgically removed, TSC patients report a seizure-free rate between 60 and 75% *(75–77)*. Despite these encouraging findings, recent studies combining high-resolution neuroimaging with electroencephalographic data suggest that adjacent perituberal, and largely non-dysplastic, tissue may be a source of seizures *(78–80)*. Data from two different conditional TSC knockout mice, neither harboring cortical tubers, further suggest that tubers may not be necessary for seizure generation *(81, 82)*. We recently analyzed brain tissue from an epileptic TSC patient and noted a significant level of synaptic hyperexcitability consistent with a pro-epileptogenic condition in tuber-free tissue samples *(83)*. Here, initial histological examinations were performed on *tsc1a* morphants (3 dpf) to determine whether tubers form. Immunostaining with the neuronal marker HuC *(84)* showed no obvious differences between control-injected and age-matched *tsc1* morphants (**Fig. 4.3A,B** and Color Plate 2, middle of book). Toluidine blue-stained semi-thin plastic sections covering the zebrafish forebrain (**Fig. 4.3C,D**), midbrain (**Fig. 4.3E,F**), and hindbrain structures (**Fig. 4.3G,J**) also failed to identify tubers, cytomegalic cells, or gross dysplasia. In the absence of tubers, we next performed electrophysiological studies to determine the physiological consequences of TSC gene knockdown in developing zebrafish. During continuous recording of midbrain activity from control-injected zebrafish larvae (3 dpf) in normal bathing medium, brief small-amplitude burst-like discharges were infrequently observed (**Fig. 4.3 M,O,P**). In contrast, large-amplitude epileptiform discharges were consistently noted in *tsc1a* MO zebrafish under the same recording conditions (**Fig. 4.3 N,O,P**). The duration of these complex field events was nearly three times that observed for spontaneous discharge in controls (**Fig. 4.3Q**). A direct implication of these studies is that reduced hamartin–tuberin expression alters neuronal or synaptic function in a manner favoring excessive excitation and the generation of spontaneous epileptiform activities. Further, these data demonstrate that a hyperexcitable condition develops in the absence of detectable tuber formation.

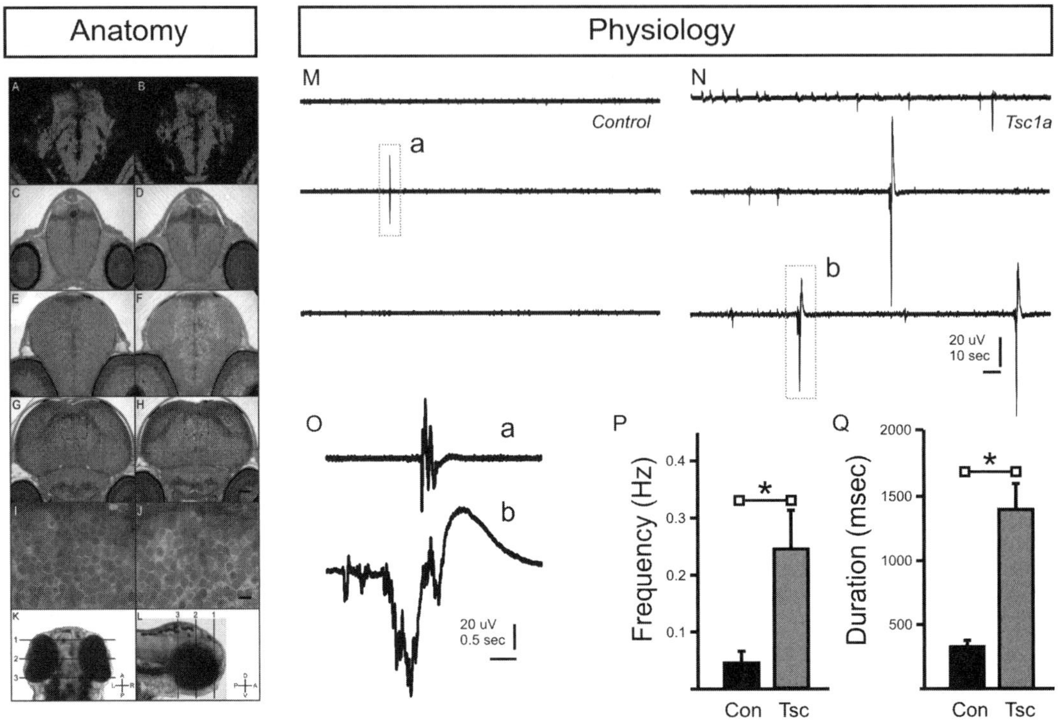

Fig. 4.3. **Tsc1 morphant zebrafish.** (*Anatomy*) Histological analysis of control and *tsc1a* morphants. Immunostaining with the neuronal marker HuC/HuD of control-injected (**A**) and *tsc1a* morphant (**B**) larvae, and semi-thin plastic sections of control-injected (**C,E,G,I**) and *tsc1a* morphant (**D,F,H,J**) larvae are shown. No obvious differences were observed between control-injected ($n = 6$) and *tsc1a* morphants ($n = 6$). Schematic representation of the location of sections [1 is shown in (**C,D**); 2 is shown in (**A,B,E,F**); 3 is shown in (**G,H,I,J**)] from a dorsal (**K**) and a lateral (**L**) perspective (A, anterior; P, posterior; L, left; R, right; D, dorsal; V, ventral). Scale bar shown in (**H**) is 50 μm in (**A–H**) and scale bar in (**J**) is 10 μm in (**I** and **J**). (*Physiology*) Gap-free recordings from control (**M**) or Tsc1a morphant (**N**) zebrafish. One brief burst-like event was observed (a, shown at high resolution in (**O**)). In contrast, *tsc1a* morphants consistently displayed complex burst-like discharge in normal bathing medium; high-resolution discharge shown in b (**O**). (**P**) Bar graph plotting the frequency of burst-like discharges during gap-free recording in control and *tsc1a* morphants. (**Q**) Same for duration. Data are presented as mean ± SEM; $^{*}p < 0.001$ using a one-way ANOVA. (*see* Color Plate 2, middle of book)

5. Conclusion

Zebrafish, like many "simple" organisms, are often overlooked in a clinically driven field such as epilepsy. Nonetheless, our studies, and complimentary research in worms (Chapter 1), flies (Chapter 2) and tadpoles (Chapter 3), clearly demonstrate seizure-like phenomena in these organisms. Here we reviewed behavioral, molecular, and electrographic correlates of seizure activity in developing zebrafish. The cellular and synaptic mechanisms underlying these events are likely similar to those responsible for seizures in mammals, and our studies also suggest innovative drug screening and genetic strategies using zebrafish that could contribute greatly to our understanding (and treatment) of epilepsy. In concluding this review, we ask not what limitations the zebrafish (or any animal model) offers, but what fundamental discoveries does it promise?

Acknowledgments

This work was supported in part by grants from the NIH (RO1 NS-053479-02) and the DOD Congressionally Directed Medical Research Program (TS050003). The author would like to thank Michael Taylor for performing morpholino injections and initial anatomical characterization of Tsc morphant zebrafish. The author would also like to thank Matthew Dinday for maintenance of the zebrafish aquarium and participation in the AED drug screening studies.

References

1. Scher MS, Hamid MY, Steppe DA, Beggarly ME, Painter MJ. Ictal and interictal electrographic seizure durations in preterm and term neonates. *Epilepsia* 1993;34(2):284–8.
2. Scher MS, Aso K, Beggarly ME, Hamid MY, Steppe DA, Painter MJ. Electrographic seizures in preterm and full-term neonates: clinical correlates, associated brain lesions, and risk for neurologic sequelae. *Pediatrics* 1993;91(1):128–34.
3. Kramer U, Nevo Y, Neufeld MY, Fatal A, Leitner Y, Harel S. Epidemiology of epilepsy in childhood: A cohort of 440 consecutive patients. *Pediatr Neurol* 1998;18(1):46–50.
4. Purpura DP, Housepian EM. Morphological and physiological properties of chronically isolated immature neocortex. *Exp Neurol* 1961;4:377–401.
5. Crain SM. Development of "organotypic" bioelectric activities in central nervous tissues during maturation in culture. *Int Rev Neurobiol* 1966;9:1–43.
6. Insel TR, Miller LP, Gelhard RE. The ontogeny of excitatory amino acid receptors in rat forebrain – I. N-methyl-D-aspartate and quisqualate receptors. *Neuroscience* 1990;35(1):31–43.
7. Casabona G, Genazzani AA, Di Stefano M, Sortino MA, Nicoletti F. Developmental changes in the modulation of cyclic AMP formation by the metabotropic glutamate receptor agonist 1S,3R-aminocyclopentane-1, 3-dicarboxylic acid in brain slices. *J Neurochem* 1992;59(3):1161–3.
8. Moshe SL, Albala BJ. Maturational changes in postictal refractoriness and seizure susceptibility in developing rats. *Ann Neurol* 1983;13(5):552–7.
9. Albala BJ, Moshe SL, Okada R. Kainic-acid-induced seizures: A developmental study. *Brain Res* 1984;315(1):139–48.
10. Baram TZ, Hirsch E, Schultz L. Short-interval amygdala kindling in neonatal rats. *Brain Res Dev Brain Res* 1993;73(1):79–83.
11. Detrich HW, 3rd, Westerfield M, Zon LI. Overview of the zebrafish system. *Methods Cell Biol* 1999;59:3–10.
12. Nasevicius A, Ekker SC. Effective targeted gene 'knockdown' in zebrafish. *Nat Genet* 2000;26(2):216–20.
13. Appel B. Zebrafish neural induction and patterning. *Dev Dyn* 2000;219(2):155–68.
14. Moens CB, Prince VE. Constructing the hindbrain: Insights from the zebrafish. *Dev Dyn* 2002;224(1):1–17.
15. Wullimann MF, Rink E. The teleostean forebrain: A comparative and developmental view based on early proliferation, Pax6 activity and catecholaminergic organization. *Brain Res Bull* 2002;57(3–4):363–70.
16. Wullimann MF, Mueller T. Teleostean and mammalian forebrains contrasted: Evidence from genes to behavior. *J Comp Neurol* 2004;475(2):143–62.

17. Ma P. Catecholaminergic systems in the zebrafish. IV. Organization and projection pattern of dopaminergic neurons in the diencephalon. *J Comp Neurol* 2003;460(1):13–37.
18. Kaslin J, Panula P. Comparative anatomy of the histaminergic and other aminergic systems in zebrafish (Danio rerio). *J Comp Neurol* 2001;440(4):342–77.
19. Higashijima S, Mandel G, Fetcho JR. Distribution of prospective glutamatergic, glycinergic, and GABAergic neurons in embryonic and larval zebrafish. *J Comp Neurol* 2004;480(1):1–18.
20. Edwards JG, Michel WC. Pharmacological characterization of ionotropic glutamate receptors in the zebrafish olfactory bulb. *Neuroscience* 2003;122(4):1037–47.
21. Kim YJ, Nam RH, Yoo YM, Lee CJ. Identification and functional evidence of GABAergic neurons in parts of the brain of adult zebrafish (Danio rerio). *Neurosci Lett* 2004;355(1–2):29–32.
22. Smear MC, Tao HW, Staub W, Orger MB, Gosse NJ, Liu Y, Takahashi K, Poo MM, Baier H. Vesicular glutamate transport at a central synapse limits the acuity of visual perception in zebrafish. *Neuron* 2007;53(1):65–77.
23. Ettl AK, Holzschuh J, Driever W. The zebrafish mutation m865 affects formation of dopaminergic neurons and neuronal survival, and maps to a genetic interval containing the sepiapterin reductase locus. *Anat Embryol (Berl)* 2006Dec;211(l) (1):73–86.
24. Scott EK, Mason L, Arrenberg AB, Ziv L, Gosse NJ, Xiao T, Chi NC, Asakawa K, Kawakami K, Baier H. Targeting neural circuitry in zebrafish using GAL4 enhancer trapping. *Nat Methods* 2007;4(4):323–6.
25. Mione M, Baldessari D, Deflorian G, Nappo G, Santoriello C. How neuronal migration contributes to the morphogenesis of the CNS: Insights from the zebrafish. *Dev Neurosci* 2008;30(1–3):65–81.
26. Baraban SC, Taylor MR, Castro PA, Baier H. Pentylenetetrazole induced changes in zebrafish behavior, neural activity and c-fos expression. *Neuroscience* 2005;131(3):759–68.
27. Kuebler D, Zhang H, Ren X, Tanouye MA. Genetic suppression of seizure susceptibility in Drosophila. *J Neurophysiol* 2001;86(3):1211–25.
28. Pavlidis P, Tanouye MA. Seizures and failures in the giant fiber pathway of Drosophila bang-sensitive paralytic mutants. *J Neurosci* 1995;15(8):5810–19.
29. Williams SN, Locke CJ, Braden AL, Caldwell KA, Caldwell GA. Epileptic-like convulsions associated with LIS-1 in the cytoskeletal control of neurotransmitter signaling in Caenorhabditis elegans. *Hum Mol Genet* 2004;13(18):2043–59.
30. Langheinrich U. Zebrafish: A new model on the pharmaceutical catwalk. *Bioessays* 2003;25(9):904–12.
31. Berghmans S, Hunt J, Roach A, Goldsmith P. Zebrafish offer the potential for a primary screen to identify a wide variety of potential anticonvulsants. *Epilepsy Res* 2007;75(1):18–28.
32. Sajovic P, Levinthal C. Inhibitory mechanism in zebrafish optic tectum: Visual response properties of tectal cells altered by picrotoxin and bicuculline. *Brain Res* 1983;271(2):227–40.
33. Kim CH, Ueshima E, Muraoka O, et al. Zebrafish elav/HuC homologue as a very early neuronal marker. *Neurosci Lett* 1996;216(2):109–12.
34. Orger MB, Smear MC, Anstis SM, Baier H. Perception of Fourier and non-Fourier motion by larval zebrafish. *Nat Neurosci* 2000;3(11):1128–33.
35. Nam RH, Kim W, Lee CJ. NMDA receptor-dependent long-term potentiation in the telencephalon of the zebrafish. *Neurosci Lett* 2004;370(2–3):248–51.
36. Edwards JG, Michel WC. Odor-stimulated glutamatergic neurotransmission in the zebrafish olfactory bulb. *J Comp Neurol* 2002;454(3):294–309.
37. Mack-Bucher JA, Li J, Friedrich RW. Early functional development of interneurons in the zebrafish olfactory bulb. *Eur J Neurosci* 2007;25(2):460–70.
38. Yamatogi Y, Ohtahara S. Early-infantile epileptic encephalopathy with suppression-bursts, Ohtahara syndrome; its overview referring to our 16 cases. *Brain Dev* 2002;24(1):13–23.
39. Al-Futaisi A, Banwell B, Ochi A, Hew J, Chu B, Oishi M, Otsubo H. Hidden focal EEG seizures during prolonged suppressions and high-amplitude bursts in early infantile epileptic encephalopathy. *Clin Neurophysiol* 2005;116(5):1113–17.
40. Baraban SC, Dinday MT, Castro PA, Chege S, Guyenet S, Taylor MR. A large-scale mutagenesis screen to identify seizure-resistant zebrafish. *Epilepsia* 2007;48(6):1151–1157.
41. Wong M, Wozniak DF, Yamada KA. An animal model of generalized nonconvulsive

status epilepticus: Immediate characteristics and long-term effects. *Exp Neurol* 2003;183(1):87–99.

42. McColl CD, Horne MK, Finkelstein DI, Wong JY, Berkovic SF, Drago J. Electroencephalographic characterisation of pentylenetetrazole-induced seizures in mice lacking the alpha 4 subunit of the neuronal nicotinic receptor. *Neuropharmacology* 2003;44(2):234–43.
43. Jutkiewicz EM, Baladi MG, Folk JE, Rice KC, Woods JH. The convulsive and electroencephalographic changes produced by nonpeptidic delta-opioid agonists in rats: Comparison with pentylenetetrazol. *J Pharmacol Exp Ther* 2006;317(3):1337–48.
44. Bains JS, Longacher JM, Staley KJ. Reciprocal interactions between CA3 network activity and strength of recurrent collateral synapses. *Nature Neurosci* 1999;2(8):720–26.
45. Dzhala VI, Staley KJ. Transition from interictal to ictal activity in limbic networks in vitro. *J. Neurosci* 2003;23(21):7873–80.
46. Khazipov R, Khalilov I, Tyzio R, Morozova E, Ben-Ari Y, Holmes GL. Developmental changes in GABAergic actions and seizure susceptibility in the rat hippocampus. *Eur J. Neurosci* 2004;19:590–600.
47. Ribera AB, Nüsslein-Volhard C. Zebrafish touch-insensitive mutants reveal an essential role for the developmental regulation of sodium current. *J Neurosci* 1998;18(22): 9181–91.
48. Drapeau P, Ali DW, Buss RR, Saint-Amant L. In vivo recording from identifiable neurons of the locomotor network in the developing zebrafish. *J Neurosci Methods* 1999;88(1): 1–13.
49. Saint-Amant L, Drapeau P. Motoneuron activity patterns related to the earliest behavior of the zebrafish embryo. *J Neurosci* 2000;20(11):3964–72.
50. Cox KJA, Fetcho JR. Labeling blastomeres with a calcium indicator: a non-invasive method of visualizing neuronal activity in zebrafish. *J Neurosci Methods* 1996;68:185–91.
51. Nicolson T, Rüsch A, Friedrich RW, Granato M, Ruppersberg JP, Nüsslein-Volhard C. Genetic analysis of vertebrate sensory hair cell mechanosensation: the zebrafish circler mutants. *Neuron* 1998;20(2): 271–83.
52. Miyawaki A, Griesbeck O, Heim R, Tsien RY. Dynamic and quantitative Ca^{2+} measurements using improved cameleons. *Proc Natl Acad Sci U S A* 1999;96:2135–40.
53. Higashijima S, Masino MA, Mandel G, Fetcho JR. Imaging neuronal activity during zebrafish behavior with a genetically encoded calcium indicator. *J Neurophysiol* 2003;90:3986–97.
54. Wachowiak M, Denk W, Friedrich RW. Functional organization of sensory input to the olfactory bulb glomerulus analyzed by two-photon calcium imaging. *Proc Natl Acad Sci U S A* 2004;101(24):9097–102.
55. Ramdya P, Reiter B, Engert F. Reverse correlation of rapid calcium signals in the zebrafish optic tectum in vivo. *J Neurosci Methods* 2006;157:230–7.
56. Merritt HH, Putnam TJ. Sodium diphenylhydantoinate in the treatment of convulsive disorders. *J Am Med Assoc* 1938;123: 1253–59.
57. Richards RK, Everett GM. Analgesic and anticonvulsive properties of 3,5,5-trimethyloxazoliidine-2,4-dione (Tridione). *Fed Proc* 1944;3:39.
58. Stilwell GE, Saraswati S, Littleton JT, Chouinard SW. Development of a Drosophila seizure model for in vivo high-throughput drug screening. *Eur J Neurosci* 2006;24:2211–22.
59. Freiman TM, Surges R, Kukolja J, Heinemeyer J, Klar M, van Velthoven V, Zentner J. K(+)-evoked [(3)H]-norepinephrine release in human brain slices from epileptic and non-epileptic patients is differentially modulated by gabapentin and pinacidil. *Neurosci Res* 2006;55(2):204–10.
60. Alzheimer C, ten Bruggencate G. Actions of BRL 34915 (Cromakalim) upon convulsive discharges in guinea pig hippocampal slices. *Naunyn Schmiedebergs Arch Pharmacol* 1988;337(4):429–34.
61. Consortium. Identification and characterization of the tuberous sclerosis gene on chromosome 16. *Cell* 1993;75(7):1305–15.
62. van Slegtenhorst M, de Hoogt R, Hermans C, et al. Identification of the tuberous sclerosis gene TSC1 on chromosome 9q34. *Science* 1997;277(5327):805–8.
63. Thomas G, Hall MN. TOR signalling and control of cell growth. *Curr Opin Cell Biol* 1997;9(6):782–7.
64. Tapon N, Ito N, Dickson BJ, Treisman JE, Hariharan IK. The Drosophila tuberous sclerosis complex gene homologs restrict cell growth and cell proliferation. *Cell* 2001;105(3):345–55.
65. Ito N, Rubin GM. gigas, a Drosophila homolog of tuberous sclerosis gene product-2, regulates the cell cycle. *Cell* 1999;96(4):529–39.
66. Gomez MR. History of the tuberous sclerosis complex. *Brain Dev* 1995;17(l):55–7.

67. Shepherd CW, Houser OW, Gomez MR. MR findings in tuberous sclerosis complex and correlation with seizure development and mental impairment. *AJNR Am J Neuroradiol* 1995;16(1): 149–55.

68. Goodman M, Lamm SH, Engel A, Shepherd CW, Houser OW, Gomez MR. Cortical tuber count: A biomarker indicating neurologic severity of tuberous sclerosis complex. *J Child Neurol* 1997;12(2): 85–90.

69. Pan D, Dong J, Zhang Y, Gao X. Tuberous sclerosis complex: From Drosophila to human disease. *Trends Cell Biol* 2004; 14(2):78–85.

70. Liu S, Lu W, Obara T, et al. A defect in a novel Nek-family kinase causes cystic kidney disease in the mouse and in zebrafish. *Development* 2002;129(24):5839–46.

71. Campbell WA, Yang H, Zetterberg H, et al. Zebrafish lacking Alzheimer presenilin enhancer 2 (Pen-2) demonstrate excessive p53-dependent apoptosis and neuronal loss. *J Neurochem* 2006;96(5):1423–40.

72. Lumsden AL, Henshall TL, Dayan S, Lardelli MT, Richards RI. Huntingtin-deficient zebrafish exhibit defects in iron utilization and development. *Hum Mol Genet* 2007;16(16):1905–20.

73. Onda H, Crino PB, Zhang H, et al. Tsc2 null murine neuroepithelial cells are a model for human tuber giant cells, and show activation of an mTOR pathway. *Mol Cell Neurosci* 2002;21(4):561–74.

74. Crino PB. Molecular pathogenesis of tuber formation in tuberous sclerosis complex. *J Child Neurol* 2004;19(9):716–25.

75. Bebin EM, Kelly PJ, Gomez MR. Surgical treatment for epilepsy in cerebral tuberous sclerosis. *Epilepsia* 1993;34(4):651–7.

76. Guerreiro MM, Andermann F, Andermann E, et al. Surgical treatment of epilepsy in tuberous sclerosis: strategies and results in 18 patients. *Neurology* 1998;51(5): 1263–9.

77. Holmes GL, Stafstrom CE. Tuberous sclerosis complex and epilepsy: recent developments and future challenges. *Epilepsia* 2007;48(4):617–30.

78. Chandra PS, Salamon N, Huang J, et al. FDG-PET/MRI coregistration and diffusion-tensor imaging distinguish epileptogenic tubers and cortex in patients with tuberous sclerosis complex: A preliminary report. *Epilepsia* 2006;47(9):1543–9.

79. Wu JY, Sutherling WW, Koh S, et al. Magnetic source imaging localizes epileptogenic zone in children with tuberous sclerosis complex. *Neurology* 2006;66(8):1270–2.

80. Madhavan D, Weiner HL, Carlson C, Devinsky O, Kuzniecky R. Local epileptogenic networks in tuberous sclerosis complex: a case review. *Epilepsy Behav* 2007;11(1):140–6.

81. Uhlmann EJ, Wong M, Baldwin RL, et al. Astrocyte-specific TSC1 conditional knockout mice exhibit abnormal neuronal organization and seizures. *Ann Neurol* 2002; 52(3):285–96.

82. Meikle L, Talos DM, Onda H, et al. A mouse model of tuberous sclerosis: neuronal loss of Tsc1 causes dysplastic and ectopic neurons, reduced myelination, seizure activity, and limited survival. *J Neurosci* 2007; 27(21):5546–58.47.

83. Wang Y, Greenwood JS, Calcagnotto ME, Kirsch HE, Barbaro NM, Baraban SC. Neocortical hyperexcitability in a human case of tuberous sclerosis complex and mice lacking neuronal expression of TSC1. *Ann Neurol* 2007;61(2):139–52.

84. Kim YJ, Nam RH, Yoo YM, Lee CJ. Identification and functional evidence of GABAergic neurons in parts of the brain of adult zebrafish (Danio rerio). *Neurosci Lett* 2004;355(1–2):29–32.

Chapter 5

Modeling Tuberous Sclerosis Complex: Brain Development and Hyperexcitability

Kevin C. Ess

Abstract

Tuberous sclerosis complex (TSC) is a very important cause of epilepsy particularly in children. Seizures are seen in up to 90% of patients with TSC and often very resistant to standard medical therapies. In addition, patients with TSC also frequently suffer from autism as well as mental retardation and various psychiatric disorders. The causative *TSC1* and *TSC2* genes were identified in the 1990s. Despite tremendous growth in our knowledge of the *TSC* genes, encoded proteins, and participation in signaling pathways, we have not yet developed breakthrough therapies that will significantly improve the quality of life of these patients. While such advances are clearly very difficult endeavors, the relative lack of appropriate animal models has impeded such translational research. Recent developments in this field are encouraging however, with several mouse models recently developed that recapitulate many neurological aspects of TSC. In this chapter I will review the clinical manifestations of TSC, molecular roles of the TSC genes, and current animal models of TSC. These and future models should facilitate the development of new and exciting discoveries that will have a meaningful impact on lives of people with TSC.

Key words: conditional mouse, cortical tuber, mTOR, rapamycin.

1. Introduction

1.1. Clinical Features

Tuberous sclerosis complex (TSC) is a devastating genetic disorder seen in approximately 1:6000 people *(1)*. Hamartomas are the defining pathologic hallmark of TSC. These focal malformations are comprised of abnormally differentiated cells and are found in multiple organ systems. TSC-associated hamartomas exhibit increased proliferation but in general represent benign lesions. True malignancies in TSC have been reported, but are fortunately quite rare *(2)*. While affecting heart, kidney, skin, lung as well as

Scott C. Baraban (ed.), *Animal Models of Epilepsy: Methods and Innovations, vol. 40*

DOI 10.1007/978-1-60327-263-6_5 Springerprotocols.com

other organs, brain involvement is the most clinically relevant due to an extremely high prevalence of neurologic features such as epilepsy and autism. In fact, seizure disorders are seen in at least 90% of patients with TSC. Seizures in TSC typically start at very young ages and are often resistant to standard anti-seizure medications. Infantile spasms, a particularly aggressive form of epilepsy, is seen in 25–50% of patients, making TSC the most common known genetic cause of this important seizure syndrome *(3)*. Fortunately, infantile spasms in children with TSC can be very effectively controlled with the anti-seizure medication vigabatrin. For these patients, rapid cessation of spasms and normalization of their EEG is often observed *(3, 4)*. This dramatic clinical response to vigabatrin suggests that in patients with TSC, abnormalities of GABAergic neurotransmission may be central to the pathogenesis of infantile spasms.

In addition to epilepsy, other neurologic features are frequently seen in TSC. These include autism, developmental delay, mental retardation, and behavioral/psychiatric problems (**Table 5.1**). The severity of these features is quite variable in any individual patient but overall contribute greatly to a poor quality of life for many patients and their families *(5)*.

1.2. Neuropathology in Human TSC

Essentially all TSC patients have extensive cortical and subcortical hamartomas known as "tubers" (**Fig. 5.1**). The disease was named by 19th century pathologists in an attempt to describe discrete brain lesions that to them were reminiscent of potatoes *(6)*. These investigators also noted the firm sclerotic nature of tubers, hence the disease name "sclerose tubereuse". Tubers are typically found in multiple locations in cortical and subcortical regions of the brain and are generally accepted as sites of seizure origination *(7)*.

Table 5.1
Neurologic features of TSC

Disorder	Prevalence
Epilepsy • Infantile spasms	90% 50
Autism and ASD	25–50
Developmental delay	25–50
Mental retardation	50–80
Behavioral problems	50–60
Psychiatric problems	50%

ASD – autism spectrum disorders.

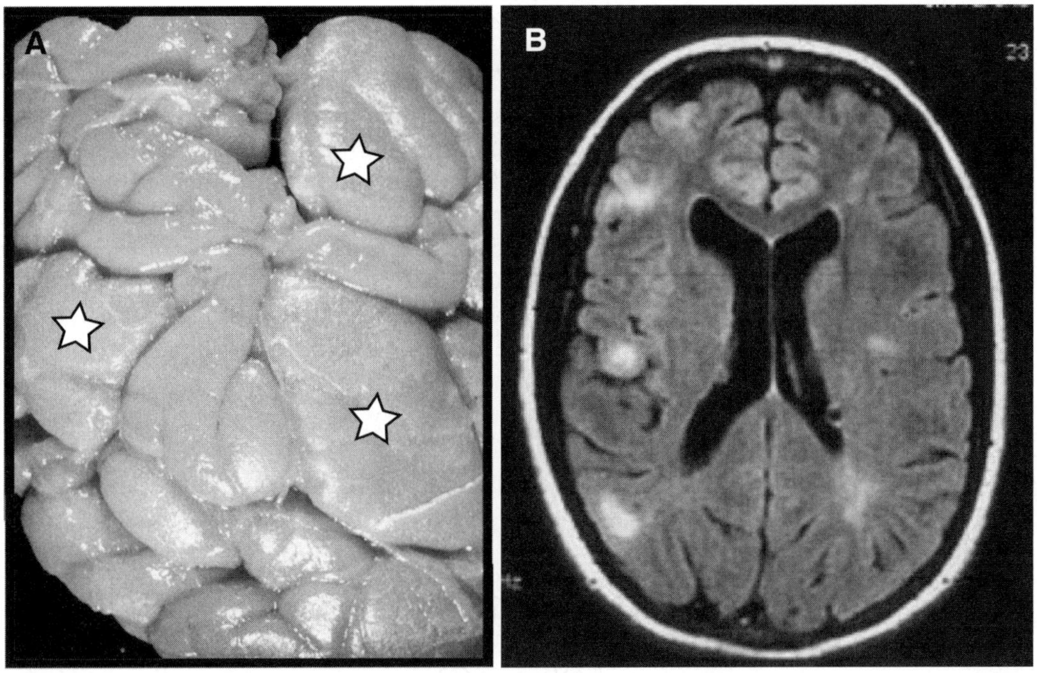

Fig. 5.1. **(A)** Gross pathology of a brain from a patient with TSC. Discrete cortical tubers (stars) are seen at the brain surface. **(B)** Brain MRI–Flair sequence of a child with TSC. Current neuroimaging allows excellent delineation of cortical and subcortical tubers in patients with TSC. Also seen in this image are subependymal nodules lining the ventricular surface.

Studies of human fetuses with TSC and current knowledge of normal human brain development suggest that tubers form quite early, prior to 20 weeks of gestation *(8–10)*. After birth, tuber number and their location are relatively static though individual tubers may become more or less epileptogenic as a function of age. In addition to tubers, patients with TSC usually have other structural brain lesions including subependymal nodules (SENs) and subependymal giant cell astrocytomas (SEGAs). These subcortical lesions are histologically benign and always found adjacent to the ventricular surface. They are clinically distinct from tubers and are not thought to participate in epileptogenesis or the phenotypes of mental retardation or autism. However, SENs and SEGAs are much more dynamic than tubers with a capacity for growth generally restricted to the first 25 years of life. SEGAs are clinically important as their propensity for continued growth near the foramen of Monro can obstruct CSF flow leading to hydrocephalus, visual impairment, and even death.

1.3. Neuropathology in TSC

Many neuropathologic studies of TSC have been published using postmortem samples as well as tubers and SEGAs resected from patients with intractable epilepsy or hydrocephalus. These studies revealed the presence of pathognomonic large, dysplastic cells ("giant cells") in tubers as well as SEGAs *(11)*. Additional immunohistochemical studies demonstrated expression of neuronal as well as glial cell markers within tubers as well as individual giant cells *(12–14)*. Examination of the entire tuber reveals a gross and dramatic loss of cortical lamination. Such a frank disruption of cortical layering points to an early developmental abnormality in brain development. These pathologic findings suggest that the *TSC* genes (see below) subserve critical functions during the specification and/or differentiation of early neural progenitor cells.

2. Genetics and Signaling Pathways

While inheritable as an autosomal dominant disorder, the majority of patients with TSC appear to have spontaneous mutations of either the *TSC1* or *TSC2* genes. Progress in defining the molecular pathogenesis of TSC has been extraordinary, with major advances reported during the past decade. Early linkage studies in the 1990s led to the cloning of the *TSC2* gene in 1993 (encodes tuberin) *(15)* and the *TSC1* gene in 1997 (encodes hamartin) *(16)*. Hamartin and tuberin physically interact and loss of either partner is sufficient to cause the full clinical spectrum of TSC. However, phenotype/genotype studies have found a trend toward increased disease severity in patients with *TSC2* gene mutations or deletions *(17)*. While found in many tissues, prominent expression of both hamartin and tuberin is seen in the CNS particularly during embryonic development *(18)*. The identification of *TSC1* and *TSC2* homologs in various species including mouse, rat, Drosophila, and yeast paved the way for multiple genetic and biochemical studies that elucidated key upstream and downstream components of the TSC signaling pathway. A key early finding was the identification of a GTPase activating protein (GAP) domain in tuberin *(19)*. While other G proteins have been identified as downstream tuberin targets, the G protein Rheb has emerged as the most important protein directly controlled by tuberin *(20)*. Through an intermediary protein FKBP38, the tuberin/hamartin complex regulates the mTOR (*m*ammalian *T*arget *Of R*apamycin) protein kinase and its downstream targets (**Fig. 5.2**) *(21)*. The discovery that mTOR kinase is closely regulated by tuberin/hamartin immediately suggested that treatment with rapamycin may be a rational therapeutic option to inhibit a constitutively active mTOR kinase and potentially ameliorate some of the clinical features of TSC.

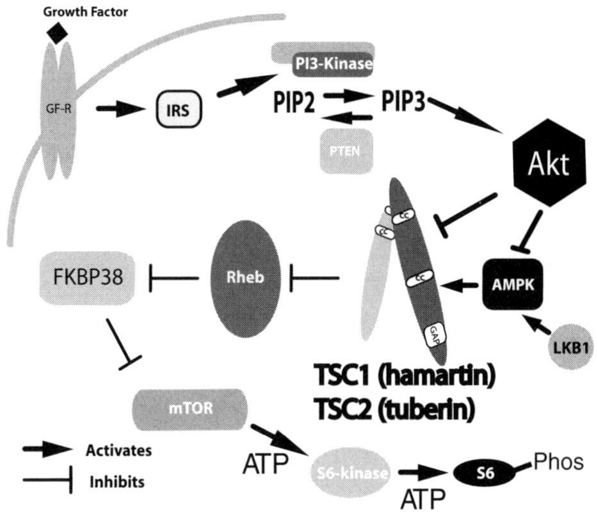

Fig. 5.2. Summary of upstream and downstream signaling pathways in TSC. The *TSC1* (hamartin) and *TSC2* (tuberin) protein products bind to each other and function within a multi-protein complex. Phosphorylation of tuberin by Akt is inactivating while AMPK phosphorylation at other sites activates tuberin function. The hamartin–tuberin complex normally inhibits Rheb, an inhibitor of FKBP38. As FKBP38 is an endogenous inhibitor of the mTOR kinase, genetic loss or inactivation of hamartin/tuberin permits constitutive activation of Rheb and the mTOR kinase. The resulting signaling changes include phosphorylation of SK-kinase, S6, and other downstream targets are currently thought to underlie much of the pathophysiology seen in TSC.

2.1. Conventional Rodent Models of TSC

After identification of the human *TSC1* and *TSC2* genes, mouse homologs were quickly identified and cloned. This allowed the molecular inactivation of these genes to further study their function. Such conventional gene "knockout" experiments were first reported for the mouse *Tsc2* gene and established its essential role during early embryogenesis as all *Tsc2*(−/−) embryos died by embryonic day 10. Heterozygote *Tsc2* (+/−) mice exhibit subtle neuropathology with some astrocytosis but exhibit a grossly normal cortex and no observed clinical phenotypes including seizures *(22, 23)*. Similar experiments were done with the mouse *Tsc1* gene with almost identical results including early lethality for *Tsc1* homozygous null embryos *(24)*. In addition to mice, the Eker rat has provided important information as well. This spontaneously arising rat model has been studied for many years as an autosomal dominant model of kidney disease *(25)*. Many years later it was discovered that the Eker phenotype results from heterozygous inactivation of the rat *Tsc2* gene *(26)*. Despite extensive evaluations and manipulations, the Eker rat has not contributed greatly to our understanding of the neuropathology of TSC *(27)*. However, there are clearly abnormalities in the brains of Eker rats. For example, increased episodic memory was reported in Eker rats as well as increased kindling after consecutive PTZ injections.

These results suggest persistent alterations in brain plasticity in rats heterozygous for the *Tsc2* gene *(28)*. These rats have very infrequent TSC-like CNS abnormalities *(29, 30)*, while similar to mice, rats homozygous for the *Eker* allele (*Tsc2−/−*) have an early embryonic lethal phenotype prior to any substantive cortical development. These complete loss-of-function experiments again confirmed the requirement of *Tsc* genes for early rodent embryogenesis. More importantly, they point out the need for alternative experimental strategies to determine the normal function of *Tsc1* and *Tsc2* genes during brain development and explore the pathogenesis of TSC.

2.2. Conditional Mouse Models

An ideal animal model of TSC would include specific clinical as well as pathologic features (**Table 5.2**). A major advance toward this goal was the creation of a conditional mouse whose *Tsc1* gene was modified through the addition of exogenous LoxP sequences flanking critical exons of this gene *(31)*. After exposure to the enzyme Cre recombinase, the genomic DNA between these LoxP sites is deleted leading to permanent inactivation of the *Tsc1* gene. By regulating the tissue and developmental expression of Cre recombinase, the *Tsc1* gene can thus be precisely inactivated in various cell types at defined stages of development. Importantly in the absence of Cre recombinase, the floxed *Tsc1* allele appears to be completely normal with no discernable effect on gene expression or function. The Cre/LoxP system was first employed in TSC research using the human *GFAP* promoter to direct Cre recombinase expression to astrocytes *(31)*. These $Tsc1^{GFAP\text{-}Cre}$ conditional knockout mice ($Tsc1^{GFAP}$ CKO) are born with normal appearing brains but exhibit an age-dependent increase in astrocyte number and brain size by 8 weeks of life. Importantly, all $Tsc1^{GFAP}$ CKO have spontaneous seizures beginning at approximately 1 month of age. The seizure frequency continues to increase as they age, likely contributing to their premature death by 3 months of age.

Table 5.2
Ideal neurologic animal model of TSC

Clinical features	Pathologic features
Epilepsy	Tubers
Behavioral/psychiatric abnormalities • Social interactions • Anxiety	SEGAs
Cognitive abnormalities	Neuronal and glia abnormalities
Postnatal viability	mTOR signaling dysregulation

Treatment of these mice with the conventional anti-seizure medications phenobarbital or phenytoin decreased seizure frequency but did not alter background EEG abnormalities or prevent premature death *(32)*. Additional hypotheses regarding the role of astrocytes in epileptogenesis were tested using this model. For example, studies using *Tsc1*^{GFAP} CKO mice demonstrated alterations in astrocyte-mediated glutamate uptake *(33, 34)*. Further investigations by the same research group also revealed abnormalities of potassium currents in astrocytes *(35)*. Alterations of both glutamate and potassium homeostasis may then contribute to epileptogenesis in patients with TSC. This concept is receiving greater support with astrocyte dysfunction even postulated to be the primary CNS abnormality in TSC *(36)*. Finally, in this model targeting the *Tsc1* gene in astrocytes, apoptotic death of neurons is seen in the cortex and hippocampus (**Fig. 5.3 A,B**) *(34)*. This result emphasizes the need to specifically target other CNS lineages to dissect the neuropathology of TSC and define the contributions of different neural cells that cause the phenotype of TSC.

While *Tsc1*^{GFAP} CKO mice provide important insights to the role of astrocytes in TSC, primary neuronal dysfunction is very likely a critical aspect of the neurologic phenotype. This is supported by extensive neuropathologic data from patients with TSC as well as from in vitro studies with *Tsc1*-deficient neurons *(37)*. This elegant study demonstrated alterations in soma size when cultured neurons were exposed to Cre recombinase. Electrophysiologic alterations in glutamate current mediated by the AMPA receptor subtype and dendrite extension were also noted. Extending these results in vivo, a conditional knockout of the *Tsc1* gene in neurons was recently reported *(38, 39)*. Again using the Cre–LoxP system, the floxed *Tsc1* gene was targeted in embryonic post-mitotic neurons using the *synapsin* promoter to direct expression of Cre recombinase (*Tsc1*^{Synapsin} CKO). These mice were reported by two independent groups to have seizures, poor growth, and premature death *(38, 39)*. The initial publication noted a grossly normal cortex but prominence of SMI311-positive (marker of neuronal neurofilaments) cells in *Tsc1*^{Synapsin} CKO mice compared to littermate controls *(38)*. Interestingly, a subsequent publication also validated the normal cortical lamination and absence of tubers but reported many large and dysmorphic neurons that were SMI311 and phospho-S6 positive in *Tsc1*^{Synapsin} CKO but not control mice *(39)* (**Fig. 5.4 A,B** and Color Plate 3, middle of book). As expected from the use of the *Synapsin* gene promoter that is expressed only in post-mitotic neurons, glutamatergic as well as GABAergic interneurons appeared to be targeted in this model as evidenced by increased expression of phospho-S6 *(39)*. Electrophysiologic evidence for decreased inhibition, however, was not found *(38)*. Glial cells were also examined in *Tsc1*^{Synapsin}

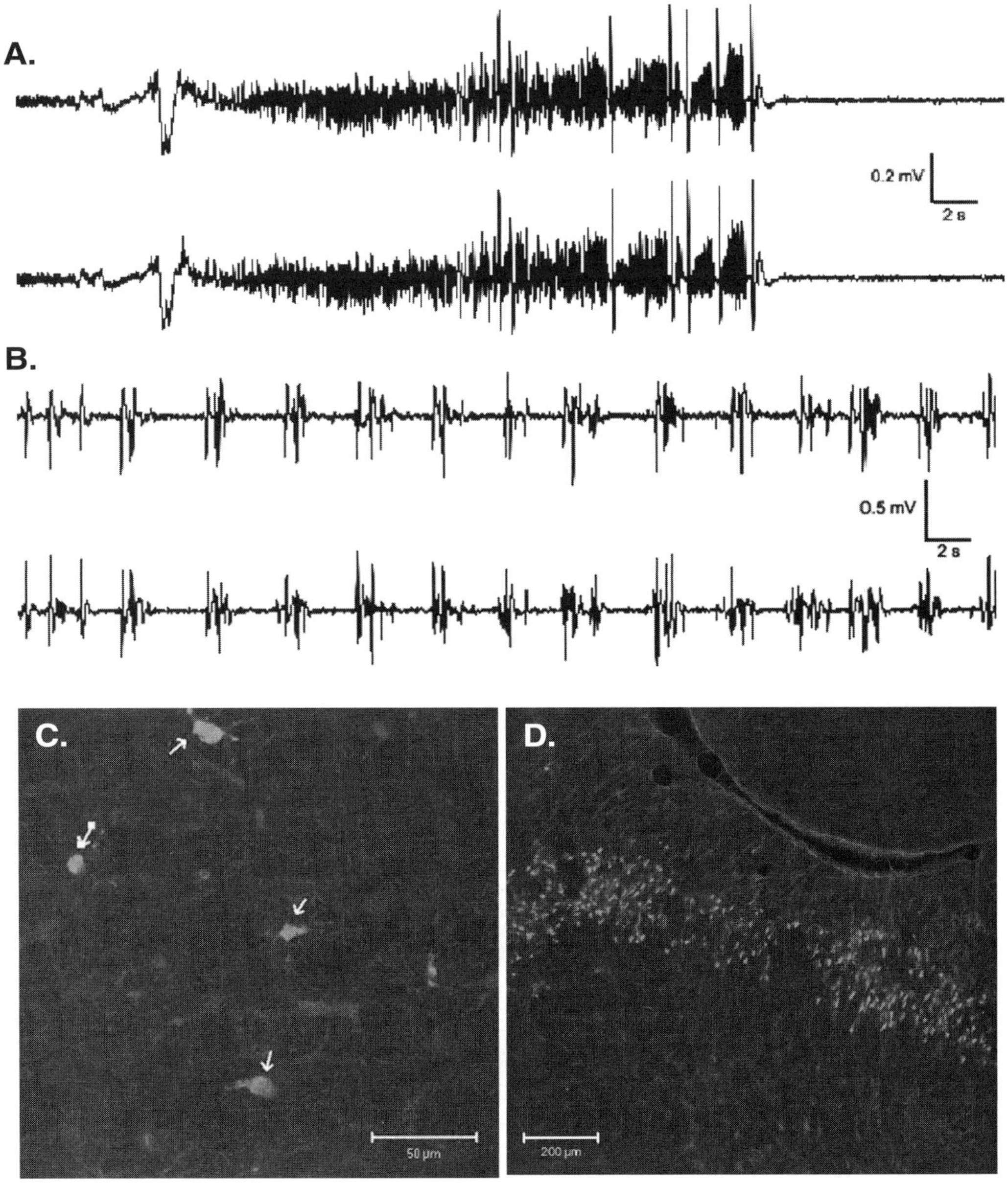

Fig. 5.3. *Tsc1*^{GFAP} *CKO mice have spontaneous seizures, abnormal EEG background, and increased neuronal death.* **(A)** Example of a seizure emanating from both hemispheres of a *Tsc1*^{GFAP} CKO mouse. These seizures increase in frequency as the animal ages. **(B)** *Tsc1*^{GFAP} CKO mice also exhibit severe interictal EEG abnormalities including burst suppression patterns. Extensive cell death is also seen in the cortex (**C**, caspase-3 immunostaining) and CA3 region of the hippocampus (**D**, Fluoro-Jade stain). Control littermates had very little to no discernable caspase-3 or Fluoro-Jade signal (data not shown).

CKO mice. Astrocytes were reported by both groups to have a normal appearance and distribution. Finally, dramatic alterations in cortical myelin expression were seen, suggesting a possible secondary effect on oligodendrocytes due to severe neuronal

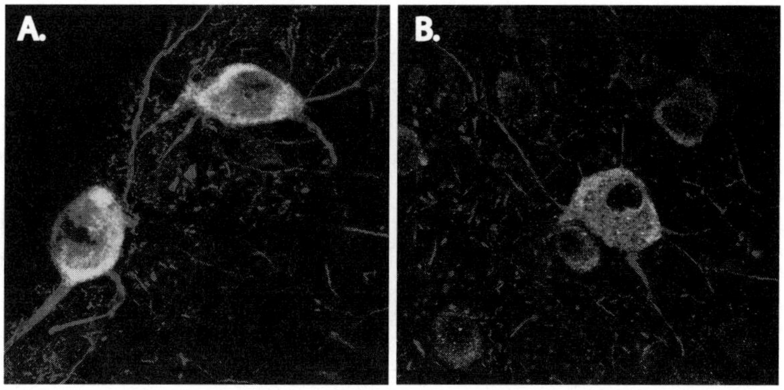

Fig. 5.4. Confocal images of brain sections from Tsc1^{Synapsin}CKO mice immunostained for SMI311 (red, neuronal marker) and phospho-S6 (green). This staining reveals large, dysplastic appearing neurons from white matter **(A)** or the cortex **(B)**. (*see* Color Plate 3, middle of book)

dysfunction *(39)*. The significance of these findings is unclear but prominent white matter abnormalities have been reported in the brains of patients with TSC *(40)*.

3. Summary and Future Directions

As is evident throughout this text, animal models have greatly facilitated advances in the basic science and function of *TSC* genes and offer much insight to the pathophysiology of this disorder. These model systems should continue to pave the way for translational research to test novel and hopefully much more effective therapies. What can these models tell us about the pathogenesis of epilepsy? The phenotype of epilepsy is almost universal in patients with TSC and as described above, perturbations of *Tsc1* gene in either neurons or astrocytes leads to a robust seizure phenotype. This suggests that the *Tsc1* gene subserves multiple critical roles during brain development and loss of function in even a subpopulation of brain cells may be sufficient for the generation of seizures. From these promising models, mechanistic insight is now emerging, for example, the identification of abnormal glutamate and potassium homeostasis in *Tsc1*^{GFAP} CKO mice. Such insight is critical to further our understanding of the pathogenesis of TSC and the continual search for improved therapeutics. First amongst potential targets are the mTOR kinase and its downstream signaling cascade. Rapamycin usage has already received much attention given the role of the *TSC2* gene to regulate the mTOR pathway and the practical fact that rapamycin is already approved by the FDA for use in the United States. In fact, rapamycin has been reported to be effective in small series of patients for the medical treatment of SEGAs as well as kidney and possibly lung manifestations in TSC *(41–43)*. It is not known whether a positive clinical response to rapamycin may also be seen for the phenotypes of epilepsy, mental retardation, or autism. While this is currently quite speculative for the treatment of patients, administration of rapamycin to either the $Tsc1^{GFAP}$ CKO or $Tsc1^{Synapsin}$ CKO mouse models has led to a clear benefit in terms of prolonged survival and seizure control *(44, 45)*. Much more work is needed to determine if these findings can be translated to a new, effective, and rational therapy for the neurologic manifestations of TSC.

Future experiments will likely build upon these existing mouse models, targeting *Tsc1* as well as *Tsc2* gene function *(46)* in other distinct CNS compartments such as neural progenitor cells, GABAergic interneurons, or the hypothalamus. In addition, "knock-in" experiments in the mouse can be performed where the mouse *Tsc1* or *Tsc2* genes are replaced with human *TSC1* or *TSC2* alleles containing disease-causing mutations. In addition to mice, other model organisms should also be considered to test novel hypotheses regarding the function of the *TSC* genes. For example, zebrafish (*Danio rerio*) is an excellent model to study developmental processes given its transparency, large numbers of offspring, and short gestational period. In addition, zebrafish offer a very potent advantage of high throughput screens as many thousands of molecules can be quickly assayed to identify potential drugs that may alter any phenotype resulting from loss of the zebrafish *Tsc1* or *Tsc2* genes. Furthermore, the study of epileptogenesis in TSC is also possible in zebrafish (also see S. Baraban, Chapter 4) including EEG recordings and treatment with anti-seizure medications *(47)*.

Acknowledgments

Thanks to Drs. Michael Wong (**Fig. 5.3**), Frances Jensen, and David Kwiatkowski (**Fig. 5.4**) for contributing original photomicrographs. K.E. was supported by the NINDS, NIH, the Tuberous Sclerosis Alliance, and an American Epilepsy Society/Milken Family Early Career Award.

References

1. Crino PB, Nathanson KL, Henske EP. The tuberous sclerosis complex. *N Engl J Med* 2006;355(13):1345–56.
2. Breysem L, Nijs E, Proesmans W, Smet MH. Tuberous sclerosis with cystic renal disease and multifocal renal cell carcinoma in a baby girl. *Pediatr Radiol* 2002;32(9):677–80.
3. Elterman RD, Shields WD, Mansfield KA, Nakagawa J. Randomized trial of vigabatrin in patients with infantile spasms. *Neurology* 2001;57(8):1416–21.

4. Parisi P, Bombardieri R, Curatolo P. Current role of vigabatrin in infantile spasms. *Eur J Paediatr Neurol* 2007;11(6):331–6.
5. Wiznitzer M. Autism and tuberous sclerosis. *J Child Neurol* 2004;19(9):675–9.
6. Bourneville DM. Tuberous sclerosis with cortical abnormalities: Mental retardation and hemiplegic epilepsy. *Archives de neurologie* 1880;1:81–91.
7. Weiner HL, Ferraris N, LaJoie J, Miles D, Devinsky O. Epilepsy surgery for children with tuberous sclerosis complex. *J Child Neurol* 2004;19(9):687–9.
8. Park SH, Pepkowitz SH, Kerfoot C, et al. Tuberous sclerosis in a 20-week gestation fetus: Immunohistochemical study. *Acta Neuropathologica* 1997;94(2):180–6.
9. Levine D, Barnes P, Korf B, Edelman R. Tuberous sclerosis in the fetus: Second-trimester diagnosis of subependymal tubers with ultrafast MR imaging. *AJR Am J Roentgenol* 2000;175(4):1067–9.
10. Wei J, Li P, Chiriboga L, et al. Tuberous sclerosis in a 19-week fetus: Immunohistochemical and molecular study of hamartin and tuberin. *Pediatr Dev Pathol* 2002;5(5):448–64.
11. Lopes MB, Altermatt HJ, Scheithauer BW, Shepherd CW, VandenBerg SR. Immunohistochemical characterization of subependymal giant cell astrocytomas. *Acta Neuropathologica* 1996;91(4):368–75.
12. Ess KC, Kamp CA, Tu BP, Gutmann DH. Developmental origin of subependymal giant cell astrocytoma in tuberous sclerosis complex. *Neurology* 2005;64(8):1446–9.
13. Sharma M, Ralte A, Arora R, Santosh V, Shankar SK, Sarkar C. Subependymal giant cell astrocytoma: A clinicopathological study of 23 cases with special emphasis on proliferative markers and expression of p53 and retinoblastoma gene proteins. *Pathology* 2004;36(2):139–44.
14. Ess KC, Uhlmann EJ, Li W, et al. Expression profiling in tuberous sclerosis complex (TSC) knockout mouse astrocytes to characterize human TSC brain pathology. *Glia* 2004;46(1):28–40.
15. Consortium ECTS. Identification and characterization of the tuberous sclerosis gene on chromosome 16. *Cell* 1993;75(7):1305–15.
16. van Slegtenhorst M, de Hoogt R, Hermans C, et al. Identification of the tuberous sclerosis gene TSC1 on chromosome 9q34. *Science* 1997;277(5327):805–8.
17. Sancak O, Nellist M, Goedbloed M, Elfferich P, Wouters C, Maat-Kievit A, Zonnenberg B, Verhoef S, Halley D, van den Ouweland A. Mutational analysis of the TSC1 and TSC2 genes in a diagnostic setting: genotype-phenotype correlations and comparison of diagnostic DNA techniques in Tuberous Sclerosis Complex. *Eur J Hum Genet* 2005; 13(6):731–41.
18. Geist RT, Reddy AJ, Zhang J, Gutmann DH. Expression of the tuberous sclerosis 2 gene product, tuberin, in adult and developing nervous system tissues. *Neurobiol Dis* 1996;3(2):111–20.
19. Inoki K, Li Y, Xu T, Guan KL. Rheb GTPase is a direct target of TSC2 GAP activity and regulates mTOR signaling. *Genes Dev* 2003; 17(15):1829–34.
20. Castro AF, Rebhun JF, Clark GJ, Quilliam LA. Rheb binds tuberous sclerosis complex 2 (TSC2) and promotes S6 kinase activation in a rapamycin- and farnesylation-dependent manner. *J Biol Chem* 2003; 278(35):32493–6.
21. Bai X, Ma D, Liu A, et al. Rheb activates mTOR by antagonizing its endogenous inhibitor, FKBP38. *Science* 2007;318(5852):977–80.
22. Kobayashi T, Minowa O, Kuno J, Mitani H, Hino O, Noda T. Renal carcinogenesis, hepatic hemangiomatosis, and embryonic lethality caused by a germ-line Tsc2 mutation in mice. *Cancer Res* 1999;59(6):1206–11.
23. Onda H, Lueck A, Marks PW, Warren HB, Kwiatkowski DJ. Tsc2(+/-) mice develop tumors in multiple sites that express gelsolin and are influenced by genetic background. *J Clin Invest* 1999;104(6):687–95.
24. Kobayashi T, Minowa O, Sugitani Y, et al. A germ-line Tsc1 mutation causes tumor development and embryonic lethality that are similar, but not identical to, those caused by Tsc2 mutation in mice. Proc Natl Acad Sci U S America 2001;98(15):8762–7.
25. Eker R. Familial renal adenomas in Wistar rats: A preliminary report. *Acta Pathol Microbiol Scand* 1954;34:554–62.
26. Kobayashi T, Hirayama Y, Kobayashi E, Kubo Y, Hino O. A germline insertion in the tuberous sclerosis (Tsc2) gene gives rise to the Eker rat model of dominantly inherited cancer. *Nat Genet* 1995;9(1):70–4.
27. Wenzel HJ, Patel LS, Robbins CA, Emmi A, Yeung RS, Schwartzkroin PA. Morphology of cerebral lesions in the Eker rat model of tuberous sclerosis. *Acta Neuropathol* 2004; 108(2):97–108.
28. Waltereit R, Welzl H, Dichgans J, Lipp HP, Schmidt WJ, Weller M. Enhanced

episodic-like memory and kindling epilepsy in a rat model of tuberous sclerosis. *J Neurochem* 2006;96(2):407–13.

29. Tschuluun N, Wenzel HJ, Schwartzkroin PA. Irradiation exacerbates cortical cytopathology in the Eker rat model of tuberous sclerosis complex, but does not induce hyperexcitability. *Epilepsy Res* 2007; 73(1):53–64.
30. Takahashi DK, Dinday MT, Barbaro NM, Baraban SC. Abnormal cortical cells and astrocytomas in the Eker rat model of tuberous sclerosis complex. *Epilepsia* 2004; 45(12):1525–30.
31. Uhlmann EJ, Wong M, Baldwin RL, et al. Astrocyte-specific Tsc1 conditional knockout mice exhibit abnormal neuronal organization and seizures. *Ann Neurol* 2002; 52(3):285–96.
32. Erbayat-Altay E, Zeng LH, Xu L, Gutmann DH, Wong M. The natural history and treatment of epilepsy in a murine model of tuberous sclerosis. *Epilepsia* 2007;48(8):1470–6.
33. Wong M, Ess KC, Uhlmann EJ, et al. Impaired glial glutamate transport in a mouse tuberous sclerosis epilepsy model. *Ann Neurol* 2003;54(2):251–6.
34. Zeng LH, Ouyang Y, Gazit V, et al. Abnormal glutamate homeostasis and impaired synaptic plasticity and learning in a mouse model of tuberous sclerosis complex. *Neurobiol Dis* 2007;28(2):184–96.
35. Jansen LA, Uhlmann EJ, Crino PB, Gutmann DH, Wong M. Epileptogenesis and reduced inward rectifier potassium current in tuberous sclerosis complex-1-deficient astrocytes. *Epilepsia* 2005;46(12):1871–80.
36. Sosunov AA, Wu X, Weiner HL, et al. Tuberous sclerosis: A primary pathology of astrocytes? *Epilepsia* 2008;49(Suppl 2):53–62.
37. Tavazoie SF, Alvarez VA, Ridenour DA, Kwiatkowski DJ, Sabatini BL. Regulation of neuronal morphology and function by the tumor suppressors Tsc1 and Tsc2. *Nat Neurosci* 2005;8(12):1727–34.
38. Wang Y, Greenwood JS, Calcagnotto ME, Kirsch HE, Barbaro NM, Baraban SC. Neocortical hyperexcitability in a human case of tuberous sclerosis complex and mice lacking neuronal expression of Tsc1. *Ann Neurol* 2007;61(2):139–52.
39. Meikle L, Talos DM, Onda H, et al. A mouse model of tuberous sclerosis: Neuronal loss of Tsc1 causes dysplastic and ectopic neurons, reduced myelination, seizure activity, and limited survival. *J Neurosci* 2007;27(21):5546–58.
40. Ridler K, Bullmore ET, Devries PJ, Suckling J, Barker GJ, Meara SJ, Williams SC, Bolton PF. 2001. Widespread anatomical abnormalities of grey and white matter structure in tuberous sclerosis. *Psychol Med* 31(8): 1437–46, 2001.
41. Davies DM, Johnson SR, Tattersfield AE, et al. Sirolimus therapy in tuberous sclerosis or sporadic lymphangioleiomyomatosis. *N Engl J Med* 2008;358(2):200–3.
42. Franz DN, Leonard J, Tudor C, et al. Rapamycin causes regression of astrocytomas in tuberous sclerosis complex. *Ann Neurol* 2006;59(3):490–8.
43. Bissler JJ, McCormack FX, Young LR, et al. Sirolimus for angiomyolipoma in tuberous sclerosis complex or lymphangioleiomyomatosis. *N Engl J Med* 2008;358(2): 140–51.
44. Kwiatkowski D. Abstract presented at "mTOR Signaling: From Cancer to CNS Function". 2008; Bethesda, MD.
45. Zeng LH, Xu L, Gutmann DH, Wong M. Rapamycin prevents epilepsy in a mouse model of tuberous sclerosis complex. *Ann Neurol* 2008;63(4):444–53.
46. Hernandez O, Way S, McKenna J, Gambello MJ. Generation of a conditional disruption of the Tsc2 Gene. *Genesis* 2007;45: 101–6.
47. Baraban SC, Dinday MT, Castro PA, Chege S, Guyenet S, Taylor MR. A large-scale mutagenesis screen to identify seizure-resistant zebrafish. *Epilepsia* 2007;48(6):1151–7.

Chapter 6

BK Potassium Channel Mutations Affecting Neuronal Function and Epilepsy

David Petrik, Qing H. Chen, and Robert Brenner

Abstract

Many experimental studies of epilepsy are based on rodent models induced by chemical or electrical insult as an instigator of seizures. Although such studies are useful in observing the physiological events that may occur during epileptogenesis, the great complexity of changes that ensue between first seizure and epilepsy makes it difficult to ascribe cause-and-effect to any single protein. In contrast, spontaneous mutations or mutations using gene targeting in mice provide a unique opportunity to evaluate a single protein, at cellular to whole organism level, for its role in epilepsy. Such approaches have uncovered a number of genes, some expected and others unexpected, which may provide novel targets for further study and drug design.

Recently, the large conductance voltage and calcium-activated potassium channel (BK channel) gene and their accessory subunits have been appended to the list of mutations causing epilepsy in humans and mice. BK channels have the specialized role of sharpening action potentials in many excitatory neurons. Surprisingly, functional analysis in heterologous expression systems indicates that some of these mutations are gain of function. This paradoxical effect of an increased BK potassium current correlating with increased excitability and seizures belies simplistic assumptions of the role of potassium channels in reducing excitability of neurons. In this review we explore genetic studies to understand how BK channel gain of function may have pro-epileptic effects. An important conclusion that may be drawn is that BK channel sharpening of action potentials may preclude recruitment of other potassium currents that otherwise reduce excitability. Accessory β subunits that either confer inactivation or slow BK channels activation are able to limit these effects. These findings suggest that BK channels and their interplay with other ion currents must be tightly regulated to serve appropriate function in neurons. Gain-of-function mutations or maladaptive changes during epilepsy may re-tailor these channels to support high-frequency action potential firing.

Key words: BK channel, MaxiK channel, calcium-activated potassium channel, $\beta 4$, beta4 accessory subunit, action potential, seizure, epilepsy.

Scott C. Baraban (ed.), *Animal Models of Epilepsy: Methods and Innovations, vol. 40*

DOI 10.1007/978-1-60327-263-6_6 Springerprotocols.com

1. Background

1.1. What Are BK Channels?

Single-channel recording from diverse cells types often demonstrate a potassium current that is both voltage dependent and calcium activated. This channel's large unitary opening makes them easy to identify resulting in their early characterization in neurons and muscle following the development of the patch clamp technique *(1–3)*. Common pseudonyms for this channel derive from their large single-channel conductance (~250 pS, roughly 10- to 20-fold larger than most other potassium currents) and therefore they are called BK ("Big K^+" current), as well as MaxiK channels *(4)*. These channels are broadly expressed in many cell types including neurons, smooth and skeletal muscle, endothelial cells, and many epithelial cell types *(5–7)*. They were originally cloned from the *Drosophila slowpoke* locus *(8)* and coined *slo* channels, and mslo *(9)* and hslo *(10, 11)* for mammalian mouse and human counterparts, respectively. More recently they have acquired the HUGO (Human Genome Organization) designation KCNMA1 for the human homologue. Unlike the great redundancy in many other functionally related voltage-gated potassium channels, the single BK channel gene encodes the pore-forming subunit underlying all BK channel currents. However, BK channels undergo extensive alternative splicing *(12, 13)* and assemble with a family of four tissue-specific accessory subunits (β1–β4, KCNMB1–4) that provide the necessary diversity to BK channel properties in different cell types *(14)*.

BK channels are formed as homomers of four α subunits *(15)*. Like their related K_v family members, BK channels have a functional voltage sensor (S4 region) embedded within six transmembrane segments (S1–S6) of each α subunit (*see* **Fig. 6.1** and Color Plate 4, middle of book) *(16)*. BK channels have two additional components that make them unique among voltage-dependent potassium channels. They have an additional transmembrane segment (S0) that places the N-terminus to the extracellular side and contributes to interactions with accessory β subunits *(17)*. In addition, they have a large carboxyl intracellular region containing proposed hydrophobic segments S7–S10 *(18)*. This region contains two putative calcium-binding domains including the RCK domain *(19)* (for "regulator conductance of potassium") that also regulates activation by magnesium *(20, 21)* and the "calcium bowl" *(22, 23)* domain (**Fig. 6.1**). BK channels are weakly activated by calcium ligand or depolarization alone under physiological conditions *(24, 25)* and often rely on the simultaneous activation of both to regulate excitability *(26–28)*. It is these properties that make BK channels an effective coincident detector of depolarization and calcium rise and gives them the specialized role to moderate excitability in response to voltage-dependent calcium channel activation.

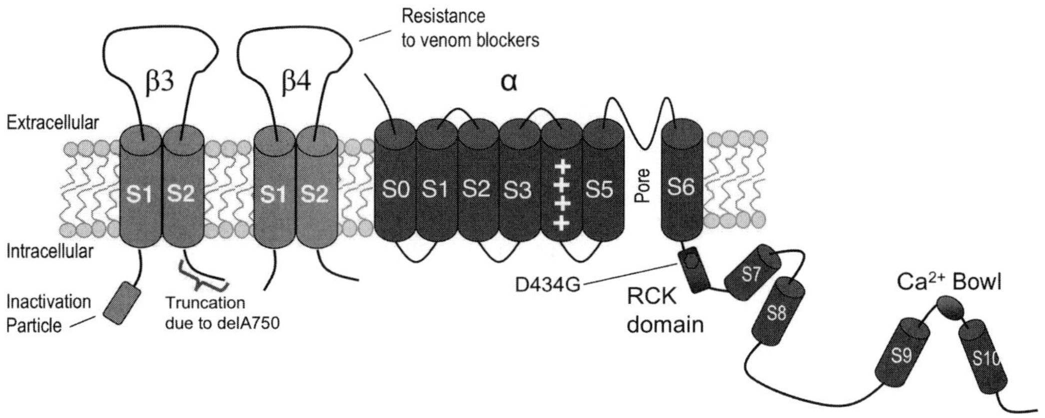

Fig. 6.1. **Diagram of BK channel α (pore-forming) subunit and accessory β subunits.** Annotations indicate relevant functional domains and sites containing epilepsy mutations (β3 delA750 and gain-of-function mutation D434G for α subunit). (*see* Color Plate 4, middle of book)

Given that BK channels require relatively high (micromolar) calcium for activation *(29)*, it is not surprising that they are frequently colocalized with calcium channels in neurons *(30)*. Although BK channels can physically co-assemble with complexes containing L, P/Q, or N-type calcium channels in the brain *(31)*, functional studies indicate that coupling with different calcium channels is cell specific. In some neurons, BK channels can be activated by calcium influx through multiple channels such as L- and Q-type calcium channels *(32)* or L- and N-type calcium channels *(33)*. In other neurons BK channels are activated exclusively by N-type *(34)* or T-type calcium channels *(35)*. In addition to extracellular sources of calcium, BK channels are functionally colocalized with calcium release from intracellular stores mediated by ryanodine receptors. This occurs at resting voltages in many smooth muscle cells *(36)* and in select neurons such as cardiac parasympathetic neurons *(37)*, auditory hair cells *(38)*, and retinal amacrine cells *(39)*. As well, calcium-induced calcium release through ryanodine receptors can modulate spike width by activation of BK channels in sympathetic neurons *(40)*.

1.2. BK Channel Effect on Action Potentials

Coincident with calcium influx, BK channels are activated during the falling phase of action potentials to narrow the spike width. In some neurons, the current is sustained transiently following repolarization (1–2 ms) to hyperpolarize the membrane below threshold (so-called fast afterhyperpolarization or fAHP) *(41)*. Because BK channels deactivate quickly following the spike, they do not generally directly affect action potential firing rates. Exceptions are vestibular nucleus neurons that can be driven to fire action potentials at very high rates (>100 Hz). In these neurons where the interpulse interval is very short, BK channels do reduce firing frequencies *(35)*. In other neurons that fire at lower rates, it is the more sustained AHPs, the medium AHP lasting up to 100 ms (contributed by SK calcium-activated potassium channels) and slow AHP that can last for several seconds *(42, 43)* (contributed by an unknown calcium-activated channel in the central nervous system, CNS), that have a greater effect on firing frequency.

It is important to note that examples exist where BK channels enhance rather than reduce excitability. The mechanism appears to be mediated by BK channels' ability to sharpen action potentials while deactivating quickly enough to allow subsequent action potentials. For example, in CA1 neurons and adrenal chromaffin cells BK channel repolarization of action potentials can increase firing frequencies by removing inactivation of sodium channels *(44, 45)*. In addition, the effect of sharpening action potentials by BK channels reduces the activation of slow-gated potassium current that otherwise can moderate firing frequency. This is seen in CA1 neurons. Block of BK channels broadens action potentials, which result in enhanced recruitment of delayed rectifier potassium current and reduced action potential firing frequency *(44)*. Finally, BK channels fast gating in pituitary somatotrophs reduces the action potential amplitude which also preclude delayed rectifier potassium channel activation. In this case, the reduced delayed rectifier current causes plateau-burst action potential firing, which enhances calcium influx and hormone secretion *(46)*. Thus, the effect of BK channels on firing rates appears to be complex. In some fast-spiking neurons such as brain stem vestibular neurons, BK channels can moderate firing frequencies *(35)*. In other neurons that fire at slower rates, BK channels either do not affect firing rates *(47)* or, through sharpening of action potentials, can have indirect effects on other currents that enhance firing rates *(44–46)*.

Given that BK channels can enhance excitability in neurons, it is not surprising that this channel is tightly regulated to limit these effects. One mechanism appears to be mediated by regulatory tissue-specific β subunits. Although BK channels in CA1 neurons enhance firing rates, they do so only at the early components of a spike train due to inactivation that is likely conferred by the accessory $\beta 2$ subunit *(48, 49)*. This is seen as spike frequency-dependent action potential broadening in CA1 and lateral amygdala pyramidal cells *(44, 49, 50)*. Thus while sharpened action potentials enhance early firing rates, inactivation of BK channels cause action potentials broadening in the later components of a spike train which reduce the firing rate *(44)*. In other neurons, the accessory $\beta 4$ subunit slows BK channel gating to rates *(51)* that likely limit BK channel contribution to action potential sharpening. As will be discussed below, knockout of $\beta 4$ creates a gain of function of BK channels that sharpens action potentials and enhances firing frequency *(52)*.

1.3. BK Channel Effect on Synaptic Transmission

Immunostaining for the channel protein indicates that they are highly enriched in synaptic regions *(53–56)*. Thus, in addition to BK channel effects on somatic action potentials, they may have a role in regulating membrane properties of synapses. Early work in *Xenopus* motor nerve terminals has shown that BK channels have the logical role of deactivating voltage-dependent calcium channels and thereby regulating neurotransmission *(26, 57)*. These findings appear to be borne out in mammalian motor *(58)* and cochlear nerve terminals *(59)*. However, examples where BK channels affect basal neurotransmission in central synapses are rare (observed in CA3–CA3 collateral synapses) *(60)*. In most CNS neurons BK channel effects on neurotransmission is minor *(61)* or absent *(62)* or are only uncovered by blocking other potassium channels *(54)* or under high-frequency stimulation *(63)*. This suggests that BK channels, despite their enrichment in synapses, may serve a secondary role in neurotransmission, perhaps as protection against overexcitability of synapses.

2. Genetics of BK Channels

Our understanding of BK channel roles in different tissues has increased significantly through the study of genetic mutations. Mutations related to BK channel function in neurons are listed in **Table 6.1** and discussed below. The pore-forming α subunit (KCNMA1) and accessory $\beta 1$ (KCNMB1) and $\beta 4$ (KCNMB4) subunits have been genetically deleted. Knockouts of the $\beta 2$ and $\beta 3$ subunits are not available; however, human mutations of the $\beta 3$ subunit (KCNMB3) have been identified. As well, recent findings have identified some mutations that act as gain of function for BK channels based on recording of the mutant channels in heterologous cells (KCNMA1 D434G mutation) and in knockout mouse tissues ($\beta 4$ knockout). An important generalization that can be made from these studies is that although BK channel knockout and deficiency is responsible for a number of neuronal defects, it is the gain-of-function mutations that appear to be more closely correlated with epilepsy phenotypes. Although a number of studies have also focused on BK α and the $\beta 1$ subunit roles in various smooth muscle and other tissues, we will confine our discussion on CNS studies for the purpose of this review.

2.1. Mutations Causing Loss of function of BK Channels

2.1.1. KCNMA1 Knockouts

Ataxia and Defects in Cerebellar Function. Initial observations of BK channel knockouts demonstrated a profound ataxia *(64, 65)*. Electrophysiological recordings from cerebellar Purkinje cells show a significant reduction of the fAHP although the action potential repolarization rate or width is unaffected. Surprisingly, knockout cells in a subpopulation of Purkinje cells had a

Table 6.1
BK channel subunit mutations affecting neuronal function

Gene	Gene alteration	Phenotype	Functional consequence	Functional alteration	Reference
Non-epileptic					
KCNMA1	Null	Cerebellar ataxia and abnormal reflex conditioning	No BK current	Reduced AHP of Purkinje neurons	Sausbier et al. *(65)*
		Ataxia and deficit in motor functions	No BK current	Neuronal mechanism not reported	Meredith et al. *(64)*
		Altered circadian rhythmicity and locomotor activity	No BK current	Increased spontaneous firing rates in the suprachiasmatic nucleus	Meredith et al. *(67)*
		High-frequency hearing loss	No BK current	Increased outer hair cell degeneration	Ruttiger et al. *(70)*
		Resistance to noise-induced hearing loss	No BK current	Loss of fast activating current in hair cells	Pyott et al. *(71)*
	Haploin sufficiency	Autism and mental retardation	Reduced BK current	Reduced expression of BK channels	Laumonnier et al. *(73)*
Epileptic					
KCNMA1	D434G	Generalized seizures with paroxysmal dyskinesia	Enhanced BK current	Enhanced activation, slower deactivation, G–V shift to more negative potentials	Du et al. *(74)*
KCNMB3	delA750	Generalized absence epilepsy	Inhibited BK current	Reduced BK activation due to enhanced $\beta 3$ inactivation	Hu et al. *(83)*; Lorenz et al. *(84)*

	Duplication	Epilepsy and mental retardation of dup(3q) syndrome	Over expression of $\beta 3$	Not known	Riazi et al. *(78)*
KCNMB4	Single nucleotide poly-morphism	Temporal lobe epilepsy	Not known	Not known	Cavalleri et al. *(96)*
	Null	Temporal lobe epilepsy	Enhanced BK current	Increased NPo, faster activation, sharpened AP peak	Brenner et al. *(52)*

significantly reduced spontaneous firing rate *(65)*. This may be due to a slight depolarization of the resting membrane potential that leads to steady-state inactivation of voltage-dependent sodium channels. Consistent with this mechanism, action potential frequencies could be recovered by hyperpolarizing current injections, which presumably remove sodium channel inactivation. The concept that BK channel knockout may affect the resting membrane potential is unique to these neurons and somewhat surprising since BK channels in most neurons do not contribute to resting membrane potentials.

Changes in circadian rhythms. Neurons of the suprachiasmatic nucleus (SCN) are regarded to be regulators of circadian rhythms. BK channels are expressed in the SCN and show day/night rhythmic changes in gene expression and ionic currents *(66)*. Knockout of the BK channel causes spontaneous electrical activity from the SCN neurons to increase during night periods *(67)*. Functional consequences of the knockout are less robust changes in behavioral patterns and temperature regulation that otherwise occurs during nightly periods, indicating perturbation of circadian behavioral rhythms *(67)*.

Auditory Hair Cells. BK channels have long been studied in auditory hair cells of non-mammalian vertebrates and shown to be critically important in electrical tuning of hair cells for discrimination of sound frequencies. In part, electrical tuning is mediated by alternative splicing of the BK channel gene to produce channels with diverse gating properties arranged along a tonotopic gradient *(12)*. Although mammalian hair cells are mechanically tuned, and do not use electrical tuning, BK channels nevertheless are expressed in these cells *(68, 69)*. Initial knockout studies demonstrate outer hair cell dysfunction and progressive high-frequency hearing loss *(70)*. In contrast, others have not observed a hearing loss and indeed saw a protection from noise-induced hearing loss in the BK channel knockout *(71)*. The discrepancy may be due to differences in genetic strains used between the two studies since mouse genetic background has been shown to affect propensity to hearing loss *(72)*.

2.1.2. Haploinsufficiency of BK Channels and Autism

In animal genetic models, loss-of-function mutations of a single copy affecting cognitive functions can sometimes appear too subtle to characterize by behavioral analysis. However, in humans deficits in cognitive function are more apparent and useful for uncovering gene functions. This is the case for the BK channel gene. Genetic mapping of an autism/mental retardation patient with a spontaneous chromosome 9/10 translocation identified a disruption of the BK channel gene that leads to haploinsufficiency of the BK channel gene *(73)*. Investigating further, the authors also found a BK channel missense mutation in an additional autism individual. The mutation results in a change at an amino acid

(A138V) that is otherwise conserved in BK channels of all organisms *(73)*. The functional consequences of the A138V change have not been studied. The relevance to epilepsy, also unexplored at this time, could relate to clinical observations of comorbidity with autism patients often suffering with seizure disorders as well.

2.2. Mutations of BK Channels Causing Epilepsy

2.2.1. D434G KCNMA1 Gain of Function of BK Channels

A mutation of the BK channel pore-forming subunit (KCNMA1) that results in epilepsy was observed in a family that had an inherited pattern of absence-type, generalized seizures with paroxysmal dyskinesia *(74)*. The mutation is an aspartate to glycine residue change (D434G) in a region of the channel (RCK domain) important for calcium sensing (*see* **Fig. 6.1**). Affected individuals have seizures at a young age (6 months to 9 years) and the mutation is partially penetrant for the epileptic phenotype (9/16 affected individuals develop epileptic seizures) although most have the paroxysmal dyskinesia (12/16). Electrophysiology studies to understand the effect of the D434G on BK channel function surprisingly showed that the mutation enhances BK channel activation in an oocyte expression system or transfected cells *(74, 75)*. Activation rate of the channel is increased, deactivation is slowed and channels open at more negative potentials. The observation that gain of function of a potassium channel causes epilepsy is somewhat confounding based on the simplistic assumption that potassium channels reduce excitability. The authors conjectured that BK channel gain of function might produce currents that are more effective at hyperpolarization and removing sodium channel inactivation. This may thereby promote increased action potential firing *(74)*. This is akin to the role that fast-gated K_v3 channels play in GABAergic inhibitory neurons to promote high-frequency action potential firing *(76)*. Indeed, knockout of $K_v3.2$ potassium channels reduces firing rate in inhibitory neurons and increases susceptibility to seizures *(77)*. Alternatively, the possibility exists that BK channels indeed reduce excitability, but do so in some critical inhibitory neurons that would promote seizures. Truly understanding the mechanism by which D434G mutation increases excitability and seizures will have to wait for construction of this mutation in mice to allow for electrophysiological recordings in neurons.

2.2.2. KCNMB3 Gene Mutations

In addition to the pore-forming subunit, modulatory accessory subunits have also shown linkage to epilepsy. After the smooth muscle-specific $\beta1$ subunit, the $\beta3$ subunit was the newest member of the BK channel β subunit family member to be identified *(78)*. Yet the functional role for the $\beta3$ subunit seems to be the least understood among the BK β subunit family members. Despite the fact that $\beta3$ shows highest expression in testis and is only weakly expressed in the CNS *(79, 80)*, there is suggestion that this subunit may have a role in epilepsy. The initial identification of the $\beta3$ subunit gene was due to its localization within a chromosome segment that is duplicated in dup(3q) syndrome *(78)*. The relatively broad symptoms include mental retardation and a high incidence (83%) of seizures *(81, 82)*. As well, a variant of the $\beta3$ (delA750) encodes a frameshift mutation that alters three amino acids and deletes the remaining 18 carboxyl terminal residues. Relative to the normal $\beta3$ variant, the mutation inhibits BK channel activation and confers inactivation to $\beta3$ splice isoforms that do not otherwise inactivate *(83)*. The delA750 polymorphism is significantly increased in German patients with idiopathic generalized epilepsy, particularly absence-type epilepsy, as compared to a control group that do not have epilepsy *(84)*. The $\beta3$ subunit contains a number of alternative splice forms that confer inactivation properties to BK channels *(85)* and outward rectification of currents *(86)* but otherwise show only relatively weak effects on steady-state open probability *(79, 80, 85, 87)*. A simplistic assumption may be that either the increased number of inactivating $\beta3$ subunits in the dup(3q) syndrome or the increased inactivation due to the delA750 polymorphism enhances excitability through an excess of BK channel inactivation. However, the affected neurons expressing the $\beta3$ subunit and the mechanisms by which $\beta3$ mutations alter their function need further investigation to understand $\beta3$ subunit relevance to epilepsy.

2.2.3. KCNMB4 Gene Knockout

In contrast to the $\beta3$ subunit, the $\beta4$ subunit is highly enriched in the CNS *(79, 80, 88)*. In heterologous expression systems, the $\beta4$ subunit dramatically slows BK channel activation and deactivation and shifts channel openings to much greater depolarizations *(79, 80)*. Thus the $\beta4$ subunit is generally assumed to be a down-regulator of BK channels *(88)*. $\beta4$ subunit also confers resistance to scorpion venoms such as iberiotoxin and charybdotoxin that otherwise block BK channels *(89)*. Thus, recording of BK channels in neurons that are insensitive to these venoms are indicative of BK channels assembled with the $\beta4$ subunit *(90–94)*. However, organic BK channel blockers such as paxilline or penitrem A can still be used to specifically block BK channels regardless of the accessory subunit *(54, 95)*.

In a large multicenter study (The Epilepsy Genetics Consortium), the $\beta4$ subunit was identified as one of only 5 genes among 279 prime candidate genes that had a single nucleotide polymorphism correlated with epilepsy *(96)*. In this case, a $\beta4$ polymorphism correlated with mesial temporal lobe epilepsy that is consistent with its enhanced expression in the hippocampus *(52)*. The polymorphism was located in the 3′ non-coding region of the $\beta4$ gene that presumably might affect expression of the gene; however, the functional consequences have not been studied as yet.

A further understanding of the role of $\beta 4$ came about with targeted knockout of the $\beta 4$ gene. Given that $\beta 4$ subunit inhibits BK channels activation *(97, 98)*, knockout of $\beta 4$ therefore would be expected to result in a gain of function. Consistent with the human D434G gain-of-function epilepsy phenotype, the $\beta 4$ knockout mice also had spontaneous seizures *(52)*. Perhaps because $\beta 4$ expression is high in the hippocampus, the seizures appear to be temporal lobe in onset as evidenced by interictal spikes that are measured in the hippocampus in the absence of cortical spikes *(52)*. Interestingly, the seizures are non-convulsive, making this knockout a unique genetic model of temporal lobe non-convulsive epilepsy.

$\beta 4$ shows greatest expression in the hippocampus dentate gyrus granule cells; therefore, firing patterns in these cells were investigated. Consistent with an increase in excitability, there is an increase in firing frequency in the $\beta 4$ knockout neurons relative to neurons from control animals (**Fig. 6.2**) *(52)*. Action potential waveforms indicate that the mechanism is similar to CA1 neurons wherein BK channels enhance firing frequency. $\beta 4$ knockout neurons have sharper action potential waveforms as expected from BK channel gain of function (**Fig. 6.3**). However, the AHP is reduced and rises to threshold more quickly than wild type (**Fig. 6.3** and Color Plate 5, middle of book). This would predict a faster approach to threshold and enhanced action potential firing. In contrast, $\beta 4$ wild-type neurons, although they have a broadened action potential, have a larger and more sustained medium AHP (**Fig. 6.3**) that may account for the lower firing frequency. By utilizing paxilline block of BK channels, a cause and effect between BK channel activation and increased excitability is easier to discern. Paxilline block in $\beta 4$ knockout indeed slows the AHP decay *(52)* and reduces firing frequencies (**Fig. 6.2**, bottom panels).

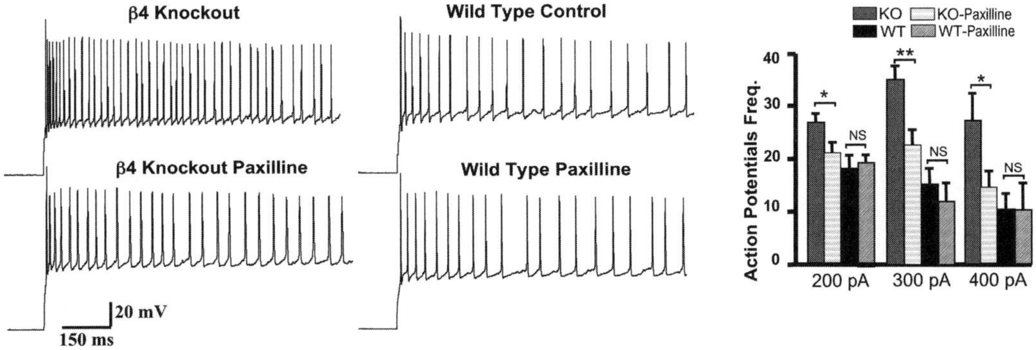

Fig. 6.2. **Firing frequency of dentate gyrus granule cells is enhanced with knockout of $\beta 4$ and reduced by BK channel block with paxilline.** Left panels show trains of action potentials evoked by 200 pA constant current injection. Average action potential frequency is shown at right. Figure was courtesy of Nature Publishing Group, modified from Brenner et al. *(52)*.

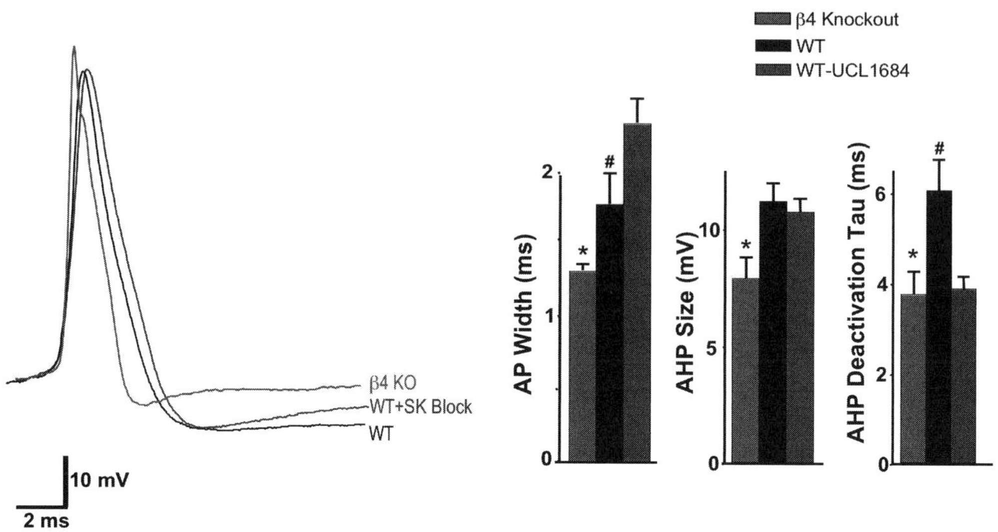

Fig. 6.3. **BK and β4 effect on action potential waveform in dentate gyrus granule cells.** Action potential was selected from 10th spike during a 300 pA current injection. Average data are plotted at right. Red trace is β4 knockout, blue trace is wild type plus SK channel blocker UCL1684, black trace is wild type. (*see* Color Plate 5, middle of book)

This indicates that gain of function of BK channels reduces recruitment of currents that can otherwise reduce membrane excitability, while wild-type neurons, where β4 inhibits BK channels, show little effect of paxilline on the AHP *(52)* or firing rates (**Fig. 6.2**).

Dentate gyrus granule cells are dependent of calcium influx during action potential firing to activate calcium-activated potassium currents that moderate firing frequencies *(52)*. Although BK channels shape action potentials, their fast deactivation after the spike generally limits their influence on subsequent action potentials or on firing rates. However, by controlling action potential shape it is possible that they may reduce calcium influx that indirectly affects other calcium-activated potassium currents that regulate firing rates. The small conductance calcium-activated (SK-type) channels are a class of calcium-activated potassium currents that have an important role in spike-frequency adaptation and moderating excitability in many neurons include dentate gyrus granule cells *(41)*. These channels can be blocked with organic blockers such as UCL1684 or the bee venom apamin *(99)*. Block of SK channels in wild-type neurons causes the action potential AHP to decay more quickly after the spike (**Fig. 6.3**). This sets the voltage on a trajectory that more quickly rises to threshold and enhances action potential firing frequency (**Fig. 6.4**). In contrast, in β4 knockout neurons, it appears that SK channels play a smaller role since SK channel block has no effect on firing frequencies (**Fig. 6.4**). Thus, BK channel gain of function, although it

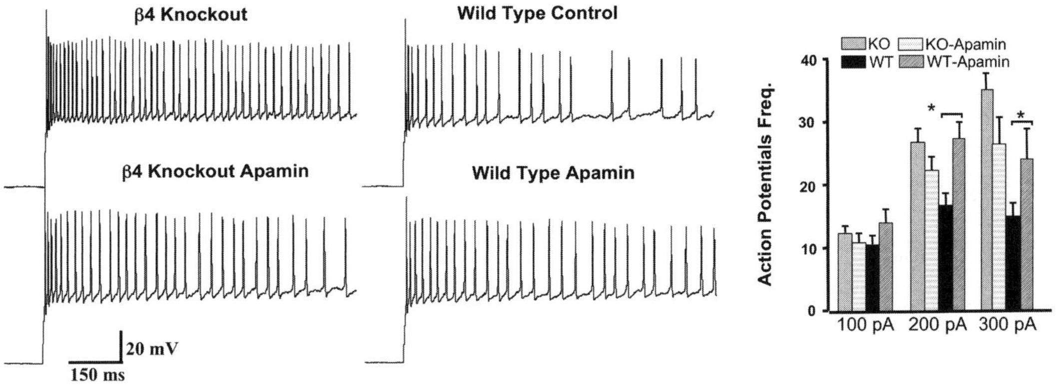

Fig. 6.4. **SK channel block enhances firing frequency in wild type but not $\beta 4$ knockout.** Left panels show trains of action potentials evoked by 200 pA constant current injection. Average action potential frequency is shown at right. Figure was courtesy of Nature Publishing Group, modified from Brenner et al. *(52)*.

sharpens action potentials, reduces contribution of SK channels and perhaps other potassium currents that would otherwise moderate excitability of neurons.

2.3. Firing Frequency and Dentate Gyrus Function

The effect of the BK channels $\beta 4$ knockout on dentate gyrus granule cells is particularly relevant to understanding epilepsy since these cells play a central role in moderating excitability and synchronization of the hippocampal circuit. Dentate gyrus granule cells are the first principle cells of a trisynaptic circuit within the hippocampus that includes downstream CA3 cells. As such these cells are regarded as a gate, interposed between excitatory input from the entorhinal cortex (perforant path) and the relatively more excitable CA3 pyramidal cells *(100)*. Much of this filtering function results from heavy inhibitory input within the dentate gyrus, and indeed there is considerable evidence that reduced inhibitory input occurs in epilepsy and contributes to breakdown of the dentate gyrus filter *(101)*. But also the intrinsic firing properties and spike-frequency adaptation may provide an important role.

How can changes in firing patterns of dentate gyrus granule cells increase excitability of the hippocampal circuit leading to seizures? Further studies are necessary to understanding how BK channels and reduced spike-frequency adaptation play a role. However, studies indicate that the ability of granule cells to drive target CA3 neurons has a frequency dependence that is affected by firing properties of granule cells *(102–104)*. In terms of excitability, dentate granule cells, through their mossy fiber axons, behave in a manner analogous to "giveth with one hand, and taketh away with the other". The "giveth" is the major excitatory pathway from the dentate gyrus that is mediated monosynaptically by mossy fiber boutons onto postsynaptic CA3 pyramidal cells *(105)*. The "taketh away" are filopodial extensions from mossy fibers that excite local inhibitory neurons. These local inhibitory neurons also synapse onto CA3 neurons to dampen the monosynaptic excitatory response *(105)*. At low-frequency excitation this results in a low probability of evoking action potentials in the CA3 postsynaptic pyramidal cells (so-called spike transmission). Firing frequency increases above 20 Hz causing depression in the inhibitory synapses, facilitation of excitatory synapses, and enhance spike transmission to CA3 neurons to high probabilities *(103)*. Spike-frequency adaptation maintains firing frequency below this critical range, whereas the β4 knockout granule cells were able to fire action potentials well above 20 Hz *(52)*. The frequency dependence for excitation of the hippocampal circuit has also been observed in vivo with coincident functional magnetic resonance imaging and electrophysiological recording of the dentate gyrus in rats *(106)*. Low-frequency stimulation (<5 Hz) of the perforant path evokes no electrophysiological response as measured by population spikes of the mossy fibers or fMRI images of the rat brain. At 5 Hz there is a spacially restricted activation of the dentate gyrus that further spreads throughout the hippocampal network at higher frequencies and even onto the contralateral hippocampus at 20 Hz *(106)*.

3. Conclusions

In retrospect, these genetic studies highlight earlier reports that the majority of dentate gyrus cells from TLE patients take on properties that allow high-frequency firing *(107)*, with a reduced AHP and a loss of spike-frequency adaptation. As well, there are several studies indicating that voltage-dependent calcium channels from dentate gyrus granule cells of TLE patients show changes that lead to reduced calcium influx *(108–111)*, consistent with an expected reduction of calcium-dependent currents that otherwise would mediate spike-frequency adaptation. The genetic studies in this review are indicative of what altered gating or expression of BK channels may do to perturb normal properties of neurons. The electrophysiological recording from dentate granule cells of epilepsy patients begs the question whether nongenetic/acquired mechanisms of epilepsy have maladaptive changes that include gain of function of BK channels. Interestingly, β4 has been identified as one of select transcripts that are downregulated in hippocampus tissues during acute and latent periods of a rat model of electrically induced epilepsy *(112)*. Consistent with the β4 gene knockout studies, one would expect that this would lead to enhanced excitability *(52)*. Certainly a number of other proteins are also likely to be altered during the progression from seizure to epilepsy. However, continuing study of the complex interplay between BK potassium channels and other ionic currents should help distinguish between adaptive and maladaptive changes affecting epileptogenesis.

Ideally, BK channels may offer a new therapeutic target for antiepileptic drugs. Recent studies suggest that this may be the case. Gain of function of BK channels was observed in a picrotoxin-induced seizure model *(113)*. This was correlated with an increase in firing frequency of layer 2/3 pyramidal neurons *(113)*. Interestingly, BK channel antagonists were able to moderate this increase in firing frequency to near-control levels *(113)*. Thus, maladaptive changes in BK subunit expression may underlie some forms of epilepsy, and BK channel antagonists may offer treatment. The next challenge will be to determine if BK channel antagonists can be useful in whole animal studies despite the side effects that might be expected from BK channels' broad expression in the nervous system. In this regard, the more restricted localization of the $\beta4$ subunit *(52)* may provide a more confined therapeutic target to antagonize BK channel activity and protect against seizures.

Acknowledgments

We thank Bin Wang for critical reading of the manuscript. Robert Brenner was supported by a National Institutes of Health Grant NS052574 and an Epilepsy Foundation of America Grant.

References

1. Lux HD, Neher E, Marty A. Single channel activity associated with the calcium dependent outward current in Helix pomatia. *Pflugers Arch* 1981;389(3):293–5.
2. Marty A. Ca-dependent K channels with large unitary conductance in chromaffin cell membranes. *Nature* 1981;291(5815):497–500.
3. Pallotta BS, Magleby KL, Barrett JN. Single channel recordings of Ca2+-activated K+ currents in rat muscle cell culture. *Nature* 1981; 293(5832):471–4.
4. Hille B. Ion channels of excitable membranes. 3rd ed. Sunderland, Mass.: Sinauer; 2001.
5. Chang CP, Dworetzky SI, Wang J, Goldstein ME. Differential expression of the alpha and beta subunits of the large-conductance calcium-activated potassium channel: Implication for channel diversity. *Brain Res Mol Brain Res* 1997;45(1):33–40.
6. Morita T, Hanaoka K, Morales MM, Montrose-Rafizadeh C, Guggino WB. Cloning and characterization of maxi K+ channel alpha-subunit in rabbit kidney. *Am J Physiol* 1997; 273(4 Pt 2): F615–24.
7. Papassotiriou J, Kohler R, Prenen J, et al. Endothelial K(+) channel lacks the Ca(2+) sensitivity-regulating beta subunit.Faseb J 2000; 14(7): 885–94.
8. Atkinson NS, Robertson GA, Ganetzky B. A component of calcium-activated potassium channels encoded by the Drosophila slo locus.*Science* 1991; 253(5019): 551–5.
9. Butler A, Tsunoda S, McCobb DP, Wei A, Salkoff L. mSlo, a complex mouse gene encoding "maxi" calcium-activated

potassium channels. *Science* 1993; 261(5118): 221–4.

10. Dworetzky SI, Trojnacki JT, Gribkoff VK. Cloning and expression of a human large-conductance calcium-activated potassium channel. *Brain Res Mol Brain Res* 1994; 27(1): 189–93.
11. Pallanck L, Ganetzky B. Cloning and characterization of human and mouse homologs of the Drosophila calcium-activated potassium channel gene, slowpoke. *Hum Mol Genet* 1994; 3(8): 1239–43.
12. Fettiplace R, Fuchs PA. Mechanisms of hair cell tuning. *Annu Rev Physiol* 1999; 61: 809–34.
13. Fury M, Marx SO, Marks AR. Molecular Biology: The study of splicing and dicing. *Sci STKE* 2002; 2002(123): PE12.
14. Orio P, Rojas P, Ferreira G, Latorre R. New disguises for an old channel: MaxiK channel beta-subunits. *News Physiol Sci* 2002; 17: 156–61.
15. Shen KZ, Lagrutta A, Davies NW, Standen NB, Adelman JP, North RA. Tetraethylammonium block of Slowpoke calcium-activated potassium channels expressed in Xenopus oocytes: Evidence for tetrameric channel formation. *Pflugers* Arch 1994; 426(5): 440–5.
16. Jan LY, Jan YN. Cloned potassium channels from eukaryotes and prokaryotes. *Annu Rev Neurosci* 1997; 20: 91–123.
17. Wallner M, Meera P, Toro L. Determinant for b-subunit regulation in high-conductance voltage-activated and Ca21-sensitive K1 channels: An additional transmembrane region at the N terminus. *Proc Natl Acad Sci U S A* 1996; 93: 14922–7.
18. Meera P, Wallner M, Song M, Toro L. Large conductance voltage- and calcium-dependent K+ channel, a distinct member of voltage-dependent ion channels with seven N-terminal transmembrane segments (S0–S6), an extracellular N terminus, and an intracellular (S9–S10) C terminus. *Proc Natl Acad Sci USA* 1997; 94(25): 14066–71.
19. Jiang Y, Pico A, Cadene M, Chait BT, MacKinnon R. Structure of the RCK domain from the E. coli K+ channel and demonstration of its presence in the human BK channel. *Neuron* 2001; 29(3): 593–601.
20. Rothberg BS, Magleby KL. Voltage and Ca2+ activation of single large-conductance Ca2+-activated K+ channels described by a two-tiered allosteric gating mechanism. *J Gen Physiol* 2000; 116(1): 75–99.
21. Xia XM, Zeng X, Lingle CJ. Multiple regulatory sites in large-conductance calcium-activated potassium channels. *Nature* 2002; 418(6900): 880–4.
22. Schreiber M, Salkoff L. A novel calcium-sensing domain in the BK channel. *Biophys J* 1997; 73(3): 1355–63.
23. Wei A, Solaro C, Lingle C, Salkoff L. Calcium sensitivity of BK-type KCa channels determined by a separable domain. *Neuron* 1994; 13(3): 671–81.
24. Prakriya M, Lingle CJ. Activation of BK channels in rat chromaffin cells requires summation of Ca(2+) influx from multiple Ca(2+) channels. *J Neurophysiol* 2000; 84(3): 1123–35.
25. Cox DH, Cui J, Aldrich RW. Allosteric gating of a large conductance Ca-activated K+ channel. *J Gen Physiol* 1997; 110(3): 257–81.
26. Robitaille R, Garcia ML, Kaczorowski GJ, Charlton MP. Functional colocalization of calcium and calcium-gated potassium channels in control of transmitter release. *Neuron* 1993; 11(4): 645–55.
27. Sah P. Ca(2+)-activated K+ currents in neurones: Types, physiological roles and modulation. *Trends Neurosci* 1996; 19(4): 150–4.
28. Storm JF. Intracellular injection of a Ca2+ chelator inhibits spike repolarization in hippocampal neurons. *Brain Res* 1987; 435(1–2): 387–92.
29. Cui J, Cox DH, Aldrich RW. Intrinsic voltage dependence and Ca2+ regulation of mslo large conductance Ca-activated K+ channels. *J Gen Physiol* 1997; 109(5): 647–73.
30. Muller A, Kukley M, Uebachs M, Beck H, Dietrich D. Nanodomains of single Ca2+ channels contribute to action potential repolarization in cortical neurons. *J Neurosci* 2007; 27(3): 483–95.
31. Grunnet M, Kaufmann WA. Coassembly of big conductance Ca2+-activated K+ channels and L-type voltage-gated Ca2+ channels in rat brain. *J Biol Chem* 2004; 279(35): 36445–53.
32. Prakriya M, Lingle CJ. BK channel activation by brief depolarizations requires Ca2+ influx through L- and Q-type Ca2+ channels in rat chromaffin cells. *J Neurophysiol* 1999; 81(5): 2267–78.
33. Sun X, Gu XQ, Haddad GG. Calcium influx via L- and N-type calcium channels activates a transient large-conductance Ca2+-activated K+ current in mouse neocortical pyramidal neurons. *J Neurosci* 2003; 23(9): 3639–48.

34. Marrion NV, Tavalin SJ. Selective activation of Ca2+-activated K+ channels by co-localized Ca2+ channels in hippocampal neurons. *Nature* 1998; 395(6705): 900–5.
35. Smith MR, Nelson AB, Du Lac S. Regulation of firing response gain by calcium-dependent mechanisms in vestibular nucleus neurons. *J Neurophysiol* 2002; 87(4): 2031–42.
36. Herrera GM, Nelson MT. Sarcoplasmic reticulum and membrane currents. *Novartis Found Symp* 2002; 246: 189–203; discussion 203–7, 221–7.
37. Parsons RL, Barstow KL, Scornik FS. Spontaneous miniature hyperpolarizations affect threshold for action potential generation in mudpuppy cardiac neurons. *J Neurophysiol* 2002; 88(3): 1119–27.
38. Marcotti W, Johnson SL, Kros CJ. Effects of intracellular stores and extracellular Ca(2+) on Ca(2+)-activated K(+) currents in mature mouse inner hair cells. *J Physiol* 2004; 557(Pt 2): 613–33.
39. Mitra P, Slaughter MM. Mechanism of generation of spontaneous miniature outward currents (SMOCs) in retinal amacrine cells. *J Gen Physiol* 2002; 119(4): 355–72.
40. Davies PJ, Ireland DR, McLachlan EM. Sources of Ca2+ for different Ca(2+)-activated K+ conductances in neurones of the rat superior cervical ganglion. *J Physiol* (Lond) 1996; 495(Pt 2): 353–66.
41. Sah P, Faber ES. Channels underlying neuronal calcium-activated potassium currents. *Prog Neurobiol* 2002; 66(5): 345–53.
42. Storm JF. Action potential repolarization and a fast after-hyperpolarization in rat hippocampal pyramidal cells. *J Physiol* 1987; 385: 733–59.
43. Storm JF. Potassium currents in hippocampal pyramidal cells. *Prog Brain Res* 1990; 83: 161–87.
44. Gu N, Vervaeke K, Storm JF. BK potassium channels facilitate high-frequency firing and cause early spike frequency adaptation in rat CA1 hippocampal pyramidal cells. *J Physiol* 2007; 580(Pt 3): 859–82.
45. Lovell PV, McCobb DP. Pituitary control of BK potassium channel function and intrinsic firing properties of adrenal chromaffin cells. *J Neurosci* 2001; 21(10): 3429–42.
46. Van Goor F, Li YX, Stojilkovic SS. Paradoxical role of large-conductance calcium-activated K+ (BK) channels in controlling action potential-driven Ca2+ entry in anterior pituitary cells. *J Neurosci* 2001; 21(16): 5902–15.
47. Faber ES, Sah P. Physiological role of calcium-activated potassium currents in the rat lateral amygdala. *J Neurosci* 2002; 22(5): 1618–28.
48. Xia XM, Ding JP, Lingle CJ. Molecular basis for the inactivation of Ca2+- and voltage-dependent BK channels in adrenal chromaffin cells and rat insulinoma tumor cells. *J Neurosci* 1999; 19(13): 5255–64.
49. McLarnon JG. Inactivation of a high conductance calcium dependent potassium current in rat hippocampal neurons. *Neurosci Lett* 1995; 193(1): 5–8.
50. Faber ES, Sah P. Ca2+-activated K+ (BK) channel inactivation contributes to spike broadening during repetitive firing in the rat lateral amygdala. *J Physiol* 2003; 552(Pt 2): 483–97.
51. Petrik D, Brenner R. Regulation of STREX exon large conductance, calcium-activated potassium channels by the beta4 accessory subunit. *Neuroscience* 2007; 149(4): 789–803.
52. Brenner R, Chen QH, Vilaythong A, Toney GM, Noebels JL, Aldrich RW. BK channel beta4 subunit reduces dentate gyrus excitability and protects against temporal lobe seizures. *Nat Neurosci*2005; 8(12): 1752–9.
53. Becker MN, Brenner R, Atkinson NS. Tissue-specific expression of a Drosophila calcium-activated potassium channel. *J Neurosci* 1995; 15(9): 6250–9.
54. Hu H, Shao LR, Chavoshy S, et al. Presynaptic Ca2+-activated K+ channels in glutamatergic hippocampal terminals and their role in spike repolarization and regulation of transmitter release. *J Neurosci* 2001; 21(24): 9585–97.
55. Knaus HG, Schwarzer C, Koch RO, et al. Distribution of high-conductance Ca(2+)-activated K+ channels in rat brain: targeting to axons and nerve terminals. *J Neurosci* 1996; 16(3): 955–63.
56. Misonou H, Menegola M, Buchwalder L, et al. Immunolocalization of the Ca2+-activated K+ channel Slo1 in axons and nerve terminals of mammalian brain and cultured neurons. *J Comp Neurol* 2006; 496(3): 289–302.
57. Robitaille R, Charlton MP. Presynaptic calcium signals and transmitter release are modulated by calcium-activated potassium channels. *J Neurosci* 1992; 12(1): 297–305.
58. Flink MT, Atchison WD. Iberiotoxin-induced block of Ca2+-activated K+ channels induces dihydropyridine sensitivity of

ACh release from mammalian motor nerve terminals. *J Pharmacol Exp Ther* 2003; 305(2): 646–52.

59. Skinner LJ, Enee V, Beurg M, et al. Contribution of BK Ca2+-activated K+ channels to auditory neurotransmission in the Guinea pig cochlea. *J Neurophysiol* 2003; 90(1): 320–32.
60. Raffaelli G, Saviane C, Mohajerani MH, Pedarzani P, Cherubini E. BK potassium channels control transmitter release at CA3-CA3 synapses in the rat hippocampus. *J Physiol* 2004; 557(Pt 1): 147–57.
61. Goldberg EM, Watanabe S, Chang SY, et al. Specific functions of synaptically localized potassium channels in synaptic transmission at the neocortical GABAergic fast-spiking cell synapse. *J Neurosci* 2005; 25(21): 5230–5.
62. Nakamura Y, Takahashi T. Developmental changes in potassium currents at the rat calyx of Held presynaptic terminal. *J Physiol* 2007; 581(Pt 3): 1101–12.
63. Bielefeldt K, Jackson MB. A calcium-activated potassium channel causes frequency-dependent action-potential failures in a mammalian nerve terminal. *J Neurophysiol* 1993; 70(1): 284–98.
64. Meredith AL, Thorneloe KS, Werner ME, Nelson MT, Aldrich RW. Overactive bladder and incontinence in the absence of the BK large conductance Ca2+-activated K+ channel. *J Biol Chem* 2004; 279(35): 36746–52.
65. Sausbier M, Hu H, Arntz C, et al. Cerebellar ataxia and Purkinje cell dysfunction caused by Ca2+-activated K+ channel deficiency. *Proc Natl Acad Sci U S A* 2004; 101(25): 9474–8.
66. Pitts GR, Ohta H, McMahon DG. Daily rhythmicity of large-conductance Ca2+ -activated K+ currents in suprachiasmatic nucleus neurons. *Brain Res* 2006; 1071(1): 54–62.
67. Meredith AL, Wiler SW, Miller BH, et al. BK calcium-activated potassium channels regulate circadian behavioral rhythms and pacemaker output. *Nat Neurosci* 2006; 9(8): 1041–9.
68. Langer P, Grunder S, Rusch A. Expression of Ca2+-activated BK channel mRNA and its splice variants in the rat cochlea. *J Comp Neurol* 2003; 455(2): 198–209.
69. Pyott SJ, Glowatzki E, Trimmer JS, Aldrich RW. Extrasynaptic localization of inactivating calcium-activated potassium channels in mouse inner hair cells. *J Neurosci* 2004; 24(43): 9469–74.
70. Ruttiger L, Sausbier M, Zimmermann U, et al. Deletion of the Ca2+-activated potassium (BK) alpha-subunit but not the BKbeta1-subunit leads to progressive hearing loss. *Proc Natl Acad Sci U S A* 2004; 101(35): 12922–7.
71. Pyott SJ, Meredith AL, Fodor AA, Vazquez AE, Yamoah EN, Aldrich RW. Cochlear function in mice lacking the BK channel alpha, beta1, or beta4 subunits. *J Biol Chem* 2007; 282(5): 3312–24.
72. Martin GK, Vazquez AE, Jimenez AM, Stagner BB, Howard MA, Lonsbury-Martin BL. Comparison of distortion product otoacoustic emissions in 28 inbred strains of mice. *Hear Res* 2007; 234(1–2): 59–72.
73. Laumonnier F, Roger S, Guerin P, et al. Association of a functional deficit of the BKCa channel, a synaptic regulator of neuronal excitability, with autism and mental retardation. *Am J Psychiatry* 2006; 163(9): 1622–9.
74. Du W, Bautista JF, Yang H, et al. Calcium-sensitive potassium channelopathy in human epilepsy and paroxysmal movement disorder. *Nat Genet* 2005; 37(7): 733–8.
75. Diez-Sampedro A, Silverman WR, Bautista JF, Richerson GB. Mechanism of increased open probability by a mutation of the BK channel*J Neurophysio*l 2006; 96(3): 1507–16.
76. Rudy B, McBain CJ. Kv3 channels: Voltage-gated K+ channels designed for high-frequency repetitive firing. *Trends Neurosci* 2001; 24(9): 517–26.
77. Lau D, Vega-Saenz de Miera EC, Contreras D, et al. Impaired fast-spiking, suppressed cortical inhibition, and increased susceptibility to seizures in mice lacking Kv3.2 K+ channel proteins. *J Neurosci* 2000; 20(24): 9071–85.
78. Riazi MA, Brinkman-Mills P, Johnson A, et al. Identification of a putative regulatory subunit of a calcium-activated potassium channel in the dup(3q) syndrome region and a related sequence on 22q11.2. *Genomics* 1999; 62(1): 90–4.
79. Behrens R, Nolting A, Reimann F, Schwarz M, Waldschutz R, Pongs O. hKCNMB3 and hKCNMB4, cloning and characterization of two members of the large-conductance calcium-activated potassium channel beta subunit family. *FEBS Lett* 2000; 474(1): 99–106.
80. Brenner R, Jegla TJ, Wickenden A, Liu Y, Aldrich RW. Cloning and functional characterization of novel large conductance calcium-activated potassium channel beta subunits, hKCNMB3 and hKCNMB4. *J Biol Chem* 2000; 275(9): 6453–61.

81. Steinbach P, Adkins WN, Jr., Caspar H, et al. The dup(3q) syndrome: Report of eight cases and review of the literature. *Am J Med Genet* 1981; 10(2): 159–77.
82. Wilson GN, Dasouki M, Barr M, Jr. Further delineation of the dup(3q) syndrome. *Am J Med Genet* 1985; 22(1): 117–23.
83. Hu S, Labuda MZ, Pandolfo M, Goss GG, McDermid HE, Ali DW. Variants of the KCNMB3 regulatory subunit of maxi BK channels affect channel inactivation. *Physiol Genomics* 2003; 15(3): 191–8.
84. Lorenz S, Heils A, Kasper JM, Sander T. Allelic association of a truncation mutation of the KCNMB3 gene with idiopathic generalized epilepsy. *Am J Med Genet B Neuropsychiatr Genet* 2007; 144(1): 10–13.
85. Xia XM, Ding JP, Zeng XH, Duan KL, Lingle CJ. Rectification and rapid activation at low Ca2+ of Ca2+-activated, voltage-dependent BK currents: Consequences of rapid inactivation by a novel beta subunit. *J Neurosci* 2000; 20(13): 4890–903.
86. Zeng XH, Xia XM, Lingle CJ. Redox-sensitive extracellular gates formed by auxiliary beta subunits of calcium-activated potassium channels. *Nat Struct Biol* 2003; 10(6): 448–54.
87. Lippiat JD, Standen NB, Harrow ID, Phillips SC, Davies NW. Properties of BK(Ca) channels formed by bicistronic expression of hSloalpha and beta1-4 subunits in HEK293 cells. *J Membr Biol* 2003; 192(2): 141–8.
88. Weiger TM, Holmqvist MH, Levitan IB, et al. A novel nervous system beta subunit that downregulates human large conductance calcium-dependent potassium channels. *J Neurosci*2000; 20(10): 3563–70.
89. Meera P, Wallner M, Toro L. A neuronal beta subunit (KCNMB4) makes the large conductance, voltage- and Ca2+-activated K+ channel resistant to charybdotoxin and iberiotoxin. *Proc Natl Acad Sci U S A* 2000; 97(10): 5562–7.
90. Dopico AM, Widmer H, Wang G, Lemos JR, Treistman SN. Rat supraoptic magnocellular neurones show distinct large conductance, Ca2+-activated K+ channel subtypes in cell bodies versus nerve endings. *J Physiol* 1999; 519(Pt 1): 101–14.
91. Reinhart PH, Chung S, Levitan IB. A family of calcium-dependent potassium channels from rat brain. *Neuron*1989; 2(1): 1031–41.
92. Reinhart PH, Chung S, Martin BL, Brautigan DL, Levitan IB. Modulation of calcium-activated potassium channels from rat brain by protein kinase A and phosphatase 2A. *J Neurosci* 1991; 11(6): 1627–35.
93. Reinhart PH, Levitan IB. Kinase and phosphatase activities intimately associated with a reconstituted calcium-dependent potassium channel. *J Neurosci* 1995; 15(6): 4572–9.
94. Wang G, Lemos JR. Tetrandrine blocks a slow, large-conductance, Ca(2+)-activated potassium channel besides inhibiting a non-inactivating Ca2+ current in isolated nerve terminals of the rat neurohypophysis. *Pflugers Arch* 1992; 421(6): 558–65.
95. Ghatta S, Nimmagadda D, Xu X, O'Rourke ST. Large-conductance, calcium-activated potassium channels: Structural and functional implications. *Pharmacol Ther* 2006; 110(1): 103–16.
96. Cavalleri GL, Weale ME, Shianna KV, et al. Multicentre search for genetic susceptibility loci in sporadic epilepsy syndrome and seizure types: A case-control study. *Lancet Neurol* 2007; 6(11): 970–80.
97. Ha TS, Heo MS, Park CS. Functional effects of auxiliary beta4-subunit on rat large-conductance Ca(2+)-activated K(+) channel. *Biophys J* 2004; 86(5): 2871–82.
98. Wang B, Rothberg BS, Brenner R. Mechanism of ß4 Subunit Modulation of BK Channels. *J Gen Physiol* 2006; 127: 449–65.
99. Stocker M. Ca(2+)-activated K(+) channels: Molecular determinants and function of the SK family. *Nat Rev Neurosci* 2004; 5 (10): 758–70.
100. Nadler JV. The recurrent mossy fiber pathway of the epileptic brain. *Neurochem Res* 2003; 28(11): 1649–58.
101. Kobayashi M, Buckmaster PS. Reduced inhibition of dentate granule cells in a model of temporal lobe epilepsy. *J Neurosci* 2003; 23(6): 2440–52.
102. Henze DA, Wittner L, Buzsaki G. Single granule cells reliably discharge targets in the hippocampal CA3 network in vivo. *Nat Neurosci* 2002; 5(8): 790–5.
103. Mori M, Abegg MH, Gahwiler BH, Gerber U. A frequency-dependent switch from inhibition to excitation in a hippocampal unitary circuit. *Nature* 2004; 431(7007): 453–6.
104. Mori M, Gahwiler BH, Gerber U. Recruitment of an inhibitory hippocampal network after bursting in a single granule cell. *Proc Natl Acad Sci U S A* 2007; 104(18): 7640–5.
105. Toth K, Suares G, Lawrence JJ, Philips-Tansey E, McBain CJ. Differential mechanisms of transmission at three types of mossy fiber synapse. *J Neurosci* 2000; 20(22): 8279–89.

106. Angenstein F, Kammerer E, Niessen HG, Frey JU, Scheich H, Frey S. Frequency-dependent activation pattern in the rat hippocampus, a simultaneous electrophysiological and fMRI study. *Neuroimage* 2007; 38(1): 150–63.
107. Dietrich D, Clusmann H, Kral T, et al. Two electrophysiologically distinct types of granule cells in epileptic human hippocampus. *Neuroscience*1999; 90(4): 1197–206.
108. Beck H, Steffens R, Heinemann U, Elger CE. Ca(2+)-dependent inactivation of high-threshold Ca(2+) currents in hippocampal granule cells of patients with chronic temporal lobe epilepsy. *J Neurophysiol* 1999; 82(2): 946–54.
109. Beck H, Steffens R, Elger CE, Heinemann U. Voltage-dependent Ca2+ currents in epilepsy. *Epilepsy Res* 1998; 32(1–2): 321–32.
110. Gorter JA, Borgdorff AJ, van Vliet EA, Lopes da Silva FH, Wadman WJ. Differential and long-lasting alterations of high-voltage activated calcium currents in CA1 and dentate granule neurons after status epilepticus. *Eur J Neurosci* 2002; 16(4): 701–12.
111. Nagerl UV, Mody I, Jeub M, Lie AA, Elger CE, Beck H. Surviving granule cells of the sclerotic human hippocampus have reduced Ca(2+) influx because of a loss of calbindin-D(28 k) in temporal lobe epilepsy. *J Neurosci* 2000; 20(5): 1831–6.
112. Gorter JA, van Vliet EA, Aronica E, et al. Potential new antiepileptogenic targets indicated by microarray analysis in a rat model for temporal lobe epilepsy. *J Neurosci* 2006; 26(43): 11083–110.
113. Shruti S, Clem RL, Barth AL. A seizure-induced gain-of-function in BK channels is associated with elevated firing activity in neocortical pyramidal neurons. *Neurobiol Dis* 2008; 30(3): 323–30.

Chapter 7

Mouse Models of Benign Familial Neonatal Convulsions (BFNC): Mutations in *KCNQ (Kv7)* Genes

Nanda A. Singh, James F. Otto, Mark F. Leppert, H. Steve White, and Karen S. Wilcox

Abstract

Benign familial neonatal convulsions (BFNC) is caused by mutations in the *KCNQ2 (Kv7.2)* or *KCNQ3 (Kv7.3)* genes. These genes encode the KCNQ2 and KCNQ3 subunits that comprise the neuronal M-type potassium channel (M channel). While numerous studies have provided evidence for the inhibitory role of normally functioning M channels in key brain structures related to seizures and epileptogenesis, the BFNC sequelae from mutation to seizure and ultimately to remission is likely very complex. In an effort to determine the role of the KCNQ genes in epilepsy, a number of mouse models with either spontaneous or transgenic mutations or targeted deletions in the *Kcnq2 or Kcnq3* gene have been described. We discuss seminal findings from the *Kcnq2* knockout, the dominant-negative *Kcnq2* G279S, the *Szt1*, and finally the recently described *Kcnq2* and *Kcnq3* knockin, mice. The approach of combining whole-animal behavior with single-cell biophysics in mouse models of BFNC has helped solidify the link between attenuated $I_{K(M)}$ function and increased seizure susceptibility that results from *Kcnq2* and *Kcnq3* mutations. The mouse models described here have all significantly contributed to our understanding of how mutations in these genes might precipitate human epilepsy and further elucidated the neurophysiologic role of $I_{K(M)}$.

Key words: benign familial neonatal convulsions, potassium channel, M-current, Kv7.2, Kv7.3, knockin.

1. Introduction

The remarkable plasticity of the neonatal brain following frequent unprovoked seizures challenged epileptologists to investigate the molecular defects underlying benign familial neonatal convulsions (BFNC). Rett and Teubel first described this early-onset autosomal dominant idiopathic epilepsy disorder in 1964 *(1)*. In otherwise thriving neonates, recurrent seizures typically

Scott C. Baraban (ed.), *Animal Models of Epilepsy: Methods and Innovations, vol. 40*

DOI 10.1007/978-1-60327-263-6_7 Springerprotocols.com

begin in the first 3 days of life and can be generalized or partial, clonic and occasionally tonic and include facial automatisms. Unremarkable interictal electroencephalographic (EEG) recordings are found in most BFNC neonates *(2, 3)*. Although the expressivity of BFNC can vary from single daily seizures to dozens of seizures per day, at approximately 3–4 months of age, seizures spontaneously remit and normal development follows in the majority of cases. Families with BFNC exhibit incomplete penetrance because approximately 15% of individuals with the disease chromosome never manifest seizures *(4)*. However, one in six individuals is diagnosed with epilepsy later in life, a rate that is considerably higher than the population prevalence of seizures *(2, 5)*. Thus, mutations underlying BFNC set the stage for enhanced seizure susceptibility.

Discovery of the genetic defects associated with BFNC was catalyzed in the pre-genome era of the late 1980s and early 1990s by more detailed genetic maps and improved genotyping techniques combined with carefully diagnosed large families described in the clinical literature *(4, 6, 7)*. Genetic linkage analysis demonstrated that mutations at either of two independent loci could result in BFNC *(4, 8)*. Although most families were too small to exhibit significant linkage, a few large families linked to the tip of the long arm of chromosome 20 and a single family linked to chromosome 8q. Positional cloning efforts on chromosome 20q revealed a novel potassium channel *KCNQ2 (Kv7.2)* that contained unique missense mutations, a large multiexonic deletion, and a frameshift mutation in unrelated families *(9, 10)*. Bioinformatic analysis of early versions of expressed sequence tag (EST) databases and molecular biology experiments that included cDNA library screening led Charlier et al., to identify an additional KCNQ gene family homolog that mapped to chromosome 8q *(11)*. This gene, *KCNQ3 (Kv7.3)*, contained a pore missense mutation in the only original family that linked to this locus. Subsequent studies in the following 10 years have shown that *KCNQ2* is the predominant BFNC gene because over 60 mutations are known (**Fig. 7.1**) whereas only 4 mutations have been reported in *KCNQ3* (**Fig. 7.2**). Missense and insertions/deletions make up the majority of *KCNQ2* mutations and only missense mutations in the pore region of *KCNQ3* have been found. More recently developed quantitative genomic detection methods have also uncovered multiexonic deletions and duplications in BFNC cases *(12)*.

As more patients with *KCNQ2* mutations are described in the literature, the BFNC phenotype is infrequently associated with a wider clinical spectrum to include myokymia *(13)*, rolandic epilepsy *(5, 14, 15, 16)*, drug-resistant epilepsy and/or mental retardation *(17, 18, 19, 20)*, and peripheral nerve

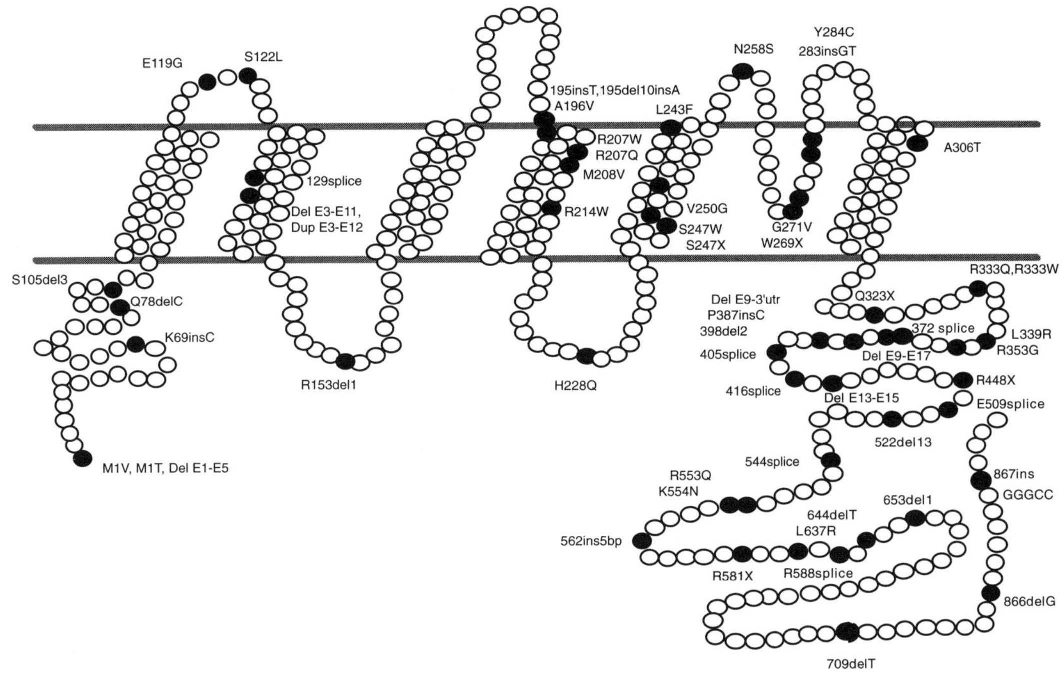

Fig. 7.1 **Location of 63 mutations found in the *KCNQ2* gene in unrelated families with benign familial neonatal convulsions.** Numbering system is based on the full-length 872 amino acid splice variant (NM 172107). Figure is modified from Singh et al. *(5)*.

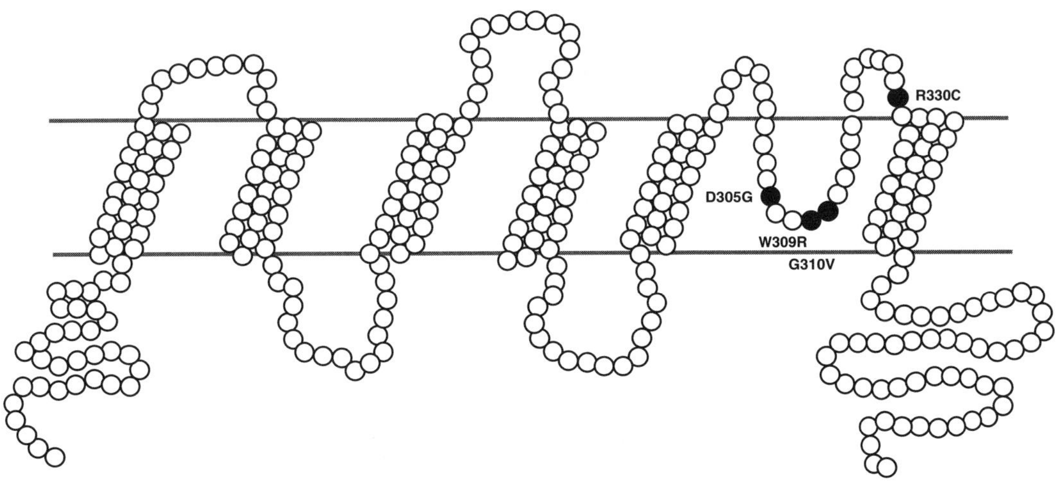

Fig. 7.2 **Location of four missense mutations found in the *KCNQ3* gene in families with benign familial neonatal convulsions.** Figure is modified from Singh et al. *(5)*.

hyperexcitability *(21)*. De novo *KCNQ2* mutations have also been uncovered in patients with benign neonatal seizures and no family history, yet their seizures remit like most familial cases *(22)*.

1.1. KCNQ2, KCNQ3, *and* KCNQ5 *Encode Subunits of the M Channel*

The *KCNQ2*, *KCNQ3*, and *KCNQ5* genes encode the KCNQ2, KCNQ3, and KCNQ5 subunits, respectively, that comprise the M-type potassium channel (M channel) *(9, 10, 23, 24)*. The current that passes through the M channel is termed the M current [$I_{K(M)}$] and is present in many neurons in the central nervous system *(25)*. The KCNQ2, KCNQ3, and KCNQ5 subunits have been shown to co-assemble in several tetrameric configurations in many different expression systems, but all M channels formed possess characteristics similar to those of the native neuronal $I_{K(M)}$. Most channel subunit expression studies to date have been conducted in the *Xenopus laevis* oocyte, Chinese hamster ovary (CHO) cell, and human embryonic kidney (HEK) cell expression systems. Such studies have established that while KCNQ2, KCNQ3, and KCNQ5 subunits can all form functional homomeric channels, the current amplitude through heteromeric channels is usually increased several fold. For example, in *Xenopus* oocytes, coexpression of KCNQ2 and KCNQ3 increases current amplitude by 10-fold relative to homomeric KCNQ2 and KCNQ3 channels *(23, 26)*. KCNQ5 homomeric channels also form in *Xenopus* oocytes, but coexpression with KCNQ3 results in a 5-fold increase in current amplitude *(24)*. There is also evidence that heteromeric channels containing all three subunits do form and are functional and that the inclusion of KCNQ5 subunits downregulates the total number of KCNQ2/KCNQ3 heteromeric channels assembled within the same cell *(27)*. Heteromeric KCNQ2/KCNQ5 channels also form, but their currents are similar in amplitude to those recorded in their respective homomeric channels.

In vitro studies designed to mimic the BFNC channelopathy show that some mutations cause a modest reduction in the heteromeric KCNQ2/KCNQ3 current in *Xenopus* oocytes *(28)*. Although the majority of BFNC mutations are splice site, nonsense, and insertion/deletion mutations predicted to cause haploinsufficiency, a variety of seizure susceptibility mechanisms have been elucidated for missense mutations using in vitro preparations. Two well-studied KCNQ2 missense mutations, Y284C and A306T, cause a heteromeric current reduction by 20–40% while having no effect on membrane surface expression in transfected *Xenopus* oocytes *(26, 29)*. Comparable effects are seen in the BFNC-causing *KCNQ3* pore mutations, D305G and G310V *(5, 11, 29)*. Atypical gating effects are seen in all six *KCNQ2* S4 transmembrane domain mutations, including dramatic effects on the voltage dependence of activation and rate of deactivation *(5, 21, 30–32)*. Finally, while it is likely that *KCNQ2* truncation mutations are subject to nonsense-mediated mRNA decay in vivo, expressed truncated KCNQ2 proteins are thought to assemble poorly and be unstable even if they reach the surface membrane, while having benign effects on biophysical properties of heteromultimeric channels *(5, 26, 33, 34)*.

In the mammalian brain, *Kcnq2*, *Kcnq3*, and *Kcnq5* mRNAs are coexpressed in many brain regions associated with seizure activity, including the hippocampus, neocortex, and striatum *(35)*; however, the expression patterns of individual subunits do display subtle differences. In the thalamus, for example, *Kcnq3* mRNA expression is qualitatively greater than that of *Kcnq2 (35)*. In hindbrain structures such as the cerebellum, it appears that only the KCNQ2 subunit mRNA is appreciably expressed. It is assumed, therefore, that many different M channel subpopulations exist in different stoichiometric configurations in various types of neurons *(27)*. Indeed, work by Cooper and colleagues has established that in CA1 pyramidal neurons, KCNQ2 subunits can be expressed with or without KCNQ3 or KCNQ5 *(36, 37)*. This finding is remarkable because it is often assumed that if a homomeric channel is expressed but produces an especially weak current in model systems (i.e., KCNQ2), then it is not likely to play a significant role in neuron physiology. The discrete location of KCNQ2 at the nodes of Ranvier, however, observed by Devaux et al. *(37)* suggests the opposite. More recent studies show that KCNQ2 is also expressed at more distal axonal surface regions where it can modulate transmission during hyperexcitable conditions *(38, 39, 40)*. Interestingly, a 50–75% reduction in this axonal surface expression is observed when the BFNC *Kcnq2* A306T and *Kcnq3* G310V patient mutations are introduced into hippocampal cultures *(39)*.

1.2. The Role of the M Current in Neuron Physiology

$I_{K(M)}$ is a potassium current that is tonically active at resting membrane potential (–65 mV), activates in response to membrane depolarization, does not inactivate, and is blocked in response to muscarinic acetylcholine receptor activation *(41, 42)*. The "M" current is thus named for its sensitivity to muscarine. $I_{K(M)}$ contributes to repolarization of the membrane and is the only voltage-gated outward current sustained in the range of action potential initiation *(25)*; moreover, it plays a major role in setting and maintaining the resting membrane potential of many neurons *(43)*. In hippocampal CA1 pyramidal neurons, KCNQ2 and KCNQ3 subunit proteins are largely coexpressed at the axon initial segment (AIS), where action potentials are generated *(37)*. Indeed, $I_{K(M)}$ provides a mechanism by which cells regulate their own excitability; in response to a prolonged membrane depolarization, activation of $I_{K(M)}$ repolarizes the membrane and attenuates action potential frequency *(44, 45)*. Pharmacologic suppression of the M channel slows membrane repolarization, delaying the decay phase of the action potential as well as the fast afterdepolarization *(46)*, and can interfere with spike frequency adaptation *(47)*.

Physiologic control of the M channel is mediated by multiple second messenger systems including Src tyrosine kinase *(48, 49)*, calmodulin *(50–52)*, A-kinase-anchoring protein/protein kinase C *(53)*, and PIP2 *(54, 55)*. Interestingly, select BFNC mutations have been shown to interrupt these second messenger modulatory interactions to impair KCNQ2 trafficking and reduce M current, thereby conferring pro-excitatory phenotypes. For example, using a yeast two-hybrid assay, the C-terminal missense mutation R353G was found to impair binding of the calmodulin calcium sensor to cause retention of the mutant protein in the endoplasmic reticulum *(56, 57)*.

2. Mouse Models of Kv7 Mutations

While numerous studies have provided evidence for the inhibitory role of normally functioning M channels in key structures related to seizures and epileptogenesis, the BFNC sequelae from mutation to seizure to remission is likely very complex. It is not surprising that ultimate proof of the role of these essential genes in hyperexcitability and the effects of inherited mutations required the whole-animal model. In an effort to determine the role of the KCNQ genes in epilepsy, a number of mouse models with either spontaneous or transgenic mutations or targeted deletions in the *Kcnq2* gene have been described. These mice will be discussed here.

2.1. Kcnq2 Knockout Mouse

A *Kcnq2* knockout mouse was first described by Watanabe et al. *(58)*. After normal in utero development and birth in the expected Mendelian ratio, homozygous knockout offspring died at postnatal day 1, indicating that the *Kcnq2* gene was essential for life. Examination of the homozygous pups suggested that they died from pulmonary atelectasis. In contrast, the heterozygous pups were fully viable, although surprisingly these animals did not exhibit spontaneous seizures during the time frame in which they were examined with EEG recordings. These authors did not record from neurons to determine if the M-current was altered. However, the heterozygous mice were found to be significantly more susceptible to pentylenetetrazol-induced seizures, suggesting that the absence of one *Kcnq2* allele confers an increase in seizure susceptibility over wild-type littermate control mice.

2.2. Dominant-Negative Kcnq2 *G279S Mutation*

It has been shown that conditional transgenic overexpression of a dominant-negative *Kcnq2* G279S mutant allele reduces $I_{K(M)}$ amplitude, increases neuronal excitability in mice, and can result in seizures *(59)*. While this particular mutation has been observed in the cardiac *KCNQ1* gene in a family with Long QT syndrome, the G279S mutation has not been found in any BFNC families to date *(60, 61)*. In this particular mouse model, only males were found to have seizures, and only if expression of the transgene occurred throughout the life of the animal. This gender difference, which is not observed in the BFNC patient population, is likely due to the fact that the transgene integrated into the murine X chromosome and was subject to random X-inactivation and mosaic expression in female mutant mice. A further caveat of this transgenic model is the unknown number of integrated and subsequently transcribed mutant gene copies that can be strikingly different from the BFNC condition. Nevertheless, in male mutant animals, M-currents recorded in hippocampal neurons were dramatically decreased. Consequently, spike frequency adaptation was attenuated in these cells, although resting membrane potential was unaltered. While being one of the first groups to describe the effect of an M-current mutation on seizure activity and M-current function, the Tet-Off system used to drive mutant *Kcnq2* over-expression significantly interfered with the expression of *both* wild-type KCNQ2 and KCNQ3 subunits. Due to the pervasive alterations in M channel subunit expression levels and the lack of this mutation in BFNC families, these results are presumably not analogous to the human condition.

2.3. Szt1 *Spontaneous C-Terminal Deletion Mutation*

The *Szt1* mouse was discovered using electroshock seizure threshold testing during a screen for ENU-induced mutations related to epilepsy. Surprisingly, one of the parent lines was found to have a heritable reduction in seizure threshold *before* treatment with ENU *(62)*. The *Szt1* mouse was subsequently found to carry a spontaneous 300 kb deletion on the C57BL/6 genetic background that includes the genomic DNA encoding the KCNQ2 C-terminus. While a *KCNQ2*-specific C-terminal deletion was previously identified as the cause of BFNC in a Czech family *(63)*, 40 out of 63 published mutations (*see* **Fig. 7.1**) could also potentially result in *KCNQ2* haploinsufficiency. Initial work with this mouse indicated that like the *Kcnq2* knockout mouse described above, the homozygous *Szt1* mutation was lethal shortly after birth, with lung atelectasis most likely the cause of death. However, the heterozygous mice are viable and although they do not appear to have spontaneous seizures, these mice have a substantial reduction in seizure threshold in several different corneal stimulation models *(62, 64)*. Interestingly, these mice also displayed a decreased sensitivity to the novel anticonvulsant retigabine (RGB) in the 6 Hz model of psychomotor seizures. The main mechanism of action for retigabine is thought to be through shifting the activation curve for I_M to a more hyperpolarized potential, which results in hyperpolarization of the resting membrane potential and a decrease in the ability to generate action potentials *(65–68)*. Interestingly, the heterozygous *Szt1* mice were also more sensitive to the M channel blocker, linopirdine (LPD), in the minimal clonic seizure paradigm *(64)*.

Electrophysiology experiments were performed on CA1 neurons in the in vitro hippocampal brain slices obtained from wild-type and *Szt1* mice to test the hypothesis that this mutation alters $I_{K(M)}$ function and pharmacology. CA1 neurons in *Szt1* mice were found to have a decreased $I_{K(M)}$ amplitude and current density compared to that of wild-type C57BL/6 littermates. Moreover, action potential accommodation was compromised in *Szt1* CA1 neurons. These results were the first to show that a BFNC-related *Kcnq2* mutation attenuates native $I_{K(M)}$ amplitude and consequently increases neuronal excitability. Differences in $I_{K(M)}$ pharmacology in *Szt1* CA1 neurons had intriguing parallels to the in vivo work and included increased sensitivity to LPD and decreased sensitivity to RGB. In addition, $I_{K(M)}$ recorded in *Szt1* CA1 neurons was insensitive to tetraethylammonium (TEA), a blocker of KCNQ2 subunit-containing M channels *(47)*. These results shed significant light on the consequences of M channel loss of function as it relates to neuroexcitability and seizure generation, and thus accentuate $I_{K(M)}$ as a therapeutic target for the treatment of epilepsy.

One of the drawbacks of the *Szt1* mouse is that the spontaneous mutation encompasses more than the *Kcnq2* gene. The 300-kb deletion in the *Szt1* mouse includes two genes, the *Chrna4* (nACh receptor α_4 subunit) that contains missense mutations in patients with an inherited partial epilepsy and *Arfgap* (guanosine triphosphatase (GTPase)-activating protein) that inactivates adenosine diphosphate (ADP)-ribosylation factor 1 *(62)*. While there is no evidence to suggest that deletion of either the *Chrna4* or *Arfgap1* genes contributes to the hyperexcitability observed, that possibility cannot be ruled out. So there is still a need to investigate mouse models that more precisely mimic the human condition.

2.4. Kcnq2 *Knockin Mice*

The ideal mouse model of BFNC would have the exact mutation that causes epilepsy in humans (**Figs. 7.1 and 7.2**) and express an identifiable seizure phenotype. Rather than a *Kcnq2* or *Kcnq3* knock *out*, the ideal model of BFNC would have a clinically and heterologously well-characterized patient mutation engineered into the mouse ortholog using homologous recombination technology, otherwise known as a knock *in*. To this end, we developed both a *Kcnq2* S6 missense knockin and a *Kcnq3* pore missense knockin and bred each to the C57BL/6 and the FVB/N genetic strains *(69)*. These first ever orthologous mouse models of BFNC display novel features consistent with the human disorder and thus represent a significant advance in our ability to understand the role of the M-current in seizure generation, seizure susceptibility, and seizure remission. The effect of a single nucleotide change is particularly striking when evaluated in two different murine genetic backgrounds and has given insight into the incomplete penetrance and variable expressivity seen in BFNC patients.

In the murine *Kcnq2* gene, we introduced the A306T S6 transmembrane domain mutation first identified in a large kindred consisting of 69 affected individuals with benign familial neonatal convulsions and a systematically documented later-onset seizure phenotype *(2)*. In this family, seizures began on day 3 in 42% of individuals and included clonic and tonic features, ocular symptoms, facial automatisms, and autonomic and respiratory features as well. Remission occurred in the first 6 weeks in 68% of BFNC individuals in this family, and 16 individuals experienced additional seizures after the neonatal period, including 6 who had epilepsy as adults. As mentioned earlier, this mutation has been extensively characterized in heterologous expression systems by many laboratories. Video or video-EEG monitoring of homozygous *Kcnq2*$^{A306T/A306T}$ mice documented spontaneous generalized tonic seizures as early as postnatal day 13 on both genetic backgrounds examined. Significant mortality in P15–P30 homozygous *Kcnq2*$^{A306T/A306T}$ mice with this mutation was noted and likely is the result of severe seizures. Yet in spite of this seizure-related mortality, hippocampal histological findings were unremarkable for cell death using apoptotic and necrotic markers *(69)*. Like the *Kcnq2* knockout model and *Szt1* mouse, adult heterozygous *Kcnq2*$^{A306T/+}$ mice exhibit a reduced threshold for induced seizures. Convulsive current curves generated from electroconvulsive threshold testing in adult B6;129-*Kcnq2*$^{A306T/+}$and littermate control B6;129*Kcnq2*$^{+/+}$ mice show that in male mice, the *Kcnq2* A306T mutation resulted in a significant reduction in the threshold of tonic hindlimb extension seizures. This finding is particularly noteworthy because 16% of the BFNC individuals with this specific mutation have seizures in adulthood, a rate that is higher than the general population *(2)* and suggests that these patients have a reduced threshold for seizure generation.

Perforated patch recording in CA1 pyramidal neurons of the acute brain slice preparation demonstrated a significant reduction in the M-current ($I_{K(M)}$) amplitude in mice homozygous for A306T on the C57Bl/6 background. Furthermore, when these currents were normalized as a function of whole cell capacitance, it was determined that there was a significant reduction of $I_{K(M)}$ density. While the amplitude of $I_{K(M)}$ was not altered in the heterozygous animals, deactivation of $I_{K(M)}$ was found to be accelerated in both heterozygous and homozygous mice *(69)*. This suggests that there is a functional alteration of $I_{K(M)}$ in neurons in these mice that could contribute significantly to hyperexcitability and enhanced susceptibility to seizures.

2.5. Kcnq3 *Knockin Mice*

The largest family characterized to date with a *KCNQ3* mutation consists of 13 affected members who exhibited the classic neonatal features of BFNC *(70)*. In multiple family members, seizures began between postnatal day 2 and 14 and spontaneously remitted in all cases by 4 months. In the proband, these seizures consisted of both partial and generalized clonic convulsions. All affected individuals appear to have normal intellectual development and none of the five individuals in the parental and grandparental generation experienced seizures past the neonatal period *(70)*. We successfully generated mice with an orthologous *Kcnq3* pore mutation that is identical to this family: the G311V mutation *(69)*. $Kcnq3^{G311V/G311V}$ mice experienced almost daily spontaneous recurrent generalized seizures into adulthood and genetic background appeared to have a robust effect on this phenotype. Interestingly, histologic assessment in $Kcnq3^{G311V/G311V}$ mice revealed no mossy fiber sprouting or obvious neuronal cell death in the hippocampal dentate gyrus and pyramidal cell layers. This is remarkably different from most adult mouse models that exhibit equally severe and frequent seizures but require status epilepticus, mossy fiber sprouting, and neuronal death for the expression of such seizures *(71, 72)*. Further investigation into the $Kcnq3^{G311V/G311V}$ model should help elucidate this protective neuroplasticity that is likely to account for the notably benign course of seizures and cognitive development seen in the majority of patients with BFNC.

To examine the effects of the *Kcnq3* G311V knockin pore mutation on adulthood seizure thresholds, electroconvulsive threshold testing was performed in B6;129-$Kcnq3^{G311V/+}$ and FVB;129-$Kcnq3^{G311V/+}$ heterozygous mutant and littermate control mice. Convulsive current values at which 50% of mice exhibit tonic hindlimb extension seizures were significantly less for the $Kcnq3^{G311V/+}$mice than for the wild-type control mice on both C57BL/6 (11.6 vs. 13.6 mA) and FVBN (7.0 vs. 8.8 mA) genetic backgrounds. Such a reduction in seizure threshold of adult heterozygous mutant mice suggests that this mutation results in mice that are at greater inherent risk of experiencing seizures.

A significant reduction compared to wild-type littermate control mice was observed in the amplitude of the M-current recorded using the perforated patch clamp technique in CA1 neurons in acute brain slices obtained from $Kcnq3^{G311V/G311V}$mice on both backgrounds. As was the case for the Kcnq2 A306T knockin mice, when these currents were normalized as a function of whole cell capacitance, it was determined that there was a significant reduction of $I_{K(M)}$ density in all CA1 cells recorded in slices obtained from $Kcnq3^{G311V/G311V}$ mice. While FVB:129 mice heterozygous for the G311V mutation did not have any detectable changes in M-current amplitude or kinetics, $I_{K(M)}$ recorded in B6;129-$Kcnq3^{G311V/+}$ mice were found to have a modest reduction in deactivation kinetics *(69)*. Additional experiments in other cell types in the heterozygous animals may reveal alterations in M-current and network function that contribute to the decreased seizure thresholds observed in these animals.

3. Conclusion

The approach of combining whole-animal behavior with single-cell biophysics in mouse models of BFNC has helped solidify the link between attenuated $I_{K(M)}$ function and increased seizure susceptibility that result from *Kcnq2* and *Kcnq3* mutations. Thus, the mouse models described here have all significantly contributed to our understanding of how *Kcnq* mutations might precipitate human epilepsy and further elucidated the neurophysiologic role of $I_{K(M)}$. In view of the fact that proper $I_{K(M)}$ function is so critically involved in regulating neuronal excitability, modulation of this current should prove useful in the management of not only BFNC but also other forms of epilepsy. Moreover, drugs designed to increase $I_{K(M)}$ activity could in fact lead to improved treatment in pathologies characterized by aberrant neuronal excitability such as chronic pain, migraine, bipolar disorder, and other neurological and psychiatric disorders. Finally, the *Kcnq3* homozygous knockin mice on the FVB genetic background provides an unambiguous, reproducible seizure phenotype that can readily be used for the evaluation of mechanistically novel antiepileptic drugs.

Acknowledgments

This work was supported in part by grants from the NIH (RO1 NS-32666 to ML, RO1 NS-44210 to KSW), the W.M. Keck Foundation (to M.L.), and the Primary Children's Medical Center Foundation (KSW).

References

1. Rett A, Teubel R.Neugeborenen Krampfe im Rahmen einer epileptisch belasten Familie. *Wiener Klinische Wochenschrift* 1964;76:609–613.
2. Ronen GM, Rosales TO, Connolly M, Anderson VE, Leppert M. Seizure characteristics in chromosome 20 benign familial neonatal convulsions. *Neurology* 1993;43 (7):1355–1360.
3. Zonana J, Silvey K, Strimling B. Familial neonatal and infantile seizures: An autosomal-dominant disorder. *Am J Med Genet* 1984;18(3):455–459.
4. Leppert M, Anderson VE, Quattlebaum T, Stauffer D, O'Connell P, Nakamura Y, et al. Benign familial neonatal convulsions linked to genetic markers on chromosome 20. *Nature* 1989;337(6208):647–648.
5. Singh NA, Westenskow P, Charlier C, Pappas C, Leslie J, Dillon J, et al. KCNQ2 and KCNQ3 potassium channel genes in benign familial neonatal convulsions: Expansion of the functional and mutation spectrum. *Brain* 2003;126(Pt 12):2726–2737.
6. Leppert M, Anderson VE, White R. The discovery of epilepsy genes by genetic linkage. *Epilepsy Res Suppl* 1991;4:181–188.
7. Leppert M, McMahon WM, Quattlebaum TG, Bjerre I, Zonana J, Shevell MI, et al. Searching for human epilepsy genes: A progress report. *Brain Pathol* 1993;3(4): 357–369.

8. Lewis TB, Leach RJ, Ward K, O'Connell P, Ryan SG. Genetic heterogeneity in benign familial neonatal convulsions: identification of a new locus on chromosome 8q. *Am J Hum Genet* 1993;53(3):670–675.
9. Biervert C, Schroeder BC, Kubisch C, Berkovic SF, Propping P, Jentsch TJ, et al. A potassium channel mutation in neonatal human epilepsy. *Science* 1998;279(5349):403–406.
10. Singh NA, Charlier C, Stauffer D, DuPont BR, Leach RJ, Melis R, et al. A novel potassium channel gene, KCNQ2, is mutated in an inherited epilepsy of newborns. *Nat Genet* 1998;18(1):25–29.
11. Charlier C, Singh NA, Ryan SG, Lewis TB, Reus BE, Leach RJ, et al. A pore mutation in a novel KQT-like potassium channel gene in an idiopathic epilepsy family. *Nat Genet* 1998;18(1):53–55.
12. Heron SE, Cox K, Grinton BE, Zuberi SM, Kivity S, Afawi Z, et al. Deletions or duplications in KCNQ2 can cause benign familial neonatal seizures. *J Med Genet* 2007; 44(12):791–796.
13. Dedek K, Kunath B, Kananura C, Reuner U, Jentsch TJ, Steinlein OK. Myokymia and neonatal epilepsy caused by a mutation in the voltage sensor of the KCNQ2 K+ channel. *Proc Natl Acad Sci USA* 2001; 98(21):12272–12277.
14. Coppola G, Castaldo P, Miraglia del Giudice E, Bellini G, Galasso F, Soldovieri MV, et al. A novel KCNQ2 K+ channel mutation in benign neonatal convulsions and centrotemporal spikes. *Neurology* 2003;61(1): 131–134.
15. Maihara T, Tsuji M, Higuchi Y, Hattori H. Benign familial neonatal convulsions followed by benign epilepsy with centrotemporal spikes in two siblings. *Epilepsia* 1999;40(1):110–113.
16. Zimprich F, Ronen GM, Stogmann W, Baumgartner C, Stogmann E, Rett B, et al. Andreas Rett and benign familial neonatal convulsions revisited. *Neurology* 2006;67 (5):864–866.
17. Bassi MT, Balottin U, Panzeri C, Piccinelli P, Castaldo P, Barrese V, et al. Functional analysis of novel KCNQ2 and KCNQ3 gene variants found in a large pedigree with benign familial neonatal convulsions (BFNC). *Neurogenetics* 2005;6(4):185–193.
18. Borgatti R, Zucca C, Cavallini A, Ferrario M, Panzeri C, Castaldo P, et al. A novel mutation in KCNQ2 associated with BFNC, drug resistant epilepsy, and mental retardation. *Neurology* 2004;63(1):57–65.
19. Dedek K, Fusco L, Teloy N, Steinlein OK. Neonatal convulsions and epileptic encephalopathy in an Italian family with a missense mutation in the fifth transmembrane region of KCNQ2. *Epilepsy Res* 2003;54(1): 21–27.
20. Schmitt B, Wohlrab G, Sander T, Steinlein OK, Hajnal BL. Neonatal seizures with tonic clonic sequences and poor developmental outcome. *Epilepsy Res* 2005;65(3): 161–168.
21. Wuttke TV, Jurkat-Rott K, Paulus W, Garncarek M, Lehmann-Horn F, Lerche H. Peripheral nerve hyperexcitability due to dominant-negative KCNQ2 mutations. *Neurology* 2007;69(22):2045–2053.
22. Claes LR, Ceulemans B, Audenaert D, Deprez L, Jansen A, Hasaerts D, et al. De novo KCNQ2 mutations in patients with benign neonatal seizures. *Neurology* 2004; 63(11):2155–2158.
23. Wang HS, Pan Z, Shi W, Brown BS, Wymore RS, Cohen IS, et al. KCNQ2 and KCNQ3 potassium channel subunits: Molecular correlates of the M-channel. *Science* 1998;282(5395):1890–1893.
24. Lerche C, Scherer CR, Seebohm G, Derst C, Wei AD, Busch AE, et al. Molecular cloning and functional expression of KCNQ5, a potassium channel subunit that may contribute to neuronal M-current diversity. *J Biol Chem* 2000;275(29):22395–22400.
25. Marrion NV. Control of M-current. *Annu Rev Physiol* 1997;59:483–504.
26. Schwake M, Pusch M, Kharkovets T, Jentsch TJ. Surface expression and single channel properties of KCNQ2/KCNQ3, M-type K+ channels involved in epilepsy. *J Biol Chem* 2000;275(18):13343–13348.
27. Schroeder BC, Hechenberger M, Weinreich F, Kubisch C, Jentsch TJ. KCNQ5, a novel potassium channel broadly expressed in brain, mediates M-type currents. *J Biol Chem* 2000;275(31):24089–24095.
28. Jentsch TJ. Neuronal KCNQ potassium channels: Physiology and role in disease. *Nat Rev Neurosci* 2000;1(1):21–30.
29. Schroeder BC, Kubisch C, Stein V, Jentsch TJ. Moderate loss of function of cyclic-AMP-modulated KCNQ2/KCNQ3 K+ channels causes epilepsy. *Nature* 1998;396 (6712):687–690.
30. Miraglia del Giudice E, Coppola G, Scuccimarra G, Cirillo G, Bellini G, Pascotto A. Benign familial neonatal convulsions (BFNC) resulting from mutation of the KCNQ2 voltage sensor. *Eur J Hum Genet* 2000;8(12):994–997.

31. Moulard B, Picard F, le Hellard S, Agulhon C, Weiland S, Favre I, et al. Ion channel variation causes epilepsies. *Brain Res Brain Res Rev* 2001;36(2–3):275–284.
32. Soldovieri MV, Cilio MR, Miceli F, Bellini G, Miraglia del Giudice E, Castaldo P, et al. Atypical gating of M-type potassium channels conferred by mutations in uncharged residues in the S4 region of KCNQ2 causing benign familial neonatal convulsions. *J Neurosci* 2007;27(18):4919–4928.
33. Schwake M, Jentsch TJ, Friedrich T. A carboxy-terminal domain determines the subunit specificity of KCNQ K(+) channel assembly. *EMBO Rep* 2003;4(1):76–81.
34. Soldovieri MV, Castaldo P, Iodice L, Miceli F, Barrese V, Bellini G, et al. Decreased subunit stability as a novel mechanism for potassium current impairment by a KCNQ2 C terminus mutation causing benign familial neonatal convulsions. *J Biol Chem* 2006;281(1):418–428.
35. Saganich MJ, Machado E, Rudy B. Differential expression of genes encoding subthreshold-operating voltage-gated K+ channels in brain. *J Neurosci* 2001;21(13):4609–4624.
36. Cooper EC, Aldape KD, Abosch A et al. Colocalization and coassembly of two human brain M-type potassium channel subunits that are mutated in epilepsy. *Proc Natl Acad Sci U S A* 2000;97:4914–4919.
37. Devaux JJ, Kleopa KA, Cooper EC, Scherer SS. KCNQ2 is a nodal K+ channel. *J Neurosci* 2004;24(5):1236–1244.
38. Weber YG, Geiger J, Kampchen K, Landwehrmeyer B, Sommer C, Lerche H. Immunohistochemical analysis of KCNQ2 potassium channels in adult and developing mouse brain. *Brain Res* 2006;1077(1):1–6.
39. Chung HJ, Jan YN, Jan LY. Polarized axonal surface expression of neuronal KCNQ channels is mediated by multiple signals in the KCNQ2 and KCNQ3 C-terminal domains. *Proc Natl Acad Sci U S A* 2006; 103(23):8870–8875.
40. Peretz A, Sheinin A, Yue C, Degani-Katzav N, Gibor G, Nachman R, et al. Pre- and postsynaptic activation of M-channels by a novel opener dampens neuronal firing and transmitter release. *J Neurophysiol* 2007; 97(1):283–295.
41. Brown DA, Adams PR. Muscarinic suppression of a novel voltage-sensitive K+ current in a vertebrate neurone. *Nature* 1980;283 (5748):673–676.
42. Constanti A, Adams PR, Brown DA. Who do barium ions imitate acetylcholine? *Brain Res* 1981;206(1):244–250.
43. Brown DA, Gahwiler BH, Griffith WH, Halliwell JV. Membrane currents in hippocampal neurons. *Prog Brain Res* 1990;83:141–160.
44. Goh JW, Pennefather PS. Pharmacological and physiological properties of the after-hyperpolarization current of bullfrog ganglion neurones. *J Physiol* 1987;394:315–330.
45. Castaldo P, del Giudice EM, Coppola G, Pascotto A, Annunziato L, Taglialatela M. Benign familial neonatal convulsions caused by altered gating of KCNQ2/KCNQ3 potassium channels. *J Neurosci* 2002;22(2):RC199.
46. Yue C, Yaari Y. KCNQ/M channels control spike afterdepolarization and burst generation in hippocampal neurons. *J Neurosci* 2004;24(19):4614–4624.
47. Otto JF, Yang Y, Frankel WN, White HS, Wilcox KS. A spontaneous mutation involving Kcnq2 (Kv7.2) reduces M-current density and spike frequency adaptation in mouse CA1 neurons. *J Neurosci* 2006;26(7): 2053–2059.
48. Gamper N, Shapiro MS. Calmodulin Mediates Ca2+-dependent Modulation of M-type K+ Channels. *J Gen Physiol* 2003.
49. Li Y, Gamper N, Shapiro MS. Single-channel analysis of KCNQ K+ channels reveals the mechanism of augmentation by a cysteine-modifying reagent. *J Neurosci* 2004;24(22):5079–5090.
50. Gamper N, Li Y, Shapiro MS. Structural requirements for differential sensitivity of KCNQ K+ channels to modulation by Ca2+/calmodulin. *Mol Biol Cell* 2005;16 (8):3538–3551.
51. Wen H, Levitan IB. Calmodulin is an auxiliary subunit of KCNQ2/3 potassium channels. *J Neurosci* 2002;22(18): 7991–8001.
52. Yus-Najera E, Santana-Castro I, Villarroel A. The identification and characterization of a noncontinuous calmodulin-binding site in noninactivating voltage-dependent KCNQ potassium channels. *J Biol Chem* 2002;277(32):28545–28553.
53. Hoshi N, Zhang JS, Omaki M, Takeuchi T, Yokoyama S, Wanaverbecq N, et al. AKAP150 signaling complex promotes suppression of the M-current by muscarinic agonists. *Nat Neurosci* 2003;6(6):564–571.
54. Suh BC, Hille B. Recovery from muscarinic modulation of M current channels requires phosphatidylinositol 4,5-bisphosphate synthesis. *Neuron* 2002;35(3):507–520.

55. Suh BC, Hille B. Electrostatic interaction of internal Mg2+ with membrane PIP2 Seen with KCNQ K+ channels. *J Gen Physiol* 2007;130(3):241–256.
56. Etxeberria A, Aivar P, Rodriguez-Alfaro JA, Alaimo A, Villace P, Gomez-Posada JC, et al. Calmodulin regulates the trafficking of KCNQ2 potassium channels. *Faseb J* 2008;22(4):1135–1143.
57. Richards MC, Heron SE, Spendlove HE, Scheffer IE, Grinton B, Berkovic SF, et al. Novel mutations in the KCNQ2 gene link epilepsy to a dysfunction of the KCNQ2-calmodulin interaction. *J Med Genet* 2004;41(3):e35.
58. Watanabe H, Nagata E, Kosakai A, Nakamura M, Yokoyama M, Tanaka K, et al. Disruption of the epilepsy KCNQ2 gene results in neural hyperexcitability. *J Neurochem* 2000;75(1):28–33.
59. Peters HC, Hu H, Pongs O, Storm JF, Isbrandt D. Conditional transgenic suppression of M channels in mouse brain reveals functions in neuronal excitability, resonance and behavior. *Nat Neurosci* 2005;8(1): 51–60.
60. Russell MW, Dick M, 2nd, Collins FS, Brody LC. KVLQT1 mutations in three families with familial or sporadic long QT syndrome. *Hum Mol Genet* 1996;5(9): 1319–1324.
61. Wollnik B, Schroeder BC, Kubisch C, Esperer HD, Wieacker P, Jentsch TJ. Pathophysiological mechanisms of dominant and recessive KVLQT1 K+ channel mutations found in inherited cardiac arrhythmias. *Hum Mol Genet* 1997;6(11):1943–1949.
62. Yang Y, Beyer BJ, Otto JF, O'Brien TP, Letts VA, White HS, et al. Spontaneous deletion of epilepsy gene orthologs in a mutant mouse with a low electroconvulsive threshold. *Hum Mol Genet* 2003;12(9):1–10.
63. Pereira S, Roll P, Krizova J, Genton P, Brazdil M, Kuba R, et al. Complete loss of the cytoplasmic carboxyl terminus of the KCNQ2 potassium channel: A novel mutation in a large Czech pedigree with benign neonatal convulsions or other epileptic phenotypes. *Epilepsia* 2004;45(4):384–390.
64. Otto JF, Yang Y, Frankel WN, Wilcox KS, White HS. Mice carrying the szt1 mutation exhibit increased seizure susceptibility and altered sensitivity to compounds acting at the m-channel. *Epilepsia* 2004;45(9): 1009–1016.
65. Otto JF, Kimball MM, Wilcox KS. Effects of the anticonvulsant retigabine on cultured cortical neurons: changes in electroresponsive properties and synaptic transmission. *Mol Pharmacol* 2002;61(4):921–927.
66. Hetka R, Rundfeldt C, Heinemann U, Schmitz D. Retigabine strongly reduces repetitive firing in rat entorhinal cortex. *Eur J Pharmacol* 1999;386(2–3):165–171.
67. Rundfeldt C, Netzer R. The novel anticonvulsant retigabine activates M-currents in Chinese hamster ovary-cells transfected with human KCNQ2/3 subunits. *Neurosci Lett* 2000;282(1–2):73–76.
68. Wickenden AD, Yu W, Zou A, Jegla T, Wagoner PK. Retigabine, a novel anti-convulsant, enhances activation of KCNQ2/Q3 potassium channels. *Mol Pharmacol* 2000;58(3):591–600.
69. Singh NA, Otto JF, Dahle EJ, Pappas C, Leslie JD, Vilaythong A, et al. Mouse models of human KCNQ2 and KCNQ3 mutations for benign familial neonatal convulsions show seizures and neuronal plasticity without synaptic reorganization. *J Physiol* 2008; 586(14):3405–3423.
70. Ryan SG, Wiznitzer M, Hollman C, Torres MC, Szekeresova M, Schneider S. Benign familial neonatal convulsions: evidence for clinical and genetic heterogeneity. *Ann Neurol* 1991;29(5):469–473.
71. Ben-Ari Y, Holmes GL. Effects of seizures on developmental processes in the immature brain. *Lancet Neurol* 2006;5(12): 1055–1063.
72. Blume WT. The progression of epilepsy. *Epilepsia* 2006;47(Suppl 1):71–78.

Chapter 8

Interneuron Loss as a Cause of Seizures: Lessons from Interneuron-Deficient Mice

Dorothy Jones-Davis, Maria-Elisa Calcagnotto, and Joy Y. Sebe

Abstract

Throughout our nervous system, excitation and inhibition are exquisitely balanced to enable a multitude of functions. When this balance is disrupted, neurons experience a surplus or a deficit in excitation, either of which can have devastating consequences. In the cortex, excitation and inhibition are mediated by glutamatergic pyramidal cells and GABAergic interneurons, respectively. The loss of GABAergic inhibition in the epileptic brain places neurons in a hyperexcitable state in which they are vulnerable to the high-frequency firing that defines seizures. The association between seizures and a loss of GABAergic transmission is supported by numerous investigations of epileptic patients and animal models of epilepsy *(1–3)*. For example, brain tissue from patients suffering from mesial temporal lobe epilepsy (MTLE), one of the most common medically intractable forms, is distinguished by a loss of specific subtypes of interneurons. Furthermore, electrophysiological studies have demonstrated that dentate gyrus granule cells from epileptic patients exhibit a functional reduction in inhibitory synaptic transmission *(4)*. These clinical findings are consistent with extensive work in pharmacologically induced animal models of epilepsy, including models of TLE and cortical dysplasia *(5–9)*. Although clinical and animal studies show a correlation between reduced inhibition and epilepsy, interneuron-deficient transgenic mice that exhibit an epileptic phenotype have recently confirmed that this link is causal. As such, interneuron-deficient transgenic mice can serve as mouse models of epilepsy and have the potential to significantly advance our understanding of interneuron development, interneuron function, and the importance of GABAergic inhibition in preventing seizure activity. In this chapter, we will review information that is directly relevant to our understanding of GABAergic interneuron-deficient mice, including interneuron origins and diversity, the characteristics of various types of interneuron-deficient mice, what we have learned from these mice, and the clinical applications of such mice in the treatment of epilepsy.

Key words: interneuron-deficient mice, interneurons, inhibition, *Nkx2.1*, *Dlx*, *Tlx*, *uPAR*, *MET*, *cD2*, *PPT1*, *Sox2*.

Scott C. Baraban (ed.), *Animal Models of Epilepsy: Methods and Innovations, vol. 40*

DOI 10.1007/978-1-60327-263-6_8 Springerprotocols.com

1. Introduction: The Utility of Interneuron-Deficient Knockout Mice

GABAergic interneuron-deficient knockout mice serve as powerful tools to (i) study interneuron development, (ii) investigate the link between a reduction in inhibition and epileptogenesis, and (iii) explore novel treatments for epilepsy. The generation of interneuron-deficient knockout mice has enabled the identification of genes, and their corresponding proteins, that are essential for normal interneuron development. Examination of these mice has also revealed the roles these proteins play in the developmental process including the specification of interneuron origin *(10)*, interneuron migration *(11, 12)*, and survival *(13)*. Furthermore, GABAergic interneuron-deficient mice demonstrate that a significant reduction in GABAergic transmission can lead to seizures. Knockout mice that exhibit a reduction in GABAergic interneurons and transmission exhibit epileptic phenotypes that are defined by spontaneous electrographic and behavioral seizures. As will be discussed later in this chapter, the association between interneuron deficiency and seizures is not purely correlative. This is evidenced by mice that exhibit normal GABAergic interneuron development during early postnatal development and only exhibit seizures when selective populations of interneurons apoptose *(13)*. To understand the importance of inhibition in suppressing seizure activity, one could replenish GABAergic interneurons and transmission in interneuron-deficient mice and determine whether seizures are alleviated or eliminated. If successful, this strategy would not only confirm that a significant loss of inhibition causes seizures, but also presents a promising new treatment for epilepsy and other neurodegenerative diseases.

2. Interneurons

2.1. GABAergic Interneuron Development

Interneuron development is a complex process that encompasses the birth of interneuron precursors, their widespread migration and differentiation into a variety of interneuron subtypes, and, finally, their survival postnatally. Most cortical interneuron precursors are born in the medial, lateral, or caudal ganglionic eminence (MGE, LGE, or CGE, respectively) of the embryonic telencephalon (**Fig. 8.1**). From these ganglionic eminences (GEs), they migrate laterally, then tangentially to the cerebral cortex, striatum, hippocampus, amygdala, nucleus accumbens, or olfactory bulb and sometimes migrate at a speed of approximately 100 μM/hour in order to reach their destinations *(15)*. This

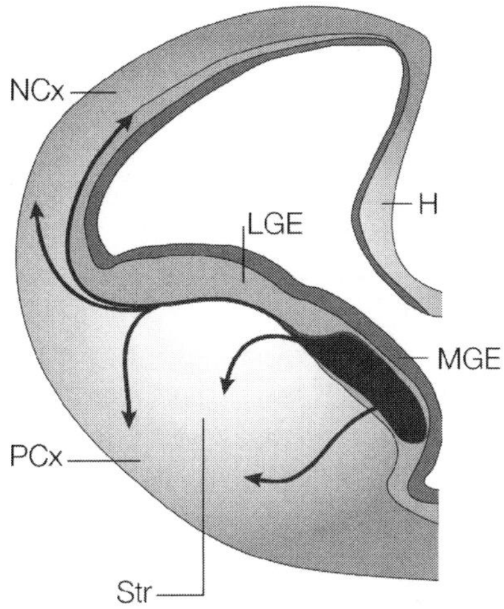

Fig. 8.1. **Schematic of a coronal section from E13.5 cortex.** The schematic shows the location of the medial and lateral ganglionic eminences (MGE and LGE, respectively), ventricular and subventricular zone (VZ and SVZ, respectively). At E13.5, interneuronal precursors migrate tangentially to the neocortex, hippocampus, and olfactory bulb. The caudal ganglionic eminence (CGE) is not shown in this schematic but is located caudal to the LGE and MGE. H, hippocampus; Ncx, neocortex; Pcx, paleocortex; Str, striatum. Modified from *(14)* with permission from J.L.R. Rubenstein.

tangential migratory pattern is in contrast to that of cortical glutamatergic neurons that migrate radially from their origin in the telencephalic ventricular zone (VZ). A number of experimental approaches have been used to elucidate the migratory pathways and cell fates of GE cells, including analysis of mutant mice, in vitro migration assays, and ultrasound-guided in utero transplantation of GE cells. These studies have demonstrated that the majority of interneuron precursors born in the MGE tangentially migrate to the cortex, striatum, and hippocampus in two distinct migratory "streams" *(10, 16, 17)*. To achieve their final location interneurons can then adopt a classically radial migratory pathway *(18)*. Although some LGE interneuron precursors also migrate to the cortex, most migrate anteriorly to the olfactory bulb *(17, 19, 20)*. In utero transplantation of CGE cells has shown that interneuron precursors of the CGE migrated caudally to a number of brain regions including the cortex, hippocampus, nucleus accumbens, and amygdala *(15, 21)*. Heterotopic transplantations of GE cells have revealed that the migratory paths and destinations of MGE and LGE cells are determined not by the external cues they receive during migration but by their origins. For example, LGE cells transplanted into the MGE migrated to the striatum, which normally receives LGE-derived projection neurons, nucleus accumbens, olfactory tubercle, and olfactory bulb *(17, 22)*. Furthermore, MGE cells transplanted into the CGE populated rostral and lateral brain regions, including the cortex, in a pattern similar to MGE cells that had been transplanted to the MGE *(21)*. The transplantation of CGE cells into the MGE has shown that CGE cells require a local CGE environment to migrate caudally *(15)*. That the migratory pathways and cell fates are somehow programmed in MGE and LGE cells is a characteristic with the potential to be exploited therapeutically.

2.2. GABAergic Interneuron Diversity

The most significant feature of interneurons is that the majority are inhibitory and thereby modulate the excitatory output of projection neurons *(23)*. Interneurons can also be distinguished from projection cells by their neurochemical markers, morphologies, and physiological properties. This suite of characteristics confers interneurons with an impressive diversity that has eluded a definitive classification system. Here, we only briefly describe the inhibitory interneuron characteristics that are relevant to our discussion of interneuron-deficient mice. Interneuron diversity is described in more detail in a number of excellent reviews (*19, 23–25)*.

Extensive immunohistochemical analysis has shown that interneurons express the Ca^{2+} binding proteins, parvalbumin (PV), calretinin (CR), and calbindin (CB) and the neuropeptides somatostatin (SOM), neuropeptide Y (NPY), vasointestinal peptide (VIP), and cholecystokinin (CCK) *(19, 24)*. Of these neurochemical markers, the majority of interneurons express PV followed by neurochemical subgroups that express SOM or CR *(19)*. Ultrasound-guided in utero transplantation and fate mapping studies have revealed that PV^+ and SOM^+ interneurons predominantly originate in the MGE whereas most CR-expressing interneurons are born in the CGE *(26, 27)*. In addition to the array of neurochemical markers that interneurons can express, histological analysis has demonstrated that GABAergic interneurons are morphologically diverse.

Within the neocortex, there are at least eight different morphological interneuron subtypes that are referred to as basket, double-bouquet, bipolar, bitufted, neurogliaform, Martinotti, Cajal-Retzius, and chandelier cells *(24)* These cells can target dendrites, somas, axons, or a combination of these structures and exhibit a variety of morphologies. For example, the axonal arborizations can be large or small, extend vertically or horizontally, or have a high or relatively low bouton density. Despite the variety of morphological subgroups, 50% of interneurons are basket cells that are further divided into three main subclasses (i.e., large, small, and nest), express PV and CB in addition to a number of neuropeptides, and are often characterized as fast-spiking (FS) interneurons *(24)*.

Electrophysiological studies of GABAergic interneurons have revealed multiple discharge patterns in response to depolarizing or hyperpolarizing current steps that can also be used to classify interneurons. For example, FS interneurons fire a fast, non-adapting train of action potentials during a depolarizing current step, burst-spiking non-pyramidal interneurons (BSNP) fire two or more rebound spikes following a hyperpolarizing current step, and regular-spiking non-pyramidal interneurons (RSNP) fire a train of action potentials during depolarization that may or may not be followed by rebound spikes. In addition to these physiological subtypes, there are the irregular-spiking (IS), delayed-spiking (DS), and late-spiking (LS) interneurons *(24, 27)*.

Interneurons are commonly classified by a combination of these neurochemical, morphological, and physiological characteristics. However, recent studies of interneuron development have revealed that interneurons may be more appropriately classified by their origins, which have already been shown to program interneurons with intrinsic properties such as their migratory pathways and destinations in the adult brain. Although interneurons derived from the MGE versus CGE share neurochemical markers, morphologies, and physiological properties, the majority of interneurons from each GE exhibit distinct characteristics. That is, most interneurons derived from E13.5 MGE are FS basket cells and express the Ca^{2+} binding proteins PV and SOM. In contrast, of the E13.5 CGE-derived interneurons that were studied (n ; 54) none were FS cells. Instead, 63% of interneurons originating in the E13.5 CGE were RSNP cells that were comprised mainly of CR^{+} bipolar or SOM^{+} multipolar interneurons *(27)*. Such characterization of interneurons derived from the MGE or CGE suggest that the GEs determine more than the migratory pathways and destinations of interneuron precursors. That is, the GEs may direct the morphological development of precursors as well as the neurochemical markers and ion channels that interneurons express. Although the mechanisms by which the GEs determine interneuron fate are unknown, one hypothesis has proposed that the transcription factors that are differentially expressed in the GEs and are required for GE development play a role *(25, 28)*. For the purposes of this review, we will often describe interneuron populations in interneuron-deficient mice using the neurochemical markers that they express.

3. Interneuron-Deficient Mice

3.1. Embryonic Lethal Interneuron-Deficient Mice

Multiple transcription factors have been identified that control the development, proliferation, and migration of GABAergic interneurons. Knockout mice that do not express these transcription factors consequently exhibit a reduction in GABAergic interneurons and emphasize the relative importance of transcription factors in interneuron development. Mice that lack the homeobox transcription factors Dlx1 and Dlx2, Dlx5, or Nkx.1 die at birth due to severe craniofacial or lung deformations and display profound deficits in cortical and hippocampal interneurons. These embryonic lethal interneuron-deficient mice have facilitated significant advances in our understanding of the homeobox genes that are essential for GABAergic interneuron migration from the subventricular zone and MGE formation. Among all of the interneuron-deficient mice including those that survive postnatally, mice lacking both the transcription factors Dlx1 and Dlx2 exhibit the greatest reduction in GABAergic interneurons (*see* **Table 8.1**). *Dlx1*

Table 8.1
Characteristics of GABAergic interneuron-deficient mice

Mouse	Class of deficient protein	Interneuron subtypes modified	Brain region affected (%INT lost, if known)	Physiological changes	Epileptic phenotype
Embryonic lethal					
Dlx1/2$^{-/-}$	Homeobox transcription factor	↓GAD67^{+}	Ncx (80%), Hipp (>95%), OB (100%)	N/A	N/A
Dlx5$^{-/-}$	Homeobox transcription factor	↓GAD65/67^{+}, ↓CR^{+}	OB	N/A	N/A
Nkx2.1$^{-/-}$	Homeobox transcription factor	↓NPY^{+}, ↓SOM^{+}	Ncx (40–50%), Hipp (50%)	N/A	N/A
Arx$^{-/-}$	Homeobox transcription factor	↓CR^{+}, ↓NPY^{+}	Striatum, globus pallidus	N/A	N/A
Non-embryonic lethal					
Dlx1$^{-/-}$	Homeobox transcription factor	↓SOM^{+}, ↓CR^{+}, ↓NPY^{+}, ↓NOS^{+},	Ctx, Hipp	↓IPSC freq. and peak amp.	Spontaneous and stress-induced seizures

(continued)

Table 8.1 (continued)

Mouse	Class of deficient protein	Interneuron subtypes modified	Brain region affected (%INT lost, if known)	Physiological changes	Epileptic phenotype
Tlx$^{-/-}$	Transcription factor	↓SOM$^+$, ↓CR$^+$	Ctx		Late onset
uPAR$^{-/-}$	Receptor that activates HGF/SF	↓PV$^+$, ↓CB$^+$,	Ctx	N/A	Spontaneous and ↑PTZ-induced seizures
MET$^{-/-}$ (conditional)	HGF/SF receptor	↑PV (DG), ↓CR$^+$ (DG, CA1, CA3), ↓PV$^+$ (CA3)	Hipp	N/A	N/A
cD2$^{-/-}$	Cell cycle protein	↓PV$^+$	Ctx (30%), Hipp (40%)	↓IPSC freq. and peak amp.	Spontaneous seizures
Ppt1$^{-/-}$	Lysosomal enzyme	↓All GABAergic interneurons			Late-onset, spontaneous seizures

Abbreviations: CR, calretinin; NPY, neuropeptide Y; SOM, somatostatin; CB, calbindin; PV, parvalbumin; NOS, nitric oxide synthase; DG, dentate gyrus; Ctx, cortex; Ncx, neocortex; Hipp, hippocampus; OB, olfactory bulb.

and −*2* are part of the Dlx family of homeobox genes that are homologous to the Drosophila *Distalless* gene and are composed of at least six genes (*Dlx1, 2, 3, 5, 6,* and *7*). Dlx1 and -2 are expressed in the developing forebrain including the subventricular zone, mature neocortex *(29)*, olfactory bulb *(30)*, developing retina *(31, 32)*, and branchial arches that give rise to head structures such as the facial skeleton *(33)*. Mice lacking both the *Dlx1* and −*2* genes suffer from severe craniofacial deformations, such as the absence of maxillary molars, and die within a few hours after birth *(34)*. In vitro migration assays using brain slices from wild-type and *Dlx1/2*$^{-/-}$ mutants have demonstrated that *Dlx1* and −*2* are also required for striatal differentiation. In *Dlx1/2* null mutants, striatal neurons born later in embryonic development accumulate within an LGE-like region, partially differentiate, and fail to migrate to the striatum. Finally, immunohistochemical studies of GAD67, an enzyme that synthesizes GABA, and GABA expression have shown that *Dlx1* and −*2* are essential for GABAergic interneuron development. *Dlx1/2*$^{-/-}$ mice lack 80% of GABA immunoreactive cells in the neocortex *(11)*, lack nearly 100% of GAD67-positive cells in the olfactory bulb *(22, 35)*, and have no detectable GABA-positive cells in the hippocampus at birth *(36)*. These deficits in GABAergic interneurons are not mirrored in *Dlx1* or *Dlx2* single mutants to the same degree (*see* discussion of *Dlx1* single mutants below), demonstrating that the mutation of both *Dlx* genes is particularly destructive for interneuron development.

Similar to *Dlx1* and *−2*, *Dlx5* is expressed throughout the brain, including in the forebrain, mature neocortex *(29)*, olfactory bulb *(37, 38)*, and developing branchial arches and inner ear *(37)*. *Dlx5* homozygous mutants exhibit a number of craniofacial deformities, including those in the olfactory and otic placodes, and die shortly after birth *(37)*. As far as we know, GABAergic interneuron loss in *Dlx5* mutants is limited to the olfactory bulb in which there is a reduction in GAD65, GAD67, and CR immunoreactivity *(38, 39)*.

Mutation of the homeobox gene *Nkx2.1*, or *T/ebp*, has revealed a number of critical roles this gene plays in embryonic development especially with regard to organogenesis and MGE formation *(10, 38)*. The importance of *Nkx2.1* is reflected in its widespread expression pattern. *Nkx2.1* is expressed in the lung, thyroid, ventral forebrain, and pituitary, all organs which are missing in *Nkx2.1* homozygous mutants. Although *Nkx2.1* null mutants survive gestation and are alive at E19.5, they die at birth. Immunohistochemical analysis of the *Nkx2.1*$^{-/-}$ telencephalon has shown that a small MGE-like structure is present at E10.5 and E11.5. However, by E12.5 the MGE and LGE cannot be anatomically distinguished due to the absence of the sulcus that normally divides these two GEs, and the MGE-like structure does not express transcription factors that are normally restricted to the MGE (i.e., *Lhx6* and *Lhx7*). By E18.5, the MGE-like structure has re-specified to form the LGE as evidenced by the expression of LGE-specific molecular markers (i.e., SCIP and GOLF). To investigate changes in the number of GABAergic interneurons in *Nkx2.1* null mutants, the neocortex, hippocampus, and olfactory bulb were immunolabeled for *Dlx2*, *GAD67*, and/or CB. The neocortex of Nkx2.1 null mutants show a 40–50% reduction in GABAergic interneurons and almost no CB staining in the neocortical intermediate zone of the subventricular zone. In the hippocampus, there is a 50% reduction in GABAergic interneurons and no expression of NPY and SOM *(36)*. In contrast, GABA immunoreactivity in the olfactory bulb is normal *(10)*. Together, these studies demonstrate that *Nkx2.1* is necessary for MGE specification and for the development GABAergic interneurons in the neocortex and hippocampus.

Similar to the *Dlx* and *Nkx2.1* homeobox genes, the Aristaless Related Homeobox gene (*Arx*) is expressed throughout development in the cortex, the striatum, the ganglionic eminences and the spinal cord. In the mature brain, expression of *Arx* persists in these areas and is specifically enriched in areas dense with GABAergic interneurons *(41)*. In 2002, Stromme et al. *(42)* identified *Arx* mutations in nine families with X-linked mental retardation (MRX), dystonia, and epilepsy (myoclonic seizures and infantile spasms). Other studies linked *Arx* mutations to Partington *(43)* and West *(44)* syndromes, and X-linked lissencephaly with abnormal genitalia *(45)*. The knowledge of the gene responsible for these conditions increased the experimental interest in designing an *Arx* null mouse. In 2002, Kitamura et al. *(46)* developed a mouse deficient in *Arx*. Male hemizygous *Arx* mutant mice die shortly after birth, exhibit abnormal testes, a shrunken brain, and display a loss of GABAergic interneurons, similar to some of the features noted in humans with *Arx* mutations. In the mutant mice, this loss of interneurons was targeted, specifically affecting the CR^+ and NPY^+ subpopulations of interneurons in the striatum and global pallidus, resulting in a smaller overall brain size in mutants. Kitamura et al. *(46)* found that this loss of GABAergic interneurons in the *Arx* mutants was due to decreased migration, specifically from the MGE to the intermediate zone, eventually leading to a buildup of interneurons in the periventricular zone. These migration defects are quite different from those noted in *Dlx* mutants, and in 2005, it was shown that *Arx* was regulated by *Dlx* genes *(47)*. Although the embryonic lethal interneuron-deficient mice have revealed the importance of *Dlx1*, *−2,* and *−5, Nkx2.1* and *Arx* in craniofacial, lung, skeletal, and GABAergic interneuron development, the behavioral outcome of interneuron loss cannot be assessed in these mice. To study the behaviors accompanied by a reduction in inhibitory transmission, interneuron-deficient mice that survive postnatally have served as a valuable resource.

3.2. Non-embryonic Lethal Interneuron-Deficient Mice

In contrast to *Dlx1/Dlx2* double mutants that die immediately after birth and only develop a subset of GABAergic interneurons, *Dlx1* single mutants are indistinguishable from wild-type mice and exhibit normal GABAergic interneuron development at birth. During postnatal development, *Dlx1* is highly expressed in the cortex, correlated with the expression of GABAergic markers such as *Gad67* (with 45% of $Gad67^+$ neurons being Dlx^+) and expressed in distinct neurochemical GABAergic interneuron subtypes (e.g., SOM^+ and CR^+ interneurons). Despite their normal appearance during early postnatal development, at 1 month of age, *Dlx1* mutants exhibited a striking 22% decrease in the number of GABAergic interneurons (as identified by *Gad67* immunoreactivity), with the specific apoptotic death of SOM^+, CR^+, NPY^+, and NOS^+ interneurons in the cortex and hippocampus. In slice electrophysiology investigations, this selective reduction of GABAergic interneurons in $Dlx1^{-/-}$ mice resulted in a reduction in inhibition as demonstrated by a decrease in the spontaneous

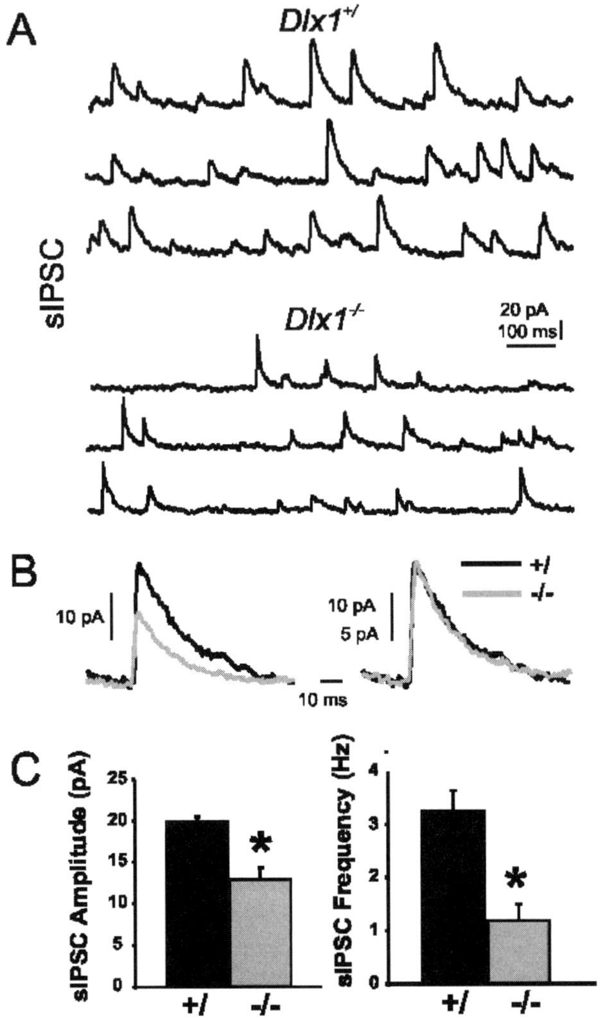

Fig. 8.2. **GABAergic synaptic transmission is reduced in neocortex of *Dlx1*$^{-/-}$ mice.** (**A**) Representative traces of sIPSCs recorded from layer II/III pyramidal cells of 2 months old *Dlx1*$^{+/}$ (*top*) and *Dlx1*$^{-/-}$ (*bottom*) mice. (**B**) Representative average traces of sIPSCs recorded from *Dlx1*$^{+/}$ (*black*) and *Dlx1*$^{-/-}$ (*gray*) cortical pyramidal cells normalized to baseline (*left*) and peak amplitude (*right*). As reflected in the average traces, sIPSC amplitude is reduced in mutants relative to controls whereas sIPSC decay is unchanged. (**C**) Average sIPSC amplitude and frequency were decreased in mutants.

inhibitory postsynaptic current (sIPSC) frequency and amplitude (**Fig. 8.2**). Not surprisingly, this decrease was observed late in development (P30^{+}) at a time when histological analysis revealed a significant drop in interneuron density, but not at earlier developmental time points. Given that the neuropeptides SOM and NPY have a potent endogenous antiepileptic effect *(48)*, Cobos et al. *(13)*, in combination with a significant reduction in both hippocampal and cortical GABA-mediated inhibitory current, proposed that the specific loss of SOM and NPY-expressing interneurons

Color Plates

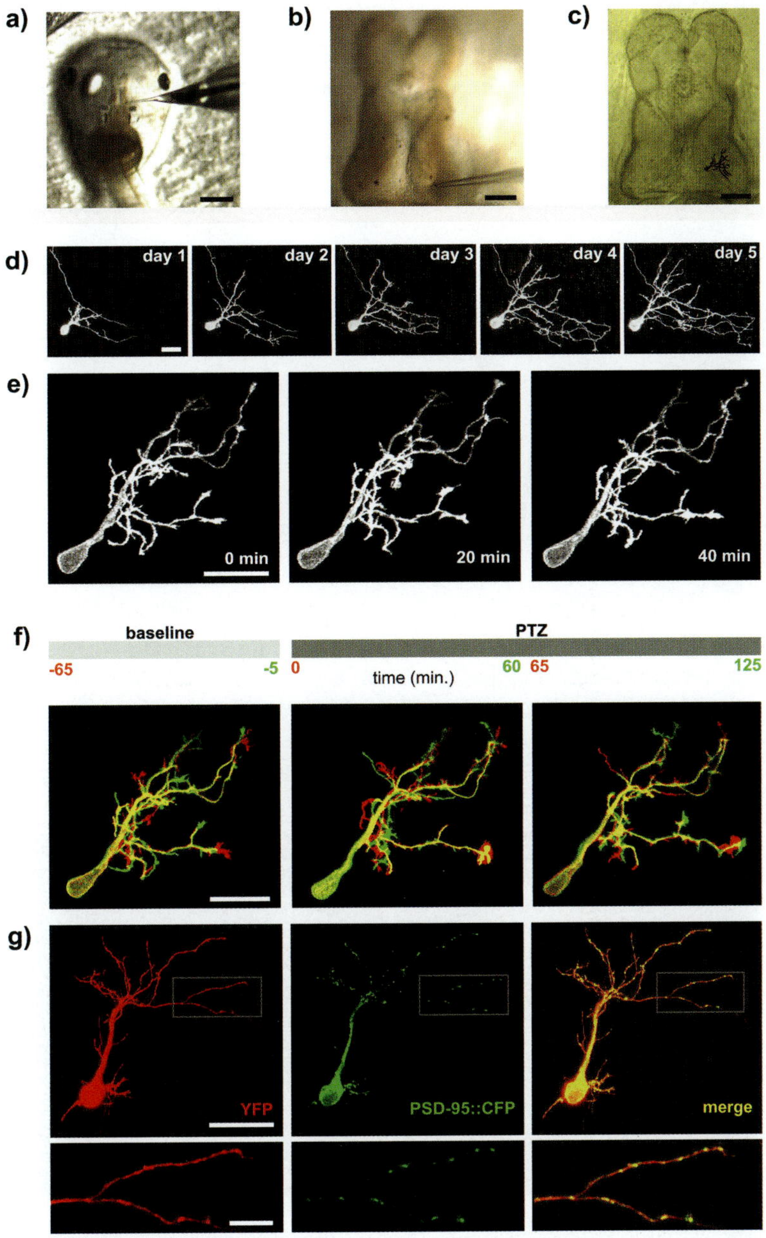

Color Plate 1. **In vivo imaging of dendritic arbor growth and synapse formation in *X. laevis* tadpoles.** (**A**) Single-cell electroporation allowing targeted neuronal labeling and transfection in vivo. Scale bar = 2 mm. (Chapter 3, Fig. 4; *see* discussion on p. 53 and complete caption on p. 54)

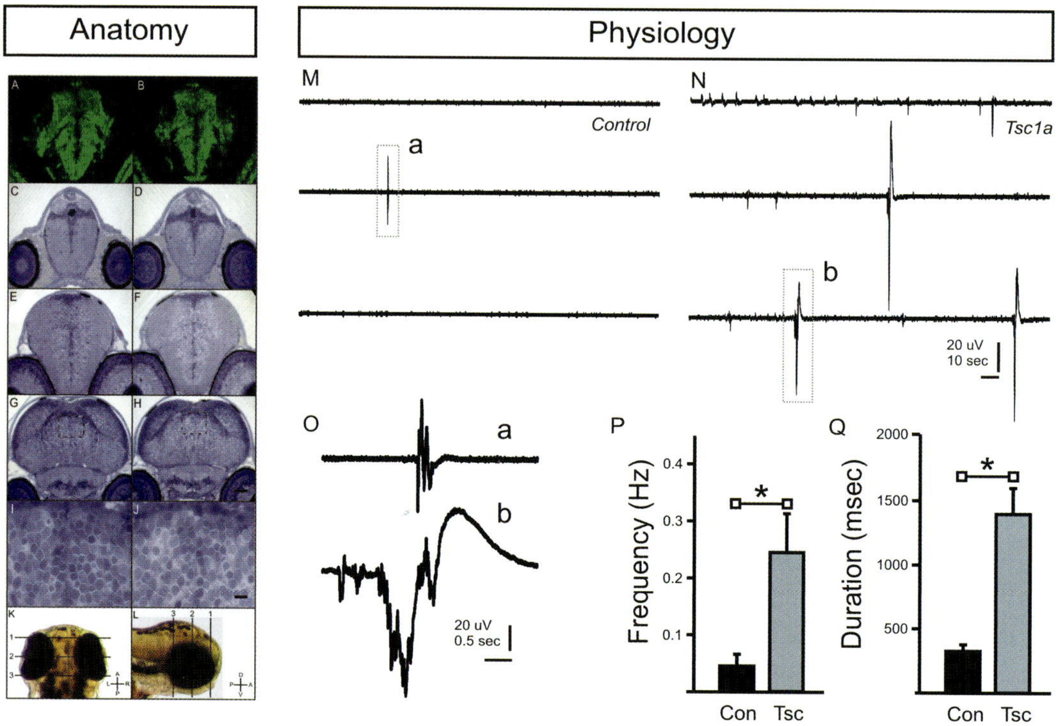

Color Plate 2. **Tsc1 morphant zebrafish.** (*Anatomy*) Histological analysis of control and *tsc1a* morphants. Immunostaining with the neuronal marker HuC/HuD of control-injected (**A**) and *tsc1a* morphant (**B**) larvae, and semi-thin plastic sections of control-injected (**C,E,G,I**) and *tsc1a* morphant (**D,F,H,J**) larvae are shown. (Chapter 4, Fig. 3; *see* discussion on p. 69 and complete caption on p. 70)

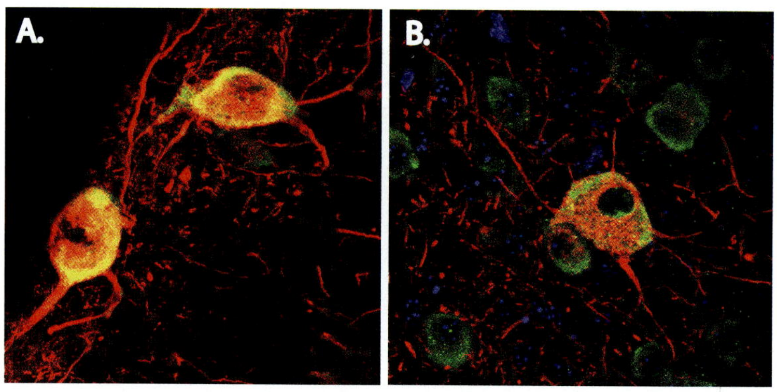

Color Plate 3. Confocal images of brain sections from Tsc1^{Synapsin}CKO mice immunostained for SMI311 (red, neuronal marker) and phospho-S6 (green). This staining reveals large, dysplastic appearing neurons from white matter **(A)** or the cortex **(B)**. (Chapter 5, Fig. 4; *see* discussion on p. 81)

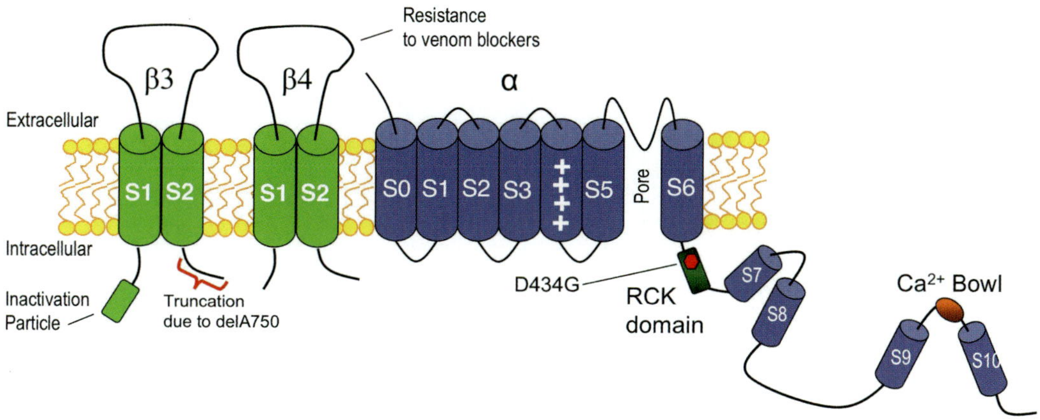

Color Plate 4. **Diagram of BK channel α (pore-forming) subunit and accessory β subunits.** Annotations indicate relevant functional domains and sites containing epilepsy mutations (β3 delA750 and gain-of-function mutation D434G for α subunit). (Chapter 6, Fig. 1; *see* discussion on p. 88)

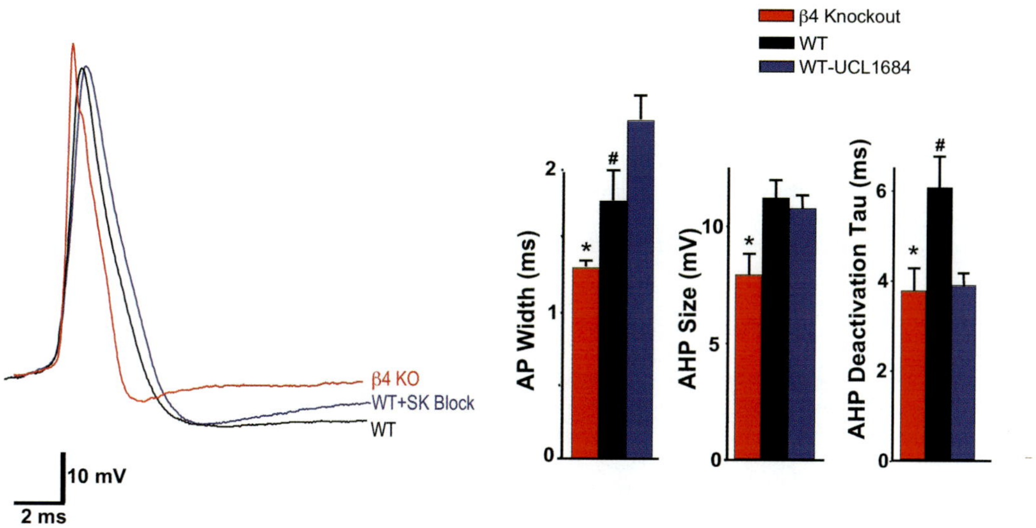

Color Plate 5. **BK and β4 effect on action potential waveform in dentate gyrus granule cells.** Action potential was selected from 10th spike during a 300 pA current injection. Average data are plotted at right. Red trace is β4 knockout, blue trace is wild type plus SK channel blocker UCL1684, black trace is wild type. (Chapter 6, Fig. 3; *see* discussion on p. 97)

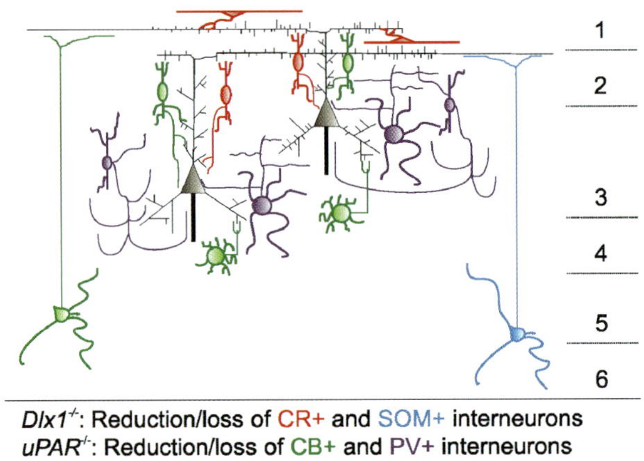

Color Plate 6. **Cortical interneuron loss in selected GABAergic interneuron-deficient mice.** Wild-type mice display normal connectivity in the cortex. (Chapter 8, Fig. 3; *see* discussion on p. 134 and complete caption on p. 135)

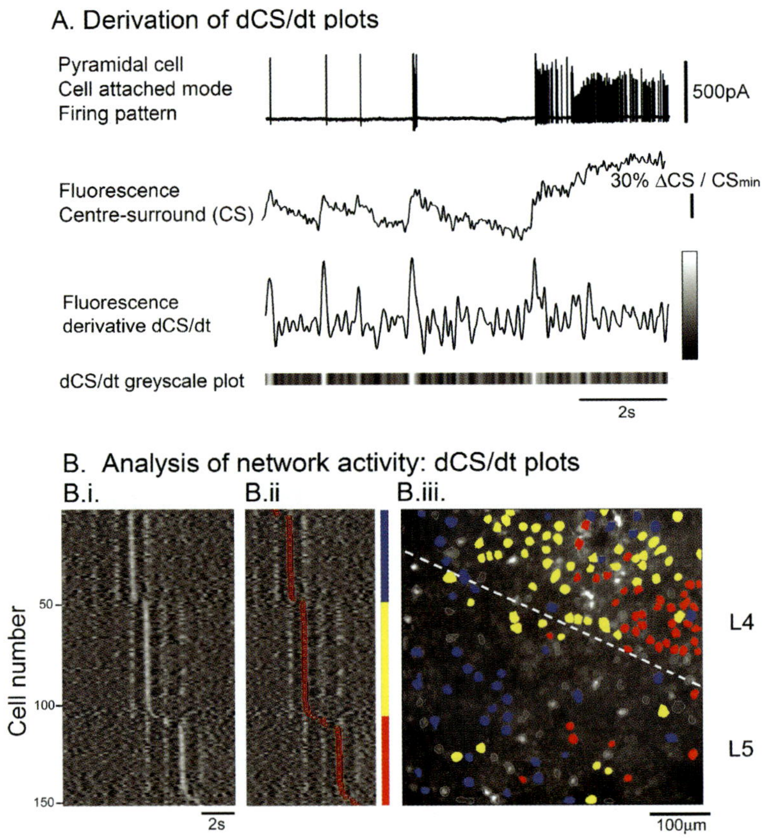

Color Plate 7. **Somatic Ca^{2+} transients reveal episodic localized recruitment. (A)** The derivation of the cellular Ca^{2+}signal. (Chapter 9, Fig. 4; *see* discussion on p. 156 and complete caption on p. 157)

would increase the likelihood of epileptic seizures in *Dlx1*$^{-/-}$ mice. Consistent with their hypothesis, *Dlx1*$^{-/-}$ mice displayed both spontaneous electrographic and behavioral seizures, as well as an increased sensitivity to seizures induced by noise or handling in parallel with the onset of interneuron loss. The increased propensity to seize as a direct consequence of the loss of specific interneuron subtypes provides an eloquent model of seizure activity. This model can be used to study the relative contribution of these particular subpopulations of interneurons to the inhibition of seizure genesis.

Mice with defects in the proliferation and differentiation of neuronal progenitor cells can also serve as informative models of late-onset epilepsy *(49, 50)*. To interpret these mice in a context of how seizures are generated it is critical to understand the role of the "timing" of interneuron loss. The transcription factor *Tlx* is evolutionarily conserved and expressed in the developing periventricular zone, where it plays a role in the proliferation and differentiation of both neuronal and glial cells in the forebrain. The absence of this transcription factor in mice causes a hypoplastic limbic system to develop as a direct result of abnormal proliferation and differentiation of these cell types. Pertinent to the present discussion, there was a selective loss of SOM^{+} and CR^{+} interneurons in the neocortex. Given these anatomical defects, *Tlx*$^{-/-}$ mice were then shown to display a wide variety of phenotypes, including late-onset epilepsy.

Until now, our discussion of interneuron-deficient mice has focused on mice lacking homeobox genes involved in neuronal differentiation, migration, and survival. Next, we turn our attention to a group of mice with mutations in molecules and proteins that are required for neuronal as well as non-neuronal development. One such mouse has a heterozygous mutation in the *Sox2* gene, which encodes a transcription factor that is essential for multipotent stem cells and cell differentiation *(51)*. Although ablation of *Sox2* leads to embryonic lethality, heterozygous *Sox2* mutants survive well into adulthood *(52, 53)*. Of the interneuron-deficient mice, the *Sox2* "knockdown" mouse is unique in that *Sox2* expression is not eliminated but reduced to 25–30% of wild-type mice levels. In comparison to the typical heterozygous mutant, *Sox2* "knockdown" mice contain one null *Sox2* allele while the expression of the remaining allele is reduced by the deletion of a *Sox2* enhancer. Investigations of *Sox2* "knockdown" mice have revealed a number of neural defects including degenerating neurons, smaller cortices and, in 40% of mutants, epileptic spikes in the EEG recordings *(53)*. In addition to disruptions in overall neural development, immunohistochemical analysis of mutant adult cortex has shown a reduction in CR^{+} interneuron number and a decrease in interneuron axonal and dendritic arborizations *(54)*. These data demonstrate that the Sox2 transcription factor is necessary for neuronal development and maintaining the balance between excitation and inhibition. The functional consequences of these changes were not reported, but acute slice electrophysiology studies should provide valuable insights into the role of CR^+ interneurons in mediating cortical circuit function. In addition, a more detailed description of interneuron loss including the percent reduction in interneuron number relative to wild type and the neurochemical subtypes affected is necessary to assess the degree and nature of interneuron deficiency, respectively, that results when *Sox2* expression is disrupted.

The interneuron-deficient mice discussed thus far demonstrate that the transcription factors (e.g., Nkx2.1, Dlx, Tlx, and Sox2) that regulate gene expression play a pivotal role in interneuron development as well as the formation of other cell types and systems. However, proteins that direct and facilitate cell migration also play critical roles during development. The importance of such migration-related proteins is evident upon investigation of mice lacking hepatocyte growth factor (HGF, also referred to as "scatter factor"), the HGF receptor MET, or the urokinase plasminogen activator receptor (uPAR), which cleaves and activates HGF. HGF has multiple important roles during embryogenesis; it inhibits tumor growth, acts as a chemoattractant for motoneurons, and serves as a growth factor for epithelial cells *(53)*. *HGF* and *MET* null mutants exhibit severe defects in liver, skeletal muscle, and placenta *(56–58)*. These mice die during mid gestation (E13.5–16.5; *56–58)* at a period when the embryos are too large to receive nutrients via diffusion and must rely on the placenta for food *(55, 57)*. In the absence of *HGF* or *MET*, these mice also die before the majority of interneuron precursors have differentiated and therefore it has not been possible to assess interneuron loss in traditional *HGF* and *MET* knockout mice. Instead, mice lacking *uPAR* have been used to examine how a reduction in HGF affects interneuron development. When bound by the urokinase plasminogen activator (uPA) or the tissue plasminogen activator (tPA), uPAR cleaves the inactive 100 kDa HGF into a 30 kDa segment and the active 70 kDA form *(59, 60)*. Mice lacking *uPAR* exhibit a reduction in active HGF/SF and MET in the embryonic forebrain and decreased scatter activity of cells seeded on the E16.5 sections of the $uPAR^{-/-}$ forebrain. Due to a disruption in migration, CB^+ cells accumulate in the corticostriatal sulcus and there is a substantial reduction (55–65%) of CB^+ neurons in the anterior regions of the P0 cortex *(12)*. Analysis of adult $uPAR^{-/-}$ cortex has demonstrated that interneuron loss is subtype specific in that there is almost a complete loss of PV^+ cells accompanied by normal numbers of SOM^+ and CR^+ neurons *(61)*. A detailed immunohistochemical analysis of GABAergic interneurons throughout the $uPAR^{-/-}$ telencephalon revealed a complex pattern of interneuron loss that is dependent on brain region (i.e., striatum, hippocampus, cortex, and amygdala) and postnatal age (i.e., P21, P45, >P90). Unlike the *HGF/SF*$^{-/-}$ mice, *uPAR*$^{-/-}$ mice survive into adulthood and it is therefore possible to characterize their epileptic phenotype. *uPAR*$^{-/-}$ mice exhibit more spontaneous seizures and seizures induced by injection of the convulsant pentylenetetrazole *(61)*. Ongoing video-EEG analysis of these mice in the Powell laboratory are likely to reveal clear evidence for electrographic seizures *(62)*; future slice electrophysiology studies similar to those described in *Dlx1* mutant mice should prove equally informative. Interestingly and suggestive of neurological deficit beyond epilepsy, in two behavioral tests *uPAR*$^{-/-}$ mice were shown to experience elevated anxiety levels. Specifically, in light–dark chambers, *uPAR*$^{-/-}$ mice spend half the time in the light side relative to WT mice. When placed on an elevated plus maze, mutants spend less time in the open arms and exhibit less exploratory behavior relative to controls *(61)*. Given that *uPAR*$^{-/-}$ mice lack a specific interneuron subtype, survive into adulthood, and are prone to seizures, they serve as valuable models through which to understand how a decrease in CB^+ and PV^+ interneurons lead to epilepsy.

Initially designed to circumvent the confounding and lethal effects of *MET* mutagenesis on non-neuronal development, two types of conditional *MET* knockout mice have recently been generated. These mice lack *MET* expression in proliferating cells of the VZ of the GE or postmitotic cells of the GE. To count the number of interneuron subtypes in the hippocampus of both types of conditional KOs, the hippocampus was immunolabeled for GABA, PV, SOM, and CR expression. The loss of *MET* expression in the GE leads to an *increase* in PV^+ cells in the dentate gyrus; a *decrease* in CR^+ cells in the dentate gyrus, CA3, and CA1; and a *decrease* in PV^+ cells in the CA3 region. In the CA3 region, some of the remaining PV^+ cells exhibited increased arborizations suggesting that surviving PV^+ cells morphologically compensate for their reduced numbers. Given that the change in interneuron number in conditional *MET* KO mice varies according to the neurochemical interneuron subtype and the hippocampal region and that interneurons in KO mice undergo morphological changes, it is difficult to interpret how the conditional loss of *MET* affects inhibition overall. Again, electrophysiological and behavioral studies are necessary to assess whether these mice exhibit a decrease or an increase in GABAergic inhibition and are prone to seizures, respectively.

Overall, immunohistochemical analysis of interneuron-deficient mice demonstrated that a mutation in one gene can disrupt the development of distinct neurochemically defined interneuron subtypes. For example, a mutation in *Dlx1* or *uPAR* leads to the selective loss of CR^+ and PV^+ cells, respectively. Like *uPAR*$^{-/-}$ mice, mice lacking the cell cycle protein cyclin D2 (cD2) are deficient in PV^+ cells *(63)*. Cyclins are activating subunits of cyclin-dependent kinases (Cdks) and together these proteins drive cells through the phases of the cell cycle – mitosis (M), first gap (G1), DNA synthesis (S), and second gap (G2). cD2, by activating Cdk4 and Cdk6, control the cells passage from G1 to S phase during which the cell undergoes chromosome duplication in preparation for mitosis *(64)*. Within the embryonic forebrain (E14.5), cD2 is more prevalent in the SVZ relative to the VZ and robustly expressed in the dorsal LGE and ventral MGE. Without cD2, it is likely that cells prematurely leave the cell cycle before they complete their normal process of proliferation. As a result, the loss of cD2 leads to a 30 and 40% reduction of PV^+ GABAergic interneurons in the neocortex and hippocampus, respectively. Voltage-clamp recordings of cortical pyramidal cells have demonstrated that the reduction in PV^+ interneurons impacts GABAergic transmission in that there is a reduction in both GABAergic mIPSC amplitude and frequency. Cortical EEG recordings of awake-behaving mice show that the spontaneous discharges of cD2 null mice occur at higher frequencies relative to wild-type mice suggesting that loss of GABAergic inhibition increases cortical excitability. To determine whether the cD2 null mouse can serve as a model for epilepsy, more extensive EEG and behavioral analysis will be valuable in detecting the high-frequency firing and seizures that define epilepsy.

Here we have discussed primary epilepsies that stem mainly from a loss or defect in the proliferation, differentiation, and/or migration of interneuron precursor cells (**Fig. 8.3** and Color Plate 6, middle of book). These epilepsies form a cohort of disorders in which the specific loss of interneurons is the primary cause of the epilepsy. However, many types of epilepsy are secondary to other causes, many of which involve organ systems outside of the nervous system. Could models of secondary epilepsies that involve interneuron loss be useful in providing additional information about seizure genesis? One such model is the $Ppt1^{-/-}$ mouse. Palmitoyl protein thioesterase 1 (Ppt1) is a lysosomal enzyme that is necessary to remove long-chain fatty acids from modified cysteine residues *(65)*. In humans, when deficient, this gene causes infantile neuronal ceroid lipofuscinosis (INCL), a disorder characterized by progressive retinal degeneration, blindness, motor and cognitive defects including epilepsy, an isoelectric EEG, and permanent vegetative state by 3 years of age, and ultimately death by 10 years of age *(66)*. When postmortem brain tissue is analyzed, it reveals vast amounts of neuronal degeneration, specifically affecting the neocortex and cerebellum. Similarly, mutants homozygous for the *Ppt1* gene display a loss of vision at 2 months of age, seizure activity at 4 months of age, and a prominent and specific loss of GABAergic interneurons at 6 months of age, secondary to inflammatory processes *(67–69)*. This model provides support for

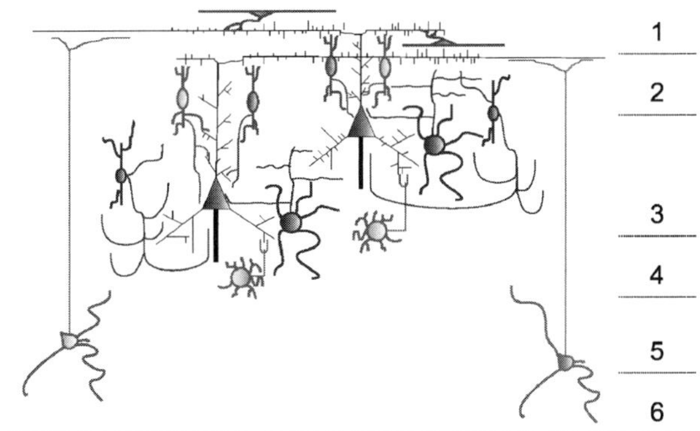

Fig. 8.3. **Cortical interneuron loss in selected GABAergic interneuron-deficient mice.** Wild-type mice display normal connectivity in the cortex. Pyramidal cells (*gray*) receive distal inputs from calbindin-positive (*green*) double-bouquet cells and calretinin-positive (*red*) double-bouquet and Cajal-Retzius cells, as well as both calbindin- and somatostatin-positive (*blue*) Martinotti cells. Additionally, pyramidal cells receive proximal inputs from parvalbumin-positive (purple) basket and chandelier cells and calbindin-positive neurogliaform cells. In contrast, *Dlx1*$^{-/-}$ mice display a selective loss of somatostatin and calretinin-positive cells, whereas *uPAR*$^{-/-}$ mice display a selective loss of parvalbumin and calbindin-expressing cells. (*see* Color Plate 6, middle of book)

the basic hypothesis that a loss of inhibition leads to epilepsy, which ultimately if uncontrolled can lead to death, regardless of the timing of or manner in which interneurons are lost.

4. Global Relevance of Interneuron-Deficient Mice to Epilepsy Research

Of the 2 million Americans who suffer from some form of epilepsy, 20% or more than 600,000 people have intractable seizures and for another 400,000 the current treatments (i.e., antiepileptic medications or surgical removal of epileptic brain tissue) provide inadequate relief *(70)*. As discussed here, emerging interneuron-deficient mouse models act as important tools with which to study the etiology of epilepsy *(4, 7, 48, 71–83)* and, perhaps, to develop more effective treatments for these patients. Published investigations have already demonstrated that a loss of GABAergic interneurons (as low as 20%) results in decreased inhibitory transmission and ultimately leads to seizure activity. While these deficits may be part of a larger circuit dysfunction that has not yet been uncovered, they suggest a critical role for interneurons in controlling overall excitation. Interestingly, data emerging from these mouse models also suggest that the severity of seizures may depend on the type of inhibition and interneuron subpopulation lost *(78)*. Additional analysis of these models, as well as the inevitable discovery of additional interneuron-deficient mice, will provide further insights and opportunities to develop new therapeutic approaches such as the transplantation of GABA progenitor cells to replace lost interneurons or novel pharmacological intervention to recover lost inhibition.

References

1. Calcagnotto ME, Paredes MF, Tihan T, Barbaro NM, Baraban SC. Dysfunction of synaptic inhibition in epilepsy associated with focal cortical dysplasia. *J Neurosci* 2005;25(42):9649–57.
2. Spreafico R, Battaglia G, Arcelli P, Andermann F, Dubeau F, Palmini A et al. Cortical dysplasia: An immunocytochemical study of three patients. *Neurology* 1998;50 (1):27–36.
3. Spreafico R, Tassi L, Colombo N, Bramerio M, Galli C, Garbelli R et al. Inhibitory circuits in human dysplastic tissue. *Epilepsia* 2000;41 (Suppl 6):168–73.
4. Williamson A, Patrylo PR, Spencer DD. Decrease in inhibition in dentate granule cells from patients with medial temporal lobe epilepsy. *Ann Neurol* 1999;45(1): 92–9.
5. Zhu WJ, Roper SN. Reduced inhibition in an animal model of cortical dysplasia. *J Neurosci* 2000;20(23):8925–31.
6. Yu FH, Mantegazza M, Westenbroek RE, Robbins CA, Kalume F, Burton KA et al. Reduced sodium current in GABAergic interneurons in a mouse model of severe myoclonic epilepsy in infancy. *Nat Neurosci* 2006;9(9):1142–49.
7. Kobayashi M, Buckmaster PS. Reduced inhibition of dentate granule cells in a model of temporal lobe epilepsy. *J Neurosci* 2003;23(6):2440–52.
8. Kumar SS, Buckmaster PS. Hyperexcitability, interneurons, and loss of GABAergic synapses in entorhinal cortex in a model of temporal lobe epilepsy. *J Neurosci* 2006; 26(17):4613–23.
9. Kumar SS, Jin X, Buckmaster PS, Huguenard JR. Recurrent circuits in layer II of medial entorhinal cortex in a model of temporal lobe epilepsy. *J Neurosci* 2007; 27(6): 1239–46.
10. Sussel L, Marin O, Kimura S, Rubenstein JL. Loss of Nkx2.1 homeobox gene function results in a ventral to dorsal molecular respecification within the basal telencephalon: Evidence for a transformation of the pallidum into the striatum. *Development* 1999; 126(15):3359–70.
11. Anderson SA, Eisenstat DD, Shi L, Rubenstein JL. Interneuron migration from basal forebrain to neocortex: Dependence on Dlx genes. *Science* 1997;78(5337):474–6.
12. Powell EM, Mars WM, Levitt P. Hepatocyte growth factor/scatter factor is a motogen for interneurons migrating from the ventral to dorsal telencephalon. *Neuron* 2001; 30(1):79–89.
13. Cobos I, Calcagnotto ME, Vilaythong AJ, Thwin MT, Noebels JL, Baraban SC et al. Mice lacking Dlx1 show subtype-specific loss of interneurons, reduced inhibition and epilepsy. *Nat Neurosci* 2005;8(8):1059–68.
14. Marin O, Rubenstein, JL. A long, remarkable journey: tangential migration in the telencephalon. *Nat Rev Neurosci* 2001; 2(11):780–790.
15. Yozu M, Tabata H, Nakajima K. The caudal migratory stream: A novel migratory stream of interneurons derived from the caudal ganglionic eminence in the developing mouse forebrain. *J Neurosci* 2005;25(31): 7268–77.
16. Marin O, Anderson SA, Rubenstein JL. Origin and molecular specification of striatal interneurons. *J Neurosci* 2000;20(16): 6063–76.
17. Wichterle H, Turnbull DH, Nery S, Fishell G, Alvarez-Buylla A. In utero fate mapping reveals distinct migratory pathways and fates of neurons born in the mammalian basal forebrain. *Development* 2001;128(19): 3759–71.
18. Rakic P. Specification of cerebral cortical areas. *Science* 1988;241(4862):170–6.
19. Wonders CP, Anderson SA. The origin and specification of cortical interneurons. *Nat Rev Neurosci* 2006;7(9):687–96.
20. Anderson SA, Marin O, Horn C, Jennings K, Rubenstein JL. Distinct cortical migrations from the medial and lateral ganglionic eminences. *Development* 2001; 128(3):353–63.

21. Nery S, Fishell G, Corbin JG. The caudal ganglionic eminence is a source of distinct cortical and subcortical cell populations. *Nat Neurosci* 2002; 5(12):1279–87.
22. Anderson SA, Qiu M, Bulfone A, Eisenstat DD, Meneses J, Pedersen R et al. Mutations of the homeobox genes Dlx-1 and Dlx-2 disrupt the striatal subventricular zone and differentiation of late born striatal neurons. *Neuron* 1997; 19(1):27–37.
23. McBain CJ, Fisahn A. Interneurons unbound. *Nat Rev Neurosci* 2001; 2(1):11–23.
24. Markram H, Toledo-Rodriguez M, Wang Y, Gupta A, Silberberg G, Wu C. Interneurons of the neocortical inhibitory system. *Nat Rev Neurosci* 2004; 5(10): 793–807.
25. Butt SJ, Cobos I, Golden J, Kessaris N, Pachnis V, Anderson S. Transcriptional regulation of cortical interneuron development. *J Neurosci* 2007; 27(44):11847–50.
26. Xu Q, Cobos I, De La Cruz E, Rubenstein JL, Anderson SA. Origins of cortical interneuron subtypes. *J Neurosci* 2004; 24(11): 2612–22.
27. Butt SJ, Fuccillo M, Nery S, Noctor S, Kriegstein A, Corbin JG et al. The temporal and spatial origins of cortical interneurons predict their physiological subtype. *Neuron* 2005; 48(4):591–604.
28. Yuste R. Origin and classification of neocortical interneurons. *Neuron* 2005; 48(4): 524–7.
29. Cobos I, Long JE, Thwin MT, Rubenstein JL. Cellular patterns of transcription factor expression in developing cortical interneurons. *Cereb Cortex* 2006; 16(Suppl 1):82–8.
30. Porteus MH, Bulfone A, Liu JK, Puelles L, Lo LC, Rubenstein JL. DLX-2, MASH-1, and MAP-2 expression and bromodeoxyuridine incorporation define molecularly distinct cell populations in the embryonic mouse forebrain. *J Neurosci* 1994; 14(11 Pt 1):6370–83.
31. Dolle P, Price M, Duboule D. Expression of the murine Dlx-1 homeobox gene during facial, ocular and limb development. *Differentiation* 1992; 49(2):93–9.
32. Eisenstat DD, Liu JK, Mione M, Zhong W, Yu G, Anderson SA et al. DLX-1, DLX-2, and DLX-5 expression define distinct stages of basal forebrain differentiation. *J Comp Neurol* 1999; 414(2):217–37.
33. Panganiban G, Rubenstein JL. Developmental functions of the Distal-less/Dlx homeobox genes. *Development* 2002; 129(19):4371–86.
34. Qiu M, Bulfone A, Ghattas I, Meneses JJ, Christensen L, Sharpe PT et al. Role of the Dlx homeobox genes in proximodistal patterning of the branchial arches: Mutations of Dlx-1, Dlx-2, and Dlx-1 and -2 alter morphogenesis of proximal skeletal and soft tissue structures derived from the first and second arches. *Dev Biol* 1997; 185(2): 165–84.
35. Bulfone A, Wang F, Hevner R, Anderson S, Cutforth T, Chen S et al. An olfactory sensory map develops in the absence of normal projection neurons or GABAergic interneurons. *Neuron* 1998; 21(6):1273–82.
36. Pleasure SJ, Anderson S, Hevner R, Bagri A, Marin O, Lowenstein DH et al. Cell migration from the ganglionic eminences is required for the development of hippocampal GABAergic interneurons. *Neuron* 2000; 28(3):727–40.
37. Depew MJ, Liu JK, Long JE, Presley R, Meneses JJ, Pedersen RA et al. Dlx5 regulates regional development of the branchial arches and sensory capsules. *Development* 1999; 126(17):3831–46.
38. Levi G, Puche AC, Mantero S, Barbieri O, Trombino S, Paleari L et al. The Dlx5 homeodomain gene is essential for olfactory development and connectivity in the mouse. *Mol Cell Neurosci* 2003; 22(4):530–43.
39. Long JE, Garel S, Alvarez-Dolado M, Yoshikawa K, Osumi N, Alvarez-Buylla A et al. Dlx-dependent and -independent regulation of olfactory bulb interneuron differentiation. *J Neurosci* 2007; 27(12):3230–43.
40. Kimura S, Hara Y, Pineau T, Fernandez-Salguero P, Fox CH, Ward JM et al. The T/ebp null mouse: Thyroid-specific enhancer-binding protein is essential for the organogenesis of the thyroid, lung, ventral forebrain, and pituitary. *Genes Dev* 1996; 10(1):60–9.
41. Poirier K, Van Esch H, Friocourt G, Saillour Y, Bahi N, Backer S et al. Neuroanatomical distribution of ARX in brain and its localisation in GABAergic neurons. *Brain Res Mol Brain Res* 2004; 122(1):35–46.
42. Stromme P, Mangelsdorf ME, Shaw MA, Lower KM, Lewis SM, Bruyere H et al. Mutations in the human ortholog of Aristaless cause X-linked mental retardation and epilepsy. *Nat Genet* 2002; 30(4):441–5.
43. Frints SG, Froyen G, Marynen P, Willekens D, Legius E, Fryns JP. Re-evaluation of MRX36 family after discovery of an ARX gene mutation reveals mild neurological features of Partington syndrome. *Am J Med Genet* 2002; 112(4):427–8.

44. Kato M, Das S, Petras K, Sawaishi Y, Dobyns WB. Polyalanine expansion of ARX associated with cryptogenic West syndrome. *Neurology* 2003;61(2):267–76.
45. Uyanik G, Aigner L, Martin P, Gross C, Neumann D, Marschner-Schafer H et al. ARX mutations in X-linked lissencephaly with abnormal genitalia. *Neurology* 2003; 61(2):232–5.
46. Kitamura K, Yanazawa M, Sugiyama N, Miura H, Iizuka-Kogo A, Kusaka M et al. Mutation of ARX causes abnormal development of forebrain and testes in mice and X-linked lissencephaly with abnormal genitalia in humans. *Nat Genet* 2002;32(3):359–69.
47. Cobos I, Broccoli V, Rubenstein JL. The vertebrate ortholog of Aristaless is regulated by Dlx genes in the developing forebrain. *J Comp Neurol* 2005;483(3):292–303.
48. Baraban SC, Tallent MK. Interneuron Diversity series: Interneuronal neuropeptides – endogenous regulators of neuronal excitability. *Trends Neurosci* 2004;27(3): 135–42.
49. Monaghan AP, Bock D, Gass P, Schwager A, Wolfer DP, Lipp HP et al. Defective limbic system in mice lacking the tailless gene. Nature 1997;390(6659):515–17.
50. Roy K, Thiels E, Monaghan AP. Loss of the tailless gene affects forebrain development and emotional behavior. *Physiol Behav* 2002;77(4–5):595–600.
51. Kamachi Y, Uchikawa M, Kondoh H. Pairing SOX off: With partners in the regulation of embryonic development. *Trends Genet* 2000;16(4):182–7.
52. Avilion AA, Nicolis SK, Pevny LH, Perez L, Vivian N, Lovell-Badge R. Multipotent cell lineages in early mouse development depend on SOX2 function. *Genes Dev* 2003;17(1):126–40.
53. Ferri AL, Cavallaro M, Braida D, Di Cristofano A, Canta A, Vezzani A et al. Sox2 deficiency causes neurodegeneration and impaired neurogenesis in the adult mouse brain. *Development* 2004;131(15):3805–19.
54. Cavallaro M, Mariani J, Lancini C, Latorre E, Caccia R, Gullo F et al. Impaired generation of mature neurons by neural stem cells from hypomorphic Sox2 mutants. *Development* 2008;135(3):541–57.
55. Birchmeier C, Gherardi E. Developmental roles of HGF/SF and its receptor, the c-Met tyrosine kinase. *Trends Cell Biol* 1998; 8(10):404–10.
56. Bladt F, Riethmacher D, Isenmann S, Aguzzi A, Birchmeier C. Essential role for the c-met receptor in the migration of myogenic precursor cells into the limb bud. *Nature* 1995;376(6543):768–71.
57. Uehara Y, Minowa O, Mori C, Shiota K, Kuno J, Noda T et al. Placental defect and embryonic lethality in mice lacking hepatocyte growth factor/scatter factor. *Nature* 1995;373(6516):702–5.
58. Schmidt C, Bladt F, Goedecke S, Brinkmann V, Zschiesche W, Sharpe M et al. Scatter factor/hepatocyte growth factor is essential for liver development. *Nature* 1995;373(6516):699–702.
59. Mars WM, Kim TH, Stolz DB, Liu ML, Michalopoulos GK. Presence of urokinase in serum-free primary rat hepatocyte cultures and its role in activating hepatocyte growth factor. *Cancer Res* 1996;56(12): 2837–43.
60. Naldini L, Tamagnone L, Vigna E, Sachs M, Hartmann G, Birchmeier W et al. Extracellular proteolytic cleavage by urokinase is required for activation of hepatocyte growth factor/scatter factor. *EMBO J* 1992; 11(13):4825–33.
61. Powell EM, Campbell DB, Stanwood GD, Davis C, Noebels JL, Levitt P. Genetic disruption of cortical interneuron development causes region- and GABA cell type-specific deficits, epilepsy, and behavioral dysfunction. *J Neurosci* 2003;23(2):622–31.
62. Bae M, Bissonette GB, Suresh T, Franz TM, Depireux DA, Powell EM. Hepatocyte growth factor (HGF) reduces seizures and behavioral deficits in a mouse model of frontal lobe epilepsy. *Epilepsia* 48[s6], 238–48. 10-27-2007. Ref Type: Abstract
63. Glickstein SB, Moore H, Slowinska B, Racchumi J, Suh M, Chuhma N et al. Selective cortical interneuron and GABA deficits in cyclin D2-null mice. *Development* 2007; 134(22):4083–93.
64. Ross ME. Cell division and the nervous system: Regulating the cycle from neural differentiation to death. *Trends Neurosci* 1996; 19(2):62–8.
65. Hofmann SL, Das AK, Lu JY, Soyombo AA. Positional candidate gene cloning of CLN1. *Adv Genet* 2001; 45:69–2.
66. Hofmann SL, Atashband A, Cho SK, Das AK, Gupta P, Lu JY. Neuronal ceroid lipofuscinoses caused by defects in soluble lysosomal enzymes (CLN1 and CLN2). *Curr Mol Med* 2002; 2(5):423–37.
67. Gupta P, Soyombo AA, Atashband A, Wisniewski KE, Shelton JM, Richardson JA et al. Disruption of PPT1 or PPT2 causes

neuronal ceroid lipofuscinosis in knockout mice. *Proc Natl Acad Sci USA* 2001; 98(24): 13566–71.

68. Jalanko A, Vesa J, Manninen T, von Schantz C, Minye H, Fabritius AL et al. Mice with Ppt1Deltaex4 mutation replicate the INCL phenotype and show an inflammation-associated loss of interneurons. *Neurobiol Dis* 2005;18(1):226–41.
69. Kielar C, Maddox L, Bible E, Pontikis CC, Macauley SL, Griffey MA et al. Successive neuron loss in the thalamus and cortex in a mouse model of infantile neuronal ceroid lipofuscinosis. *Neurobiol Dis* 2007;25(1): 150–62.
70. National Institute of Neurological Disorders and Stroke. Seizures and Epilepsy: Hope Through Research. http://www.ninds.nih.gov/disorders/epilepsy/detail_epilepsy.htm. 2-13-2008. 3-11-2008.
71. Treiman DM. GABAergic mechanisms in epilepsy. *Epilepsia* 2001; 42(Suppl 3):8–12.
72. Magloczky Z, Freund TF. Impaired and repaired inhibitory circuits in the epileptic human hippocampus. *Trends Neurosci* 2005;28(6):334–40.
73. Ben Ari Y. Seizures beget seizures: The quest for GABA as a key player. *Crit Rev Neurobiol* 2006;18(1–2):135–44.
74. Fritschy JM, Kiener T, Bouilleret V, Loup F. GABAergic neurons and GABA(A)-receptors in temporal lobe epilepsy. *Neurochem Int* 1999;34(5):435–45.
75. Martin JL, Sloviter RS. Focal inhibitory interneuron loss and principal cell hyperexcitability in the rat hippocampus after microinjection of a neurotoxic conjugate of saporin and a peptidase-resistant analog of Substance P. *J Comp Neurol* 2001;436(2): 127–52.
76. Sloviter RS. Decreased hippocampal inhibition and a selective loss of interneurons in experimental epilepsy. *Science* 1987; 235(4784):73–6.
77. Sloviter RS. The functional organization of the hippocampal dentate gyrus and its relevance to the pathogenesis of temporal lobe epilepsy. *Ann Neurol* 1994;35(6): 640–54.
78. Cossart R, Dinocourt C, Hirsch JC, Merchan-Perez A, De Felipe J, Ben Ari Y et al. Dendritic but not somatic GABAergic inhibition is decreased in experimental epilepsy. *Nat Neurosci* 2001;4(1):52–62.
79. Dinocourt C, Petanjek Z, Freund TF, Ben Ari Y, Esclapez M. Loss of interneurons innervating pyramidal cell dendrites and axon initial segments in the CA1 region of the hippocampus following pilocarpine-induced seizures. *J Comp Neurol* 2003; 459(4):407–25.
80. Sayin U, Osting S, Hagen J, Rutecki P, Sutula T. Spontaneous seizures and loss of axo-axonic and axo-somatic inhibition induced by repeated brief seizures in kindled rats. *J Neurosci* 2003;23(7):2759–68.
81. Kobayashi M, Wen X, Buckmaster PS. Reduced inhibition and increased output of layer II neurons in the medial entorhinal cortex in a model of temporal lobe epilepsy. *J Neurosci* 2003;23(24):8471–9.
82. de Lanerolle NC, Kim JH, Robbins RJ, Spencer DD. Hippocampal interneuron loss and plasticity in human temporal lobe epilepsy. *Brain Res* 1989;495(2):387–95.
83. Magloczky Z, Wittner L, Borhegyi Z, Halasz P, Vajda J, Czirjak S et al. Changes in the distribution and connectivity of interneurons in the epileptic human dentate gyrus. *Neuroscience* 2000;96(1):7–25.

Chapter 9

Imaging Seizure Propagation In Vitro

Andrew J. Trevelyan and Rafael Yuste

Abstract

This is perhaps the most beautiful time in human history; it is really pregnant with all kinds of creative possibilities made possible by science and technology. Jonas Salk's quotation seems particularly pertinent to recent developments in imaging technology, which have provided both beauty and insight in equal measure. We can now manipulate biological systems, both genetically and otherwise, to introduce fluorescent markers, literally adding colour to our preparations. These advances have occurred in parallel with remarkable developments in microscopy technology, with novel means of illumination and light detection allowing imaging to be done in ever more inaccessible places, with ever improving temporal and spatial resolution. A critical step in the application of these new technologies though is to characterize their relationship to older ways of measuring phenomena. In this chapter, we describe some of our efforts to use Ca^{2+} dyes to follow network activity, and in particular the progress of epileptiform events through cortical networks. The conventional means of recording epileptiform events has always been, and likely will continue to be, electrophysiological, but a careful calibration of imaging signals with respect to electrophysiological recordings can extend our data set immeasurably, providing new insights into old problems.

Key words: calcium imaging, voltage-sensitive dye, intrinsic optical imaging, pyramidal cell, interneuron, inhibitory surround, astrocyte, glial cell, propagation, ictal, interictal.

1. Introduction: The Importance of Cortical Maps

The pattern of the flow of activity through cortical networks defines their function. Information is represented within topographic maps, and these maps demarcate the various cortical areas. For instance, the basic organizing principal of the lower cortical visual areas (primary and secondary visual cortices, or V1 and V2) is visuotopic, meaning that the map reflects visual space. Thus, a particular sensory stimulus, such as a bar moving across the

Scott C. Baraban (ed.), *Animal Models of Epilepsy: Methods and Innovations, vol. 40*

DOI 10.1007/978-1-60327-263-6_9 Springerprotocols.com

visual field, activates a limited set of neurons located at the visuotopic location of the stimulus. This focal activity in V1 and V2 is then relayed on to higher visual areas, where the representations are increasingly abstract, ceasing to have a clear visuotopic basis, but instead starting to represent other features of the stimulus such as the nature of the movement, colour of the bar or other visual attributes. Consistently though, in all these areas, activation of a local territory is associated with suppression of the cortical areas immediately adjacent. Thus, the processing of information flows through cortical areas in a very focal fashion, activating subsequent cortical areas, but generally not spreading laterally within cortical areas.

Why begin a chapter on epilepsy with a discussion of topographic maps? Well, there are two reasons. First, the two phenomena are linked in one of the truly great neuroscientific insights of the pre-Cajal era. The English neurologist, Hughlings Jackson, intuited that the gradual progressive spread of convulsions he observed in some patients, first involving the fingers of one hand, then the whole hand, next the forearm and so on until the whole body was convulsing, strongly implied the existence of a map of the body within the cortex. He proposed that the spread of convulsions through the limbs reflected the spread of pathological activity across the body map. This pattern of epileptic recruitment, subsequently referred to as a Jacksonian march, must have exercised Jackson greatly, because the patient he was observing was his own wife.

The second reason is to introduce the idea that the same inhibitory surround mechanism which keeps normal physiological activity focussed may also be involved in restricting epileptic activity from spreading. This is a natural consequence of the topographic organization of the cortex and one that has profound implications for epilepsy research. The significance of this inhibitory surround theory, first proposed by Prince and Wilder in 1967 *(1)*, is that it immediately suggests a cause of seizures arising from a deficit in this mechanism. It also raises an interesting therapeutic avenue, seeking to bolster the inhibitory surround mechanism to counter epileptiform activity. We will return to these topics later in the chapter.

One might speculate from these introductory remarks that one could make a broad distinction between physiological activity which passes through cortical areas while retaining focality and pathological activity which spreads progressively across cortical areas in a contiguous fashion. This distinction is blurred though, because there are clearly physiological ripples of activity that progress across the cortex during sleep and also during early development (variously called early network oscillations *(2)* or giant depolarizing potentials *(3)*). In addition, epileptiform spread can undoubtedly spread upwards and downwards through the hierarchically arranged cortical areas, as well as the more contiguous pattern of activation seen in Jacksonian marches. Most in vitro studies, however, have for obvious reasons, only explored contiguous spread, and this is what we will concentrate on in this chapter.

We will discuss first the relevance of in vitro studies in the current neuroscientific era, which is increasingly characterized by heroically difficult in vivo experiments. As implied in this introduction, our angle will be to explore how inhibition restricts the spread of epileptiform seizures. We will then provide some details of the methodology of our in vitro imaging experiments and some of the main findings of our and others' research in this field. And finally, we will discuss briefly where these studies might lead next.

2. The Usefulness of Epilepsy Models

Epilepsy is a serious, chronic condition, and large numbers of sufferers live their lives without proper seizure control. We need better therapies, but how to develop these? Our efforts to understand the disease are continually thwarted by the sheer complexity of epileptic activity patterns, the most intense of any neural activities. A particular difficulty lies in adequately sampling the various neural firing patterns. Some researchers, most notably Mircea Steriade and his colleagues, have made a lifetime's work building up a database of single cell in vivo recordings in various epileptic states *(4, 5)*. This work is of course an invaluable resource. There remains, however, a great need for simplifying models of the disease, which focus on particular patterns of cellular activity that are pivotal in both the model and the patient.

The ideal model would allow one to control for extraneous variables and study a particular process in isolation. To this end, researchers have shown great ingenuity over the years, developing numerous models of epilepsy. Our problems now arise because many different activity patterns are termed "epileptiform", and there is a good deal of confusion about how different models relate to each other. In many cases, we have only scratched the surface in our attempts to resolve this confusion. The key issue is not whether the model replicates all the in vivo activity, but whether we can identify which part it does replicate and provide validation that the activity patterns in vitro and in vivo do indeed represent the same phenomenon. Many in vitro models appear to capture facets of the in vivo activity very well. For instance, the zero magnesium (0 Mg^{2+}) in vitro model manifests a wide range of epileptiform activity patterns, including interictal activity *(6, 7)*, slow and rapid patterns of generalization *(8–10)*, tonic–clonic transitions *(6, 7)* and status epilepticus *(11–13)*. Another approach has been to try to mimic specific epilepsy phenotypes, looking to develop a range of genetic and pharmacological manipulations that create similar phenotypes in rodents *(14–19)*.

To confirm the equivalence of these activity patterns in the different models, however, requires a deep knowledge of these myriad activity patterns, from the molecular activity up through the cellular to the network behaviour. These are extreme requirements, but essential. Without this understanding, the models remain obscure, but with it, they can elucidate genetic causes of epilepsy, aid drug development by identifying molecular targets and indicate appropriate assays to test new drugs.

2.1. The Usefulness of In Vitro Models

What research is possible then using in vitro preparations? Most people accept that such preparations are appropriate for examining subcellular issues (e.g. protein expression, synaptic physiology). In these studies, it is presumed that the processes being investigated are relatively unaffected by the fact that the preparation destroys parts of the network. As such, the benefits in terms of access to the tissue and the huge gain in control of the experimental conditions greatly outweigh the loss of part of the network. The concern with studying network dynamics is whether this "net gain" in benefits over drawbacks is still the case.

Dealing first with the drawbacks of in vitro preparations, the most important of these regard the loss of long neuronal connections and also the lifetime of various molecular and ionic species within the extracellular milieu. Long-range axonal pathways are severed by sectioning the brain, although there are ways of preparing the tissue that preserve particular tracts such as the interhemispheric, callosal pathways or the thalamo-cortical pathways *(20, 21)*. It remains an open question how much the activity patterns observed in in vitro preparations are altered by loss of these long-range connections. It is extremely pertinent, however, that the vast majority of cellular interactions in cortical networks are local *(22, 23)*, and this particularly applies to the ephaptic interactions *(24)*, glial–neuronal interactions *(25–27)*, gap-junction coupling *(28–33)* and the distribution of inhibitory synaptic connections. Indeed, these issues can be used to the researcher's advantage if a particular activity pattern survives the sectioning process, because it provides strong circumstantial evidence that the behaviour is mediated largely by local rather than long-range connections: a case in point is the presence of a very powerful inhibitory restraint that is still readily apparent in brain slices (*see* **Section 4**).

Another issue in vitro concerns the modulation of activity by substances normally present in the external milieu. In vitro and in vivo preparations are likely to differ in the delivery and lifetime of various neuromodulators (acetylcholine, noradrenaline, 5HT, etc.) or ion species (particularly K^+ in the aftermath of epileptiform bursts). The extreme cooling of the brain during slicing may also have consequences for the network, since cooling causes the collapse of dendritic spines, although these quickly recover on rewarming *(34)*.

In spite of these various concerns regarding in vitro preparations, the intrinsic conductances of neurons and their local connectivity patterns are thought to be preserved in brain slices. Furthermore, the fact remains that many epileptiform and physiological activity patterns (e.g. UP states *(35, 36)*, stable models of various brain rhythms in brain slices *(37)*) are remarkably similar in vivo and in vitro (**Fig. 9.1**). This similarity suggests that many of the most important determinants of local network behaviour are still present in the slices.

What, then, are the benefits of in vitro experimentation? Well, there are many. The preparation is straightforward and permits great control over and reproducibility of the experimental conditions. Indeed this facilitated the development of Ca^{2+} imaging of network dynamics by allowing a rapid exploration of various means of loading cells with Ca^{2+} dyes *(28)*.

A second major benefit over in vivo preparations is the greatly enhanced visualization and access of the tissue in vitro. For instance it is possible to identify individual cells either by their distinctive somatic morphology (using differential interference contrast (DIC) microscopy) or their molecular expression pattern (using a co-expressed fluorescent marker such as green fluorescent protein (GFP)) or their activity patterns (assessed using a bulk-loaded

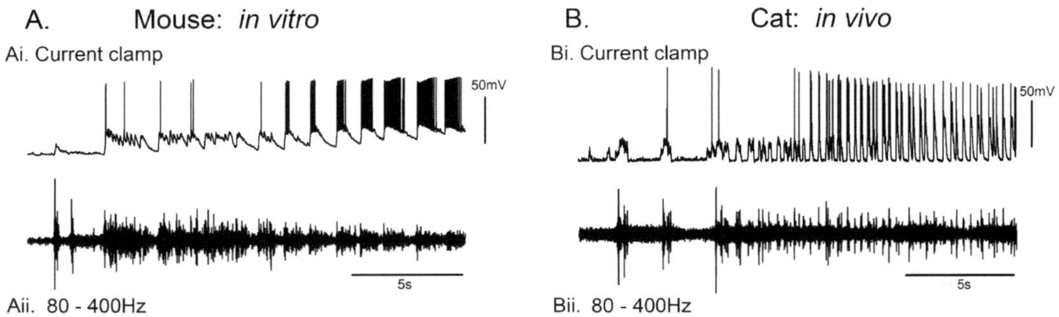

Fig. 9.1. **Comparable firing patterns and high-frequency oscillations (HFOs) in vitro and in vivo. (A)** Interictal to ictal transition in a mouse brain slice bathed in 0 Mg^{2+}solution. Current clamp recording from a layer 5 pyramidal cell (Ai) is shown together with the bandpass filtered (80–400 Hz) V_{clamp} trace from an adjacent pyramidal cell (Aii). **(B)** Equivalent traces showing the recruitment of a putative pyramidal cell recorded in vivo in cortical area 7 in an adult cat (recording kindly provided by Igor Timofeev).

Ca^{2+} dye) and then target these cells for patching. Furthermore, one can do this for multiple closely apposed cells. Henry Markram's laboratory has developed this to an extreme art form, patching on to as many as 12 juxtaposed cells simultaneously. Even patching on to just two adjacent cells is currently beyond our capabilities in vivo, yet this technique is the basis for much of our knowledge of CNS synaptic physiology. And as we will show in **Section 4**, such recordings from adjacent pyramidal cells have also proved incredibly informative in our attempts to understand in vitro epileptiform activity *(8)*.

It is also worth mentioning temperature at this juncture. Cooling the tissue to room temperature causes many functional changes in neuronal behaviour including depolarizing the cells *(38, 39)*, reducing the probability of vesicular release following an action potential *(40)*, but increasing the probability of the postsynaptic response triggering an action potential (cooling causes a drop in membrane resistivity, insulating the dendrites better, and thus reducing the electrotonic length of the dendritic arbour, resulting in larger postsynaptic potentials at the soma *(39)*). The net effect is to change the network from a more deterministic network to a more probabilistic one, thus introducing uncontrolled variability instead of reducing it. With this caveat in mind, temperature is, of course, an important topic of epilepsy research. Febrile seizures are the most common form of childhood seizure, and there is evidence that cooling may be a means of terminating seizures *(41)*, perhaps by inducing a depolarizing block ahead of the ictal wavefront. Unless though one is studying these specific temperature issues, it seems sensible to study network behaviours at physiological temperatures.

To summarize this section on in vitro experimentation, it has been argued that such preparations have misled us as to the nature of in vivo activity. We would counter, as others have also argued, that there are indeed clear differences between the in vivo and in vitro states, and researchers need to exercise caution. But to disregard in vitro experimentation completely seems unreasonable, because it provides technical possibilities that do not exist in other preparations, and these have illuminated our understanding of the brain immeasurably.

3. A Double-Barrel Shotgun: Ca^{2+} Imaging and Electrophysiology

The critical task facing us is to describe the different models in terms of the behaviour of key populations of neurons. Which are those "key" neurons though, and how to identify them? Conventional approaches have used various electrophysiological techniques, but these present serious sampling issues. Intracellular recordings provide a detailed view of single neurons, but only a handful can be recorded simultaneously. In contrast, extracellular recordings reflect the activity levels of the entire network, but lack cellular resolution.

If a process can be visualized though, a third option exists: to use imaging to record at multiple locations simultaneously, with a high resolution limited only by the optics of the system (essentially determined by the wavelength of the light, the quality of the lenses and the specifications of the detectors). A very useful experimental tool in this regard is the fast network imaging of Ca^{2+} transients, a technique pioneered in the laboratory of the late Larry Katz in the late 1980 s *(42, 43)*. When neurons fire an action potential, they experience a sharp rise in cytosolic Ca^{2+} level *(44)*. Thus, by bulk loading the tissue with a Ca^{2+} dye and monitoring these Ca^{2+} transients, one can chart the activity of hundreds of neurons simultaneously with single-cell resolution and with fine temporal resolution *(28)*.

How might this technique be used and what are its limitations? Is it possible for instance to merely video the tissue and derive the neural code? The first question to ask is how fast one can image the tissue? Seizures, after all, can propagate extremely quickly. Can imaging realistically capture the dynamics of these intense and rapidly evolving network events?

With high numerical aperture (NA) microscope objectives, focussing on subcellular compartments, it is quite possible to detect significant Ca^{2+} increases in less than a millisecond (i.e. imaging at speeds in excess of 1 kHz). When one is interested in network activity, however, one sacrifices sensitivity for an increase in the field of view. As with all scientific measurement, one then needs to determine the sampling protocol that allows an appropriate signal to noise, and this has to be assessed empirically for each rig. In our experiments, we found that using a 20 × objective gave a reasonable balance between having an extended field of view and a reduced photon capture, while still producing an image on the CCD camera in which the cell soma extends over 10 s of pixels (important for cell soma identification and averaging of the signal). This trade-off has been helped recently by several microscope manufacturers developing high NA 20 × lens specifically for electrophysiologists and in vivo 2-photon microscopists (these are water-dipped lenses with long working distance to allow an electrode to be placed underneath the objective or to permit imaging deep to the pial surface). For our recent epilepsy research, we used a spinning disc confocal scanner (there are a number of suppliers; Solamere Technology Group and Visitech International provided the ones in our respective laboratories) with a fast CCD camera (again there are several options; we use Stanford Photonics or Hamamatsu cameras), which allowed us to image successfully at 30–60 Hz.

This imaging speed sets some limitations to what might usefully be studied using Ca^{2+} dyes. Consider, for example, the propagation of epileptiform events in the disinhibited slice. The various estimates of propagation speeds for these events range from 10 to 100 mm/s. Taking an approximate average to be about 40 mm/s and an average field of view to be about 400 μm across (imaging at 20 ×), a back-of-the-envelope calculation shows that the event would traverse the entire field of view in about 10 ms, or within a single time frame when imaging at 60 Hz. Clearly, Ca^{2+} imaging is not going to capture much of the dynamics of such events. Fortunately, when inhibition is relatively intact within the slice, the epileptiform dynamics are far slower and captured well by Ca^{2+} imaging (*see* **Section 4**).

Keeping with the same topic, the asymmetric kinetics of Ca^{2+} binding to the dye also influence the temporal resolution of Ca^{2+} imaging: the onset of the signal is extremely rapid after Ca^{2+} enters the cellular compartment containing the dye, but the signal then only gradually decays. The time constant of the decay can be several hundreds of milliseconds or even seconds, depending on the exact dye concentration and K_d *(45)*. Thus the signal from successive action potentials summates at short intervals, and the termination of firing can be difficult to assess from the slow fluorescence decay. In terms of monitoring action potential trains then, one can readily detect step increases in firing, but the timing of drops in firing is less precise. Furthermore, sustained high-frequency firing may also present detection problems. Unless one has an independent measure of the fluorescence intensity when the cell is completely quiescent, then high-frequency activity and complete quiescence can result in small rapidly fluctuating fluorescent signals that are impossible to distinguish. In contrast to these extremes (quiescence and persistent intense firing), we and others have shown that intermittent, low-level firing can be followed with great accuracy (*8*, *21*, *28*, *35*, *44*, *46–51)*.

To exemplify these issues, say one wanted to visualize cellular activity patterns during γ-rhythms. Basket cells are known to fire on every cycle of the rhythm and their Ca^{2+} dye signal would be high relative to baseline (inactive) and fluctuating rapidly. The onset of the γ-frequency rhythm would be visualized as a sudden change in fluorescence and its offset as a more gradual decay back to baseline. If, however, the rhythm were present throughout the entire imaging time and one did not know the baseline fluorescence, then the basket cell signal may be indistinguishable from that of a quiescent cell with a high resting $[Ca^{2+}]$, where the fluctuations in this latter case are merely noise. Pyramidal cells on the other hand, which fire only intermittently during the γ-frequency rhythm, may be followed rather well. Assessment of basket cell activity may be compromised in other ways too: fast-spiking basket cells have shorter action potentials and one class of basket cells has high levels of an intrinsic Ca^{2+} buffer, parvalbumin, in its cytosol. For these reasons, the Ca^{2+} signal in parvalbumin-positive, fast-spiking interneurons is likely to be qualitatively different to that in pyramidal cells.

Our assessment of Ca^{2+} dye signals has been carefully worked out using combined imaging and electrophysiological recordings. The key experiment is to record the action potentials in a prior dye-loaded cell, using a cell-attached patch electrode *(8)*. This recording mode preserves the internal milieu of the cell (including its dye concentration) and allows one to correlate directly the activity pattern and the Ca^{2+} signal. These experiments also flagged an important source of imaging artefact, due to the Ca^{2+} dye being distributed in all parts of the cell. Not only does signal arise from the somata, but also from dendritic branches and axonal terminals. This extrasomatic signal can give rise to false-positives when trying to derive the firing pattern from the Ca^{2+} fluorescence, especially when the tissue is very active. We discovered from our cell-attached recordings that prior to the recruitment of pyramidal cells to an ictal event, they regularly experience intense barrages of activity without firing, and at these times, the local neuropil lights up (when loaded with Ca^{2+} dye). With the relatively low NA objectives used for network imaging, even if one is focussed on the cell soma, the fluorescence measures also include photons arising from the neuropil either above or below the soma. Thus one detects fluorescence changes from the neuropil activity and not the somatic activity. We found though that simply subtracting the signal from the surrounding area produces a cleaned-up signal which does accurately reflect the firing pattern of the cell.

These kinds of imaging-electrophysiology experiments have told us how to interpret the Ca^{2+} signals and why, in heterogeneous populations, we cannot as yet derive precise firing patterns from imaging alone (although see *(50)*). In the next two sections, we discuss two other important reasons for combining electrophysiology and imaging: first to control the imaging and second to provide a cross-reference for other epilepsy databases.

3.1. Using Electrophysiology to Guide Imaging

An important difference between electrophysiology and imaging is the duration of recordings. In the clinic, EEG recordings can be monitored for days on end. Other electrophysiological recordings are not as robust, but all can easily outlast imaging experiments. The problem with imaging in this regard is that the dyes are eventually inactivated by light (photobleaching) and in the process can release toxic elements, usually in the form of free radicals (phototoxicity). It is therefore of utmost importance to limit the imaging time to the most interesting periods of activity. One way is to time-lock the imaging to a stimulus. For instance, we have analysed activation patterns in cortex subsequent to thalamic stimulation *(21)*, and others have imaged visually induced activity in cortex in whole animals *(51)* and local neuronal activity subsequent to Ca^{2+} uncaging events in glia *(52)*.

Capturing spontaneous activity though presents a problem, particularly if events occur only infrequently and are unpredictable. For this reason, we have not yet addressed ictogenesis using imaging, but we found that the propagation of ictal events in the 0 Mg^{2+} model could be captured relatively easily because the arrival of the wavefront was heralded by a very particular pattern of voltage clamp (V_{clamp}) trace *(8)* (*see* **Section 4**). A similar approach used current clamp patch recordings to trigger image capture when studying spontaneous UP states in barrel cortex *(21)*.

3.2. Using Electrophysiology to Interpret Imaging

A second important reason to combine electrophysiology and imaging is that most of our understanding of epileptic discharges comes from electrophysiological recordings. And for obvious practical reasons, extracellular recordings (indeed "extraskullular" in the form of most EEG recordings) will continue to be the recording mode of choice in the clinical setting. Thus, electrophysiology provides the means of cross-referencing our newly acquired imaging data. An added incentive in this regard comes because currently our Ca^{2+} imaging protocols remain rather limited. For reasons that are not completely clear, our ability to load tissue with the dye drops off markedly with the increasing age of the animal. Thus, almost all our Ca^{2+} imaging is limited to neonatal or young adolescent mice, and even then, there is variability in different brain regions. Consequently, while Ca^{2+} imaging provides an unprecedented detail about the network activity patterns, imaging can only be performed on a restricted range of preparations. One must then generalize the imaging data using particular features of the electrophysiological recordings that correlate with particular imaging findings. To give an example from our own work, we described in juvenile animals a particular pattern of intermittent, stepwise ictal progression associated with a particular transition from rhythmic inhibitory to rhythmic excitatory barrages seen in V_{clamp} recordings of pyramidal cells *(8)*. The same pattern of V_{clamp} recordings is seen in adult animals *(8)*, and furthermore, the distribution of propagation speeds in young and old animals is the same *(8, 10)*, strongly suggesting that our other imaging-specific findings can also be extrapolated from young to old animals too.

Neocortex can be loaded well, but again for obscure reasons, hippocampal and thalamic preparations are not labelled so well with Ca^{2+} dyes. These problems are likely to be overcome in the future. One promising avenue is that age appears not to be such a limitation when loading cells in vivo *(53)*, thus presenting the possibility of preloading cells prior to sacrificing the animal. A second avenue is the development of genetically expressed

Ca^{2+} sensors *(54)*. The latest of these use various members of the GFP family of fluorescent proteins tagged to a Ca^{2+} sensor, and thus provide a relatively photostable dye which lasts the entire lifetime of the animal. The challenge now is to get these proteins expressed in the right cell classes at sufficient concentration to be useful.

3.3. Using Imaging to Guide Electrophysiology

Perhaps the simplest experimental paradigm combining imaging and electrophysiology, though, is the converse of what we have described in the previous sections: using imaging to identify "interesting" cells. In this way, we use imaging as a screening tool, using it to highlight cell classes by their distinctive activity patterns. A prerequisite is that the activity being imaged can either be triggered or occurs sufficiently frequently that one can capture events just by turning the camera on. To capture spontaneous activity in this way, our chances of success are obviously boosted if we can image for longer durations, and consequently multiphoton imaging is the best option, followed by spinning disc confocal imaging, while epifluorescence imaging is a poor third. Another prerequisite is that the analysis of the movies can be done quickly, so as to be able to return to the slice while it is still alive and perform targeted patch clamp recordings.

As a technical aside, Fura-2 lends itself particularly well to targeted patching experiments. Fura-2 traditionally has been used as a ratiometric Ca^{2+} dye, making use of its unusual fluorescence emission at different Ca^{2+} concentrations: an increase in Ca^{2+} causes an increased fluorescence emission to illumination below 360 nm and a drop in fluorescence to illumination with light above 360 nm (the inflexion, referred to as the isosbestic point of the dye, occurs at ~360 nm). By collecting alternate frames imaging either side of the isosbestic point, one can derive the exact Ca^{2+} concentration, independent of the concentration of the dye (and hence independent of bleaching). Fura-2 can also be used at a single wavelength illumination. Typically, the preferred wavelength then is the higher, which shows a drop in emitted light when Ca^{2+} is bound. The benefit for patching comes because in the resting state, Ca^{2+} is low, and so the cells emit fluorescence strongly. As a result, one sees a kind of Nissl-like labelling appearance, with every cell visible, giving a direct comparison with the DIC image down the scope. Particular cells showing interesting patterns of fluorescence flux can thus be identified under DIC very easily. In contrast, other commonly used Ca^{2+} dyes, such as Oregon Green BAPTA 1, have very low fluorescence at rest, and so most cells are almost invisible at rest. As a result, the correspondence between the fluorescence view and the DIC view is a lot less clear, making targeted patching more difficult, but not impossible.

These targeted patching experiments have been used regularly by our and other laboratories to identify cells active during UP states *(35)*, cells that are rhythmically active after all neurotransmission is blocked and cells showing distinct activity patterns in early neonatal life *(55)*.

4. Inhibitory Restraints on Epileptiform Spread: In Vitro Evidence

Having so far given a general discourse on Ca^{2+} imaging in in vitro slices, in this next section we will describe in rather more detail our own studies on epileptiform propagation using one particular in vitro model, induced by bathing slices in ACSF devoid of Mg^{2+} (the 0 Mg^{2+} model). The attraction of this model was that, in contrast to many other models, epileptiform activity occurs in a network in which inhibition is relatively preserved.A further appeal was that a previous study of this model had shown that propagation can proceed extremely slowly, thus presenting a more tractable imaging proposition *(10)*.

It was apparent though that the epileptiform activity in the 0 Mg^{2+} model was a moving target in more ways than one. Epileptiform activity in this model clearly evolves. While the immediate effect of removing Mg^{2+} ions from the extracellular space is to relieve NMDA receptors of their voltage-dependent gating mechanism, there is in addition a second epileptogenic development, because over a period of several hours there evolves a gradual reduction in the amplitude of evoked GABAergic synaptic events *(56)*. Whittington et al. proposed the following cascade: reducing extracellular Mg^{2+} causes the intracellular levels to drop, albeit much more slowly, which in turn causes a drop in Mg-ATP levels and a reduction in GABA receptor phosphorylation *(56)*. These dephosphorylated channels show a reduced conductance, and so consequently, by this convoluted cascade, reducing the extracellular Mg^{2+} levels also disinhibits the slice. This is just one documented way in which the network evolves, but there are likely to be other pro-epileptic changes in the 0 Mg^{2+} slices. To give just one example, given what we know about how Ca^{2+} entry through NMDA receptors influences synaptic plasticity, it can be presumed with some certainty that ictal events will induce changes in synaptic strength, especially in the 0 Mg^{2+} model.

4.1. Rapidly Propagating Epileptiform Activity

Our studies of propagation speeds in this model confirmed the progressive nature of epileptiform activity, with each successive ictal event propagating faster than the one before *(9)*. This finding provided insight into why Wong and Prince had observed both slow and fast propagation in slices that had been pre-incubated in

0 Mg^{2+}: those slices displaying fast propagation had already experienced many ictal events, whereas the slowly propagating events were the first such events to occur in those slices. In contrast to the increasing speed of the wavefront, the afterdischarge propagation speed does not change in successive events *(9)*. Interestingly, the ictal wavefront speed appears to converge with the afterdischarge speed, which coincidentally is also within the published range of speeds for disinhibited slices (**Fig. 9.2**).

This latter model, the disinhibited slice, is the best studied of the in vitro models. It is also the topic of an excellent and extensive monograph by Traub and Miles *(57)*, in which they sought to understand the network behaviour by modelling the activity patterns computationally. They chose the disinhibited slice for their initial attempts at modelling network activity because, at a stroke, it massively reduced the computational task by removing the inhibitory components of the network. These modelling studies indicated that in the disinhibited slice, ictal propagation is over an order of magnitude slower than axonal propagation because the rate-limiting step is postsynaptic, with the charging up of newly recruited neurons. Based on the admittedly circumstantial evidence of the propagation speeds, could it

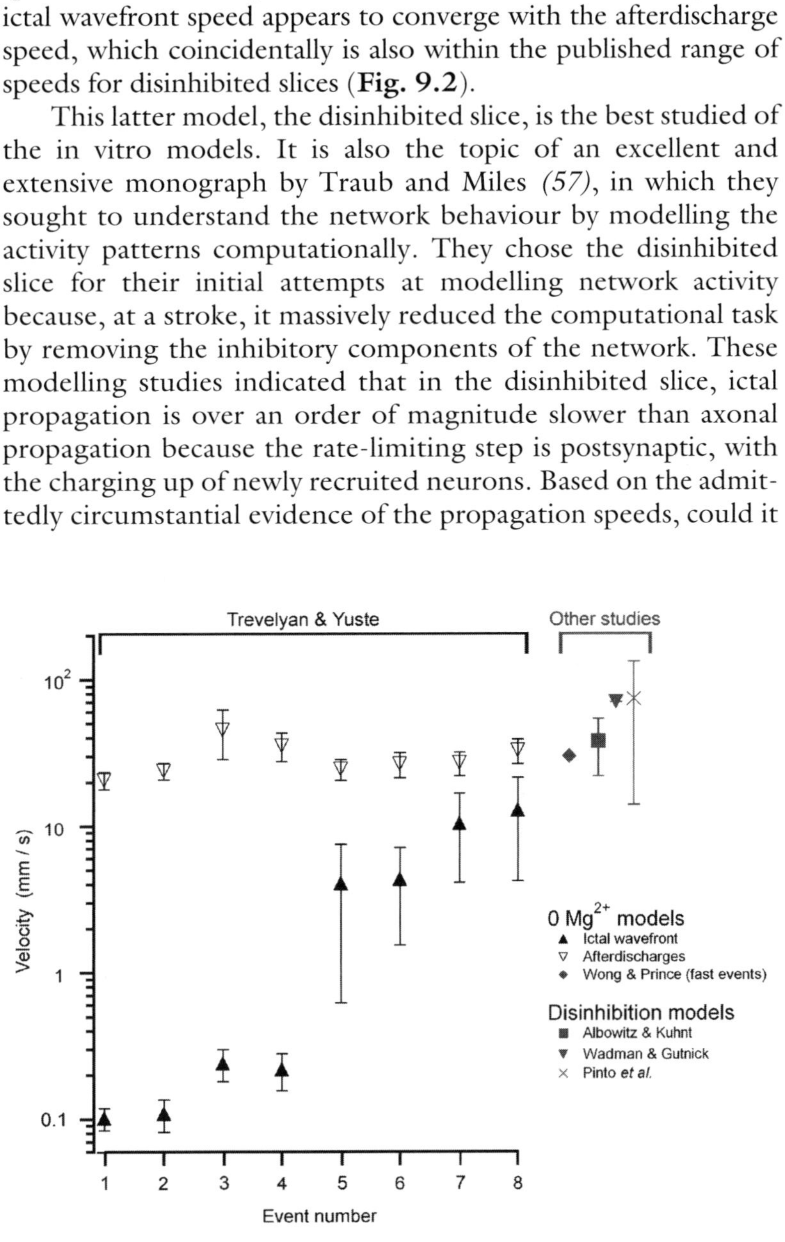

Fig. 9.2. **The velocity of the ictal wavefront and the afterdischarge is independently set**. The speed of afterdischarges does not change with successive ictal events, unlike the speed of the ictal wavefront which increases with each successive event (data are from *(9, 58)*). For comparison, we also plot published velocity measures from another study of fast $0Mg^{2+}$ ictal events *(10)* and also from several studies of disinhibition model *(59–61)*.

be that the propagation of afterdischarges in the 0 Mg^{2+} model and also of ictal wavefronts after many ictal events have occurred (late ictal events) are also best understood in terms of Traub and Miles' model, being limited primarily by how rapidly neurons are charged up to fire?

4.2. Slowly Propagating Epileptiform Activity

These considerations suggest that we may have a coherent view of fast propagation, as occurs in a variety of conditions (disinhibition, late 0 Mg^{2+} events, afterdischarges), but what about the slowly propagating events? When we were careful to record the very first ictal events experienced by a slice (we patched on to pyramidal cells prior to washing out Mg^{2+} ions), consistently we recorded propagation speeds of between two and three orders of magnitude slower than seen either in disinhibited slices or our late 0 Mg^{2+} events *(8, 9)*. Thus, a single slice could exhibit over a 1000-fold range of propagation speeds if bathed in 0 Mg^{2+} for a protracted period, a remarkably similar range to that seen in human patients. What is different about the early 0 Mg^{2+} events? While we do not yet have the whole picture, our imaging and dual patch clamp recordings have been extremely illuminating.

When one records pyramidal cells in current clamp mode, the cell displays rhythmic, δ-frequency depolarizations prior to the first full ictal paroxysmal depolarizations. An identical progression is also seen in in vivo recordings *(5)*. The key to interpreting this activity pattern came from paired recordings of adjacent pyramidal cells *(8)*. The critical observation was that when we recorded adjacent pyramids in voltage clamp mode, during ictal barrages they experienced almost identical synaptic barrages. This was a statistical effect reflecting the sheer intensity of the synaptic barrages, because during quiescent periods, the association of individual postsynaptic currents between the two cells was quite weak. During epileptiform barrages, the parallel synaptic barrages presented us with a fantastic experimental resource: to record the two cells in different recording modes to allow us to dissect out the synaptic forces and see how they set the pyramidal cell firing. The most informative recording mode was to voltage clamp cells at –30 mV, roughly half way between the reversal potentials of GABA and glutamate receptors. Our paired recordings of adjacent pyramidal cells allowed us to understand the relationship of these –30 mV V_{clamp} recordings to our other recordings. These –30 mV recordings thus became our "reference recordings", giving a coherent view of all our other electrophysiology and imaging recordings.

In this way, we showed that the rhythmic, δ-frequency depolarizations started when the ictal wavefront was still several hundred microns away from the recorded cell (**Fig. 9.3**). This pattern of activity of course leant itself beautifully to imaging because it

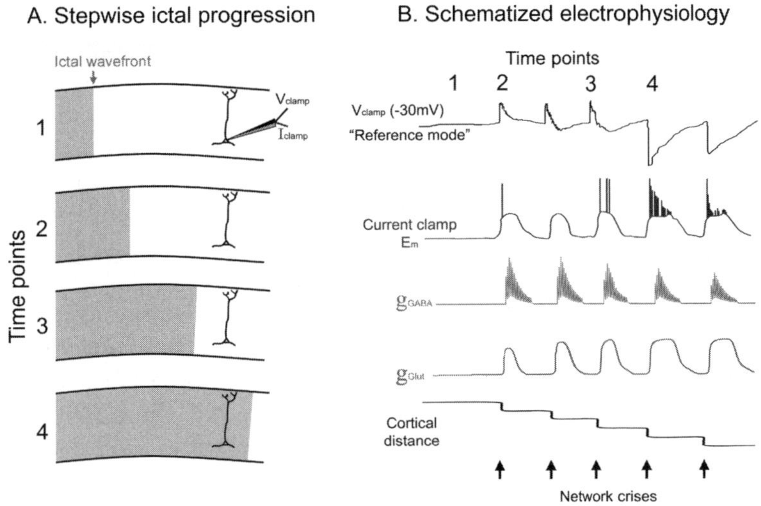

Fig. 9.3. **Schematic illustration relating the spread of the ictal event to the typical electrophysiological recording patterns. (A)** Schematic view of a cortical slice showing four time points as an ictal event progresses across the tissue towards a layer 5 pyramidal cell that is being recorded electrophysiologically. **(B)** Schematized electrophysiological recordings to show the relationship between the "reference mode" V_{clamp} recordings (at -30mV, halfway between $E_{glutamate}$ and E_{GABA}; top trace) and current clamp (I_{clamp}) recordings (2nd trace) at the times depicted in panel (A). Below are estimates of the glutamatergic and GABAergic synaptic conductances (3rd and 4th traces respectively). Note particularly that the earliest synaptic barrages, when the ictal wavefront is still several 100 μm away, comprise large amplitude inhibitory and excitatory drives. At this time, the inhibition is dominant, and so, although the cell becomes quite depolarized, it does not generally fire. The start of each barrage is a crisis time for the network, and it is at these times that new cells are recruited to the event. Thus, the ictal wavefront takes steps across the network at these crisis moments, as depicted in the bottom trace.

generally provided a few seconds prior warning before the ictal wavefront entered the field of view. It proved a very useful trigger for us to start the imaging acquisition. The rhythmicity of the depolarizations matched the rhythmicity of activity at the ictal wavefront, and thus represented a feedforward synaptic drive from the wavefront itself. This feedforward synaptic barrage included an extremely powerful excitatory drive which ordinarily would have greatly exceeded threshold. The cells failed to fire at this time however, because of an even more powerful feedforward inhibition which thus provides a restraint of pyramidal cells ahead of the wavefront. The recruitment of pyramidal cells to the ictal event occurs then, NOT when the excitatory barrage begins but rather much later, when the inhibitory restraint fails. This model thus suggests itself as the first in vitro model of the "inhibitory surround" mechanism *(1)*, which was hypothesized to preserve normal cortical function by preventing the lateral spread of seizure activity.

The power of the inhibitory restraint is indeed remarkable. When we recorded the excitatory drive by holding pyramidal cells at the GABA reversal potential, the inward (depolarizing) current generally was in excess of a nanoamp – single EPSCs are generally around 10 pA at this holding potential. In the absence of any inhibitory drive, we estimated action potential threshold to be between 5 and 10% of the epileptiform barrage, and if the full current was injected, the firing pattern resembled a full ictal depolarizing shift *(8)*.

These experiments suggested a completely new view of epileptiform spread, with each ictal barrage representing a crisis for the still-to-be-recruited cortical territories. Neurons are either recruited to the event at these times of crisis or they resist by dint of their inhibitory restraint. With each crisis, a few more neurons are generally recruited, and the activity of others ratchets up. Remarkably, when we imaged these recruitment steps, we found that local clusters of cells were recruited together, suggestive of a modular propagation pattern (**Fig. 9.4** and Color Plate 7, middle of book). Our electrophysiology analyses had indicated that the final recruitment was governed more by inhibition than by excitation, suggesting that these modules are defined by a common inhibitory restraint. It is interesting to speculate that these represent functional modules, and, given the tantalizing link between topography and epilepsy mentioned in our introduction, whether this suggests a particular physiological role for the same population for interneurons that also restrain epileptiform spread. Having discussed at length how in vitro experimentation can illuminate in vivo activity patterns, this correspondence between functional modules and epileptiform modules is one obvious instance where the opposite may be the case, which we will only truly understand with careful in vivo work.

This concept of episodic network crises appears to be a novel one with regard to epileptiform propagation. Its significance, though, is that it can explain the huge range of propagation speeds seen both in vitro and in vivo. If there is no effective inhibitory restraint, then the network succumbs to the first crisis it experiences. The propagation speed in this case is extremely rapid, being limited primarily by the time to charge up newly recruited neurons to firing, as elucidated by Traub and Miles' modelling work. If, on the other hand, the inhibitory restraint is manifest, then repeated crises can be resisted, with the ictal wavefront taking only short steps across the tissue. Our V_{clamp} recordings allowed us to make exactly this correlation. Thus, we could count how many crises a given cell experienced before being recruited, and by recording at a distant location in the slice we could also monitor the speed of the wavefront. These two measures, the number of crises and the speed of ictal propagation, proved to have a

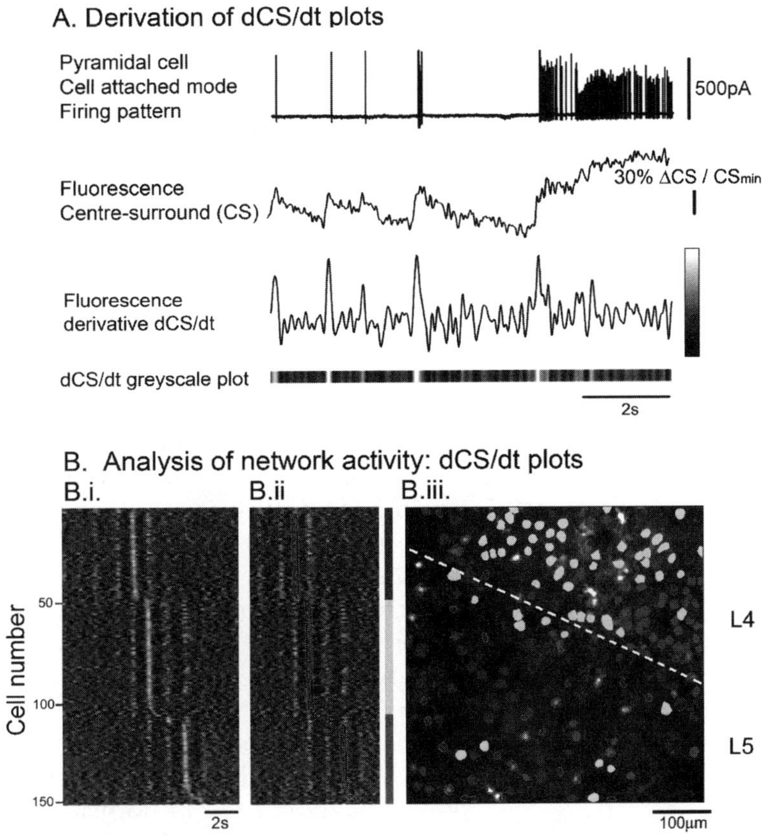

Fig. 9.4. **Somatic Ca^{2+} transients reveal episodic localized recruitment. (A)** The derivation of the cellular Ca^{2+}signal. We recorded the firing pattern of a pyramidal cell (top trace) that had been bulk loaded with the Ca^{2+}dye, Oregon Green BAPTA 1. The second trace shows the somatic Ca^{2+}fluxes – note the sharp rises concomitant with each action potential. The derivative of the Ca^{2+}signal highlights the sudden increases in fluorescent signal, which reflect the sudden recruitment of neurons. The bottom panel shows how this derivative trace is transformed into the greyscale plots shown in Fig. (**B**) which shows greyscale representations from 152 identified neurons. (**Bi**) "Greyscale plot" of the temporal derivatives (dCS/dt) of the centre surround signals from 152 neurons ordered by the time of the maximum for each trace. Each row of pixels represents a dCS/dt trace: high rate of change of fluorescence is depicted as lighter grey. Each trace has several peaks. (**Bii**) The same plot, with the single maximum rate of change for each cell marked (*red*). This ordering clearly shows three main clusters of cells. They are colour coded as shown, and their positions are indicated in the photomicrograph in (**Biii**). Note the tight and largely non-overlapping spatial distributions of the *yellow* and *red* groups in layer 4. The lamina borders were determined from the location of the filled cells following post hoc staining for biocytin and matching this up to the electrode location in the field of view (the faint triangular shadow of one of the electrodes can just be seen at the bottom right corner of the field of view). Movie taken using a 20 × objective. Figure used with permission from *(8)*. (*see* Color Plate 7, middle of book)

strong inverse correlation, confirming that the inhibitory restraint ahead of the wavefront is indeed the prime determinant of propagation speeds in this model. The comparable range of propagation speeds estimated from subdural recordings in humans suggests that this same pattern of episodic, modular spread also occurs in vivo.

5. Future Research Lines

The development of new technology frequently provides new insights into old problems, and network Ca^{2+} imaging is no different in this regard. The concept of the inhibitory restraint was first proposed 40 years ago and offers a coherent view of many epileptiform activities. In the disinhibited slice, activity of a single neuron can trigger an epileptiform event. Furthermore, there are many recognized triggers of synchronized neuronal activity, from sensory stimuli to release of glutamate from glia. Clearly such potential ictogenic activity occurs continuously within us all. Without some kind of intrinsic protection mechanism, we would all experience repeated seizures. The onus then is not to describe more types of epileptogenic triggers, but rather to understand how they are resisted in normal brains. These, though, remain speculations since the cellular basis of the restraint has eluded us. Our recent work now provides a means of examining this important hypothesis.

An important first step is to identify which cells provide the restraint. This may be achieved by examining interictal events in the 0 Mg^{2+} model since our preliminary data suggest that such events involve the same neurons as are active during the period of restraint immediately prior to recruitment of a new module of cortex to an ictal event. Theoretical considerations suggest that the fast-spiking basket cell population are likely to play a pivotal role.

We believe that the same experimental protocols, combining fast network imaging of somatic Ca^{2+} signals allied with patch clamp recordings, will provide important new insights if applied to other in vitro models. For instance, one can record transient network events lasting about 1s in slices bathed in either 0 Mg^{2+}, GABAergic antagonists or in the K^+ channel blocker, 4-aminopyridine. These transient network events in the three different models are all frequently referred to by the same name as "interictal events", simply because of their time course: interictal events in humans also last about 1 s on the EEG traces. Ca^{2+} imaging provides a new way of looking at these activity patterns to see whether they do correspond in the three models. Our expectation is that they will not and that we can anticipate new terms being required for the three models. But new terminology in this case will reflect an improved understanding of cortical pathological activity patterns, which ultimately we hope will lead to new ways of thinking about epilepsy and new therapies.

References

1. Prince DA, Wilder BJ. Control mechanisms in cortical epileptogenic foci. "Surround" inhibition. *Arch Neurol* 1967;16(2): 194–202.
2. Garaschuk O, Linn J, Eilers J, Konnerth A. Large-scale oscillatory calcium waves in the immature cortex. *Nat Neurosci* 2000;3(5): 452–9.
3. Ben-Ari Y, Cherubini E, Corradetti R, Gaiarsa JL. Giant synaptic potentials in immature rat CA3 hippocampal neurones. *J Physiol* 1989;416:303–25.
4. Steriade M. Neuronal substrates of sleep and epilepsy. Cambridge, UK: Cambridge University Press; 2003.
5. Timofeev I, Steriade M. Neocortical seizures: Initiation, development and cessation. *Neuroscience* 2004;123(2):299–336.
6. Anderson WW, Lewis DV, Swartzwelder HS, Wilson WA. Magnesium-free medium activates seizure-like events in the rat hippocampal slice. *Brain Res* 1986;398(1): 215–19.
7. Flint AC, Maisch US, Kriegstein AR. Postnatal development of low [Mg2+] oscillations in neocortex. *J Neurophysiol* 1997; 78(4):1990–6.
8. Trevelyan AJ, Sussillo D, Watson BO, Yuste R. Modular propagation of epileptiform activity: Evidence for an inhibitory veto in neocortex. *J Neurosci* 2006;26(48):12447–55.
9. Trevelyan AJ, Sussillo D, Yuste R. Feedforward inhibition contributes to the control of epileptiform propagation speed. *J Neurosci* 2007;27(13):3383–7.
10. Wong BY, Prince DA. The lateral spread of ictal discharges in neocortical brain slices. *Epilepsy Res* 1990;7(1):29–39.
11. Dreier JP, Heinemann U. Late low magnesium-induced epileptiform activity in rat entorhinal cortex slices becomes insensitive to the anticonvulsant valproic acid. *Neurosci Lett* 1990;119(1):68–70.
12. Dreier JP, Zhang CL, Heinemann U. Phenytoin, phenobarbital, and midazolam fail to stop status epilepticus-like activityinduced by low magnesium in rat entorhinal slices, but can prevent its development. *Acta Neurol Scand* 1998;98(3):154–60.
13. Pfeiffer M, Draguhn A, Meierkord H, Heinemann U. Effects of gamma-aminobutyric acid (GABA) agonists and GABA uptake inhibitors on pharmacosensitive and pharmacoresistant epileptiform activity in vitro. *Br J Pharmacol* 1996;119(3):569–77.
14. Glasscock E, Qian J, Yoo JW, Noebels JL. Masking epilepsy by combining two epilepsy genes. *Nat Neurosci* 2007;10(12):1554–8.
15. Noebels JL. The biology of epilepsy genes. *Annu Rev Neurosci* 2003;26:599–625.
16. Noebels JL, Sidman RL. Inherited epilepsy: Spike-wave and focal motor seizures in the mutant mouse tottering. *Science* 1979;204 (4399):1334–6.
17. Ludwig A, Budde T, Stieber J, et al. Absence epilepsy and sinus dysrhythmia in mice lacking the pacemaker channel HCN2. *Embo J* 2003;22(2):216–24.
18. van Luijtelaar EL, Coenen AM. Two types of electrocortical paroxysms in an inbred strain of rats. *Neurosci Lett* 1986;70(3): 393–7.
19. Bertaso F, De Bock F, Bockaert J, Fagni L, Lerner-Natoli M. Disruption of mGluR7a C-terminal protein interactions triggers absence-like epileptic seizures. In Society for Neuroscience Abstracts; 2007, p. 333.11.
20. MacLean JN, Fenstermaker V, Watson BO, Yuste R. A visual thalamocortical slice. *Nat Methods* 2006;3(2):129–34.
21. MacLean JN, Watson BO, Aaron GB, Yuste R. Internal dynamics determine the cortical response to thalamic stimulation. *Neuron* 2005;48(5):811–23.
22. Douglas RJ, Markram H, Martin KAC. Neocortex. In: Shepherd GM, ed. *The Synaptic Organisation of the Brain*. 5th edn. Oxford: Oxford University Press; 2004:499–558.
23. Douglas RJ, Martin KA. Neuronal circuits of the neocortex. *Annu Rev Neurosci* 2004;27:419–51.

24. Jefferys JG. Nonsynaptic modulation of neuronal activity in the brain: Electric currents and extracellular ions. *Physiol Rev* 1995;75(4):689–723.

25. Angulo MC, Kozlov AS, Charpak S, Audinat E. Glutamate released from glial cells synchronizes neuronal activity in the hippocampus. *J Neurosci* 2004;24(31):6920–7.

26. Parri HR, Gould TM, Crunelli V. Spontaneous astrocytic Ca2+ oscillations in situ drive NMDAR-mediated neuronal excitation. *Nat Neurosci* 2001;4(8):803–12.

27. Parpura V, Basarsky TA, Liu F, Jeftinija K, Jeftinija S, Haydon PG. Glutamate-mediated astrocyte-neuron signalling. *Nature* 1994;369(6483):744–7.

28. Yuste R, Peinado A, Katz LC. Neuronal domains in developing neocortex. *Science* 1992;257(5070):665–9.

29. Draguhn A, Traub RD, Schmitz D, Jefferys JG. Electrical coupling underlies high-frequency oscillations in the hippocampus in vitro. *Nature* 1998;394(6689):189–92.

30. Gibson JR, Beierlein M, Connors BW. Two networks of electrically coupled inhibitory neurons in neocortex. *Nature* 1999;402 (6757):75–9.

31. Perez Velazquez JL, Carlen PL. Gap junctions, synchrony and seizures. *Trends Neurosci* 2000;23(2):68–74.

32. Hamzei-Sichani F, Kamasawa N, Janssen WG, et al. Gap junctions on hippocampal mossy fiber axons demonstrated by thin-section electron microscopy and freeze fracture replica immunogold labeling. *Proc Natl Acad Sci U S A* 2007;104(30):12548–53.

33. Galarreta M, Hestrin S. A network of fast-spiking cells in the neocortex connected by electrical synapses. *Nature* 1999;402 (6757):72–5.

34. Kirov SA, Petrak LJ, Fiala JC, Harris KM. Dendritic spines disappear with chilling but proliferate excessively upon rewarming of mature hippocampus. *Neuroscience* 2004; 127(1):69–80.

35. Cossart R, Aronov D, Yuste R. Attractor dynamics of network UP states in the neocortex. *Nature* 2003;423(6937):283–8.

36. Sanchez-Vives MV, McCormick DA. Cellular and network mechanisms of rhythmic recurrent activity in neocortex. *Nat Neurosci* 2000;3(10):1027–34.

37. Whittington MA, Traub RD, Jefferys JG. Synchronized oscillations in interneuron networks driven by metabotropic glutamate receptor activation. *Nature* 1995;373(6515): 612–15.

38. Volgushev M, Vidyasagar TR, Chistiakova M, Yousef T, Eysel UT. Membrane properties and spike generation in rat visual cortical cells during reversible cooling. *J Physiol* 2000;522(Pt 1):59–76.

39. Trevelyan AJ, Jack J. Detailed passive cable models of layer 2/3 pyramidal cells in rat visual cortex at different temperatures. *J Physiol* 2002;539(Pt 2):623–36.

40. Hardingham NR, Larkman AU. The reliability of excitatory synaptic transmission in slices of rat visual cortex in vitro is temperature dependent. *J Physiol* 1998;507(Pt 1): 249–56.

41. Hill MW, Wong M, Amarakone A, Rothman SM. Rapid cooling aborts seizure-like activity in rodent hippocampal-entorhinal slices. *Epilepsia* 2000;41(10):1241–8.

42. Yuste R, Katz LC. Transmitter-induced changes in intracellular free calcium in brain slice of developing neocortex. In Abstracts *Society for Neuroscience Abstracts* 1989;4,5(15):2.

43. Yuste R, Katz LC. Control of postsynaptic Ca2+ influx in developing neocortex by excitatory and inhibitory neurotransmitters. *Neuron* 1991;6(3):333–44.

44. Smetters D, Majewska A, Yuste R. Detecting action potentials in neuronal populations with calcium imaging. *Methods* 1999;18(2):215–21.

45. The Handbook — A Guide to Fluorescent Probes and Labeling Technologies. 10th edn: Invitrogen.

46. Badea T, Goldberg J, Mao B, Yuste R. Calcium imaging of epileptiform events with single-cell resolution. *J Neurobiol* 2001; 48(3):215–27.

47. Kerr JN, Greenberg D, Helmchen F. Imaging input and output of neocortical networks in vivo. *Proc Natl Acad Sci USA* 2005;102(39):14063–8.

48. Sullivan MR, Nimmerjahn A, Sarkisov DV, Helmchen F, Wang SS. In vivo calcium imaging of circuit activity in cerebellar cortex. *J Neurophysiol* 2005;94(2):1636–44.

49. Murayama M, Miyazaki K, Kudo Y, Miyakawa H, Inoue M. Optical monitoring of progressive synchronization in dentate granule cells during population burst activities. *Eur J Neurosci* 2005;21(12):3349–60.

50. Yaksi E, Friedrich RW. Reconstruction of firing rate changes across neuronal populations by temporally deconvolved Ca2+ imaging. *Nat Methods* 2006;3(5):377–83.

51. Ohki K, Chung S, Ch'ng YH, Kara P, Reid RC. Functional imaging with

cellularresolution reveals precise micro-architecture in visual cortex. *Nature* 2005;433 (7026):597–603.

52. Tian GF, Azmi H, Takano T, et al. An astrocytic basis of epilepsy. *Nat Med* 2005;11(9): 973–81.
53. Stosiek C, Garaschuk O, Holthoff K, Konnerth A. In vivo two-photon calcium imaging of neuronal networks. *Proc Natl Acad Sci USA* 2003;100(12):7319–24.
54. Heim N, Garaschuk O, Friedrich MW, et al. Improved calcium imaging in transgenic mice expressing a troponin C-based biosensor. *Nat Methods* 2007;4(2):127–9.
55. Crepel V, Aronov D, Jorquera I, Represa A, Ben-Ari Y, Cossart R. A parturition-associated nonsynaptic coherent activity pattern in the developing hippocampus. *Neuron* 2007;54(1):105–20.
56. Whittington MA, Traub RD, Jefferys JG. Erosion of inhibition contributes to the progression of low magnesium bursts in rat hippocampal slices. *J Physiol* 1995;486 (Pt 3):723–34.
57. Traub RD, Miles R. Neuronal networks of the hippocampus. Cambridge: Cambridge University Press; 1991.
58. Trevelyan AJ, Baldeweg T, van Drongelen W, Yuste R, Whittington M. The source of afterdischarge activity in neocortical tonic-clonic epilepsy. *J Neurosci* 2007;27(49): 13513–19.
59. Albowitz B, Kuhnt U. Epileptiform activity in the guinea-pig neocortical slice spreads preferentially along supragranular layers–recordings with voltage-sensitive dyes. *Eur J Neurosci* 1995;7 (6):1273–84.
60. Wadman WJ, Gutnick MJ. Non-uniform propagation of epileptiform discharge in brain slices of rat neocortex. *Neuroscience* 1993;52(2):255–62.
61. Pinto DJ, Patrick SL, Huang WC, Connors BW. Initiation, propagation, and termination of epileptiform activity in rodent neocortex in vitro involve distinct mechanisms. *J Neurosci* 2005;25(36): 8131–40.

Chapter 10

Complexity Untangled: Large-Scale Realistic Computational Models in Epilepsy

Robert J. Morgan and Ivan Soltesz

Abstract

In the epilepsy field, the creation of large-scale data-driven models that incorporate decades worth of experimental data has led to substantial innovations over current methodologies. Such models and sophisticated visualization software that makes the models truly come to life have brought computational neuroscience closer to reality for all epilepsy researchers. In this chapter, we discuss detailed, data-driven models that have resulted in significant, testable, theoretical advances that have contributed to our knowledge of how large-scale biological neuronal networks interact to promote hyperexcitability and hypersynchrony in epilepsy syndromes. Additionally we elaborate on how computational advances in software infrastructure have greatly increased the accessibility and applicability of computational modeling, especially for biologists who have previously employed only experimental techniques in their research.

Key words: synchrony, dentate gyrus, granule cell, network.

1. Introduction

In the field of epilepsy, attempts at rational development of novel therapeutic regimens have been spectacularly unsuccessful largely because of a lack of understanding of how large, complex networks of cells interact to create hyperexcitability and hypersynchrony. The best experimental efforts to understand these interactions have been impressive both technically and in the amount of new information which they have yielded, but they have also been limited. It is possible to record from individual neurons (and sometimes even tens of neurons) in a particular brain region and even in multiple regions at once, or to record the simultaneous bulk activity of many hundreds of thousands of cells via EEG.

Scott C. Baraban (ed.), *Animal Models of Epilepsy: Methods and Innovations, vol. 40*

DOI 10.1007/978-1-60327-263-6_10 Springerprotocols.com

However, technical limitations continue to prevent the recording of the precise activity patterns of tens of thousands of single cells simultaneously either in vitro or more importantly in vivo. While the innovative animal models of epilepsy syndromes discussed in this text will undoubtedly bring the field closer to this goal, the understanding of large-scale network interactions will likely remain incomplete without the use of highly data-driven, reliable computational models of complex brain anatomy and physiology.

Computational neuroscientists are constantly developing new and innovative ways of visualizing and analyzing the myriad complex processes occurring in single cells and cellular networks in order to understand the structure and function of the mammalian nervous system. In particular, computational models of pathological phenomena have been instrumental in furthering our understanding of the molecular and cellular mechanisms of such diseases as schizophrenia, Parkinson's disease, stroke, and epilepsy, among many others *(1–4)*. The diversity of computational modeling approaches to the questions arising from a given neurological disorder is at least as great as the diversity of the pathological alterations involved in the disorder itself. Computational models of epilepsy syndromes suffer from the desire to balance physiological accuracy with computational feasibility. This leads to a wide selection of models which includes small, integrate-and-fire networks of neurons *(5, 6)* up to large-scale data-driven networks of physiologically realistic multi-compartmental single-cell models *(7–9)*.

While a complete survey of all modeling techniques in the epilepsy field would require at least an entire book, and has in fact resulted in such a book *(3)*, it is important to focus on some of the most recent and innovative techniques in the field of computational modeling of epilepsy. In particular, the ability to create large-scale data-driven models that incorporate decades worth of experimental data has led to significant innovations over current methodologies. Additionally, the incorporation of such models in sophisticated visualization software that is able to interface with multiple simulation platforms *(10)* has brought computational neuroscience closer to reality for experimental labs that have not previously focused on it.

In this chapter, we will discuss detailed, data-driven models that have resulted in significant, testable, theoretical advances that have contributed to our knowledge of how large-scale biological neuronal networks interact to promote hyperexcitability and hypersynchrony in epilepsy syndromes. Additionally we will elaborate on how computational advances in software infrastructure have begun to allow computational modeling to become accessible to biologists who have previously employed only experimental techniques in their research.

2. Computational Modeling Approaches in Epilepsy

As stated above, computational modeling of epilepsy has struggled with trying to balance physiological realism with computational reality. That is, models must be sufficiently detailed to accurately reproduce both physiological and anatomical properties yet must at the same time be sufficiently simple to be simulated on a computer with limited capacity. Some solutions to this problem have been to utilize either small networks of detailed cells *(11, 12)* or large networks of relatively simple cells with few compartments or even leaky integrate-and-fire neurons *(5, 6)*. Other solutions have focused on macroscopic modeling which utilizes the average activity of massive subpopulations of neurons without explicitly modeling the physiological mechanisms in single cells *(13–15)*. These models have been particularly useful for interpreting physiological data based on similar types of recordings such as EEG.

While all of the aforementioned modeling strategies have greatly contributed to our understanding of epilepsy, and each has aided the progression of computational neuroscience toward where it stands today, they will not be discussed in detail in this chapter. To truly appreciate their contributions, the interested reader should refer to two excellent books *(3, 16)* for an extensive review of such models. Here, we will focus on models which incorporate highly detailed single cells into complex, large-scale physiologically realistic networks, complete with anatomically accurate topographies and massive numbers of synaptic connections. These networks have only recently become feasible due to the rapidly increasing computational power available to modelers which has allowed large-scale simulations to be carried out without the need for large supercomputers. We will focus on a particular large-scale network which exemplifies recent innovations in computational neuroscience, the dentate gyrus network first implemented by Santhakumar et al. *(9)*.

3. Development of a Large-Scale Dentate Gyrus Model

Work on development of this model first began many decades ago (unbeknownst to anyone) as the anatomical and physiological characteristics of the dentate gyrus were meticulously and laboriously characterized. Over the years, numerous cell types were enumerated, their physiological properties examined, and their synaptic connections explored *(17–21)*. The tremendous amount of information about the dentate that was gathered was then collected, organized, and incorporated into a database which characterized the anatomical structures of dentate gyrus cells, their passive parameters and channel conductances, their physiological properties, and their synaptic parameters *(9, 20)*. Detailed channel mechanisms were developed and incorporated into complex single-cell models of four major dentate cell types. These cells were then tuned to reproduce physiologically accurate input–output characteristics (**Fig. 10.1**). The single cells were cloned to produce a 1:2000 scale model of the dentate with 500 granule cells (excitatory output cells of the dentate; **Fig. 10.1A**), 15 mossy cells (excitatory cells in the hilus with a high degree of connectivity; **Fig. 10.1B**), 6 basket cells (perisomatically projecting inhibitory cells; **Fig. 10.1C**), and 6 hilar interneurons with axonal projections to the perforant path (dendritically projecting inhibitory cells; **Fig. 10.1D**). Finally, synaptic connections were made between cells on a cell type-specific basis, resulting in a fully functional (albeit somewhat small and anatomically unrealistic – due to the small number of cells, overall connectivity was exaggerated and anatomically realistic connection patterns and probabilities were impossible to implement) model of the dentate gyrus (**Fig. 10.2**). The process just described is simply an outline of the construction of a very detailed model and does not do justice to the amount of effort put forth either by the experimental neuroscientists who studied the anatomical and physiological properties of the dentate or the modelers who constructed the database and the final model. However, a full description of the construction of such a model requires an entire chapter or manuscript *(9, 22)*.

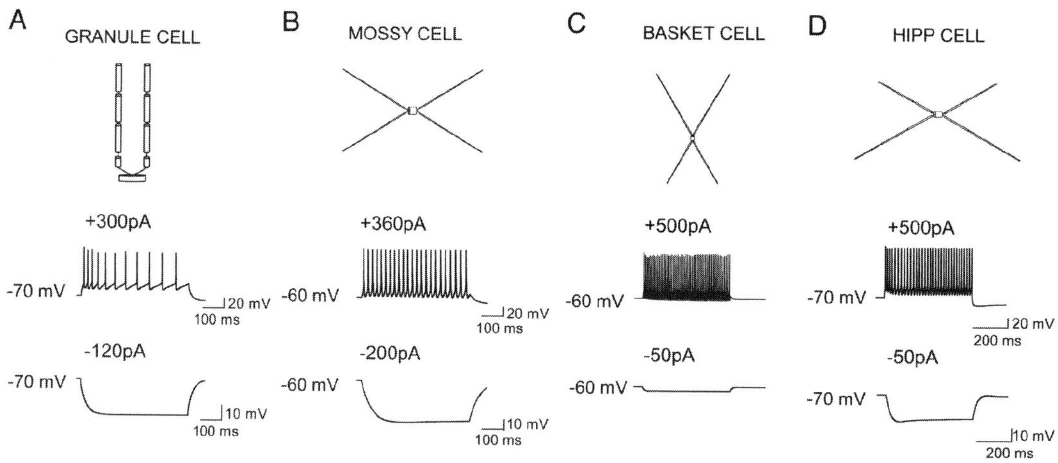

Fig. 10.1. **Structure and intrinsic excitability of model cells.** (**A**) Schematic representation of the structure (*top*) of the granule cell model and membrane voltage traces of the granule cell in response to 300 pA (*middle*) and 120 pA (*bottom*) current injections. (**B**) Illustration of the structure of the model mossy cell (*top*) and the membrane voltage responses to 360 pA (*middle*) and 200 pA (*bottom*) current injections. (**C**) The structure of the model basket cell (*top*) and responses to 500 pA (*middle*) and 50 pA (*bottom*) current injections. (**D**) Structure (*top*) and responses to 500 pA (*middle*) and 50 pA (*bottom*) of the model hilar perforant path-associated cells (HIPP cells). Used with permission from Santhakumar et al., *J Neurophysiol* 2005 Jan;93(1):437–53.

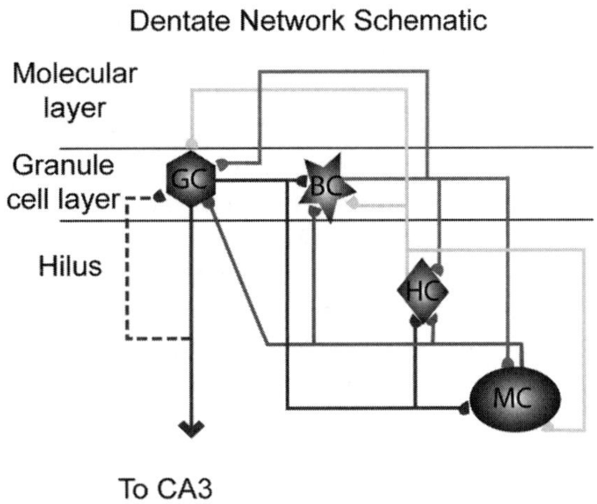

Fig. 10.2. **Dentate network schematic.** Schematic of the dentate gyrus model showing the four cell types implemented, the layers in which they reside, and their cell type-specific connectivity. Cell types from left to right: excitatory granule cells (GC, hexagon), inhibitory somatically projecting basket cells (BC, star), inhibitory dendritically projecting hilar interneurons with axonal projections to the perforant path (HIPP cells, HC, diamond), and excitatory mossy cells (MC, oval). Dashed line: recurrent GC connections from sprouted mossy fibers following dentate injury. Used with permission from Morgan and Soltesz, *Proc Natl Acad Sci U S A* 2008;105:6179–84.

4. Utility of a Large-Scale Dentate Gyrus Model for Epilepsy Research

Having discussed the construction of a model dentate network, one is left to wonder what advantages a realistic yet admittedly imperfect model can provide over the actual biological specimen. The answer to this question lies in the many alterations that the dentate gyrus undergoes following insults that result in epileptogenesis in both humans and animals. Following repetitive seizures, head trauma, and ischemia, the dentate gyrus undergoes dramatic structural reorganization *(23–25)*, thereby providing an excellent experimental model to study the effects of structural alterations during epileptogenesis. However, the two primary structural changes that occur, loss of hilar cells and mossy fiber (granule cell axon) sprouting, create a fundamentally transformed dentate gyrus microcircuit that is superimposed on a background of intrinsic cellular and synaptic alterations. The biological dentate gyrus in an animal model is an excellent tool for studying the effects of all anatomical and physiological alterations together, but it is only by utilizing computational modeling that a particular network alteration can be studied in isolation.

The 1:2000 scale model of the dentate gyrus provided the perfect tool for studying the roles of mossy fiber sprouting and hilar cell loss on dentate excitability both together and separately in order to more fully understand the biological basis for hyper-excitability after traumatic brain injury. The healthy dentate gyrus contains no granule cell-to-granule cell connections. It is therefore relatively simple to simulate experimentally observed post-injury changes in the dentate gyrus by first determining the maximum number of new granule cell-to-granule cell connections that occur under the most extreme circumstances (approximately 275 new connections per granule cell following pilocarpine-induced status epilepticus in rats *(26)*) and then implementing a percentage of that maximum for more moderate injuries. Likewise, hilar interneuron and mossy cell loss can be implemented based on percentages of cells lost at different levels of injury. By using these methods, Santhakumar et al. *(9)* were able to show that mossy fiber sprouting substantially increases dentate excitability (**Fig. 10.3**). Additionally, loss of hilar mossy cells was shown to decrease excitability (**Fig. 10.4**), providing supporting evidence for the in vitro findings of Ratzliff et al. *(27)* and contrasting with an alternative theory *(28)* that hypothesized that decreased excitation from mossy cells onto inhibitory basket cells should result in a net increase in granule cell activity (the dormant basket cell hypothesis).

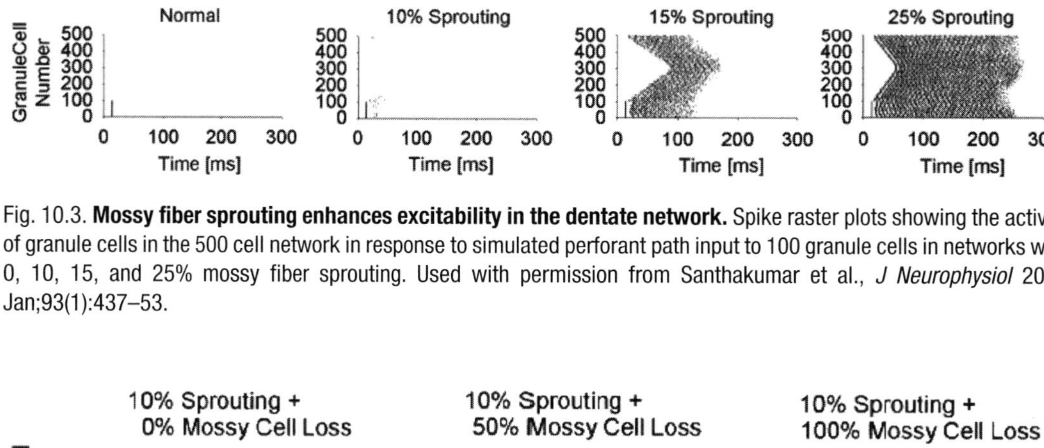

Fig. 10.3. **Mossy fiber sprouting enhances excitability in the dentate network.** Spike raster plots showing the activity of granule cells in the 500 cell network in response to simulated perforant path input to 100 granule cells in networks with 0, 10, 15, and 25% mossy fiber sprouting. Used with permission from Santhakumar et al., *J Neurophysiol* 2005 Jan;93(1):437–53.

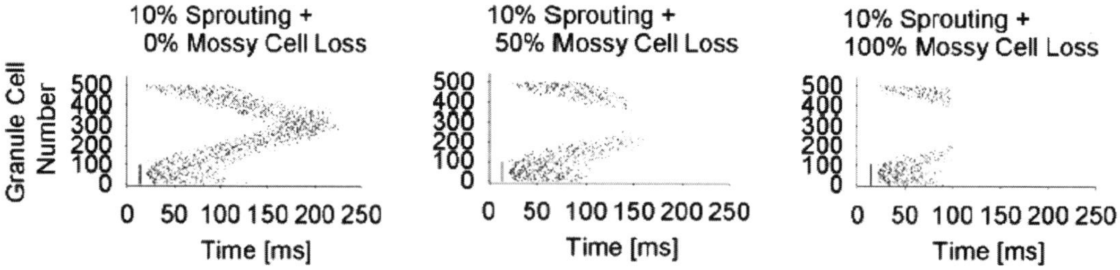

Fig. 10.4. **Effect of mossy cell loss on dentate excitability.** Spike raster plots showing the activity of granule cells in the 500 cell network in response to simulated perforant path input to 100 granule cells in networks with 10% mossy fiber sprouting and 0, 50, or 100% mossy cell loss. Note the decreased excitability with progressive mossy cell loss at a constant percentage of mossy fiber sprouting. Used with permission from Santhakumar et al., *J Neurophysiol* 2005 Jan;93(1):437–53.

5. Expansion of the Dentate Gyrus Model and Use of Large-Scale Structural Models to Explore Topological Determinants of Epileptogenesis

A model network with approximately 500 cells is quite large, especially when the cells are detailed multi-compartmental neurons. A typical simulation on a desktop computer takes several minutes to simulate approximately 1 s of network activity. Five hundred cells is really nothing, however, when faced with the complexity of the dentate gyrus that contains over one million cells in the rat. In order to be able to accurately model synaptic connectivity numbers, strengths, and patterns, as well as the topography of cellular distributions, the network must be substantially larger. Indeed, to provide the most realistic estimation of the true topography of the dentate gyrus in model form, a 1:1 scale model is ideal. Clearly, this cannot be implemented as a functional model with detailed single cells except on the most powerful supercomputers in the world, but it is certainly a possibility for a model in which the desire is to study only the structural topography of the dentate. Dyhrfjeld-Johnsen et al. *(7)* undertook this daunting task of creating a 1:1 scale structural model of the dentate gyrus in order to understand the basic structure of the healthy dentate gyrus and how that structure changes during epileptogenesis, potentially resulting in hyperexcitability. This 1:1 scale model was constructed such that each cell was simply a node in the computer (nodes have no functional properties and are therefore much less computationally demanding than the multi-compartmental cell models described previously) and synapses between cells were represented as non-functioning directed links *(7)*. Connections between cells were made probabilistically based on cell type-specific connection probabilities derived from the literature as well as from Gaussian axonal distributions for the various cell types derived from in vivo cell fills (**Fig. 10.5**).

Network topology was determined using tools available from graph theory which were previously employed to assess the structure of the nervous system of the worm *Caenorhabditis elegans (29)*. The two primary measures used to characterize the dentate gyrus were the average path length, *L*, and the average clustering coefficient, *C*. The average path length is defined as the average number of steps required to move from any node to any other node in the network, and it is therefore a measure of how well connected the network is globally. The average clustering coefficient, on the other hand, is a measure of local connectivity. It is defined as the fraction of all possible connections between "post-synaptic" nodes of a given node that are actually formed. The average clustering coefficient for an entire network is simply the average of the clustering coefficients for each node in the network.

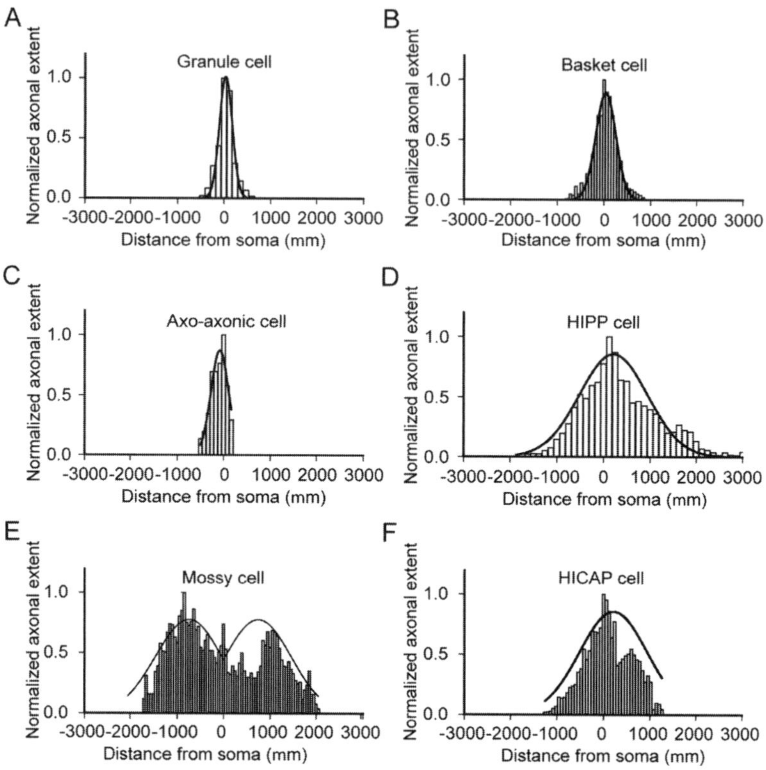

Fig. 10.5. **Gaussian fits to experimentally determined distributions of axonal branch length used in construction of the structural and functional models of the dentate gyrus.** (**A**) Plot shows the averaged axonal distribution of 13 granule cells *(24)* and the corresponding Gaussian fit. (**B**) Fit to the septotemporal distribution of axonal lengths of a filled and reconstructed basket cell *(57)*. (**C**) Fit to the axonal distribution of a CA1 axo-axonic cell *(58)*. (**D**) Gaussian fit to the averaged axonal distributions of 3 HIPP cells from gerbil *(59)*. (**E**) Fit to averaged axonal distributions of three mossy cells illustrates the characteristic bimodal pattern of distribution *(60)*. (**F**) Histogram of the axonal lengths of a HICAP cell along the long axis of the dentate gyrus *(57)* and the Gaussian fit to the distribution. All distributions were based on axonal reconstruction of cells filled in vivo. In all plots, the septal end of the dentate gyrus is on the *left* (indicated by negative coordinates) and the soma is located at zero. Used with permission from Dyhrfjeld-Johnsen et al., *J Neurophysiol* 2007;97:1566–87.

Based on these two topological measures, networks can be divided into three classes: (1) *Regular*, with a long path length and large clustering coefficient; (2) *Random*, with a short path length and small clustering coefficient; and (3) *Small World*, with a short path length and large clustering coefficient *(30)*. **Figure 10.6** shows examples of these types of networks.

As shown, the regular network (**Fig. 10.6A**) is characterized by a high local connectivity, but the distance (via connected nodes) between any two random nodes, especially nodes at the ends of the graph, can be quite large. Contrarily, the random network (**Fig. 10.6B**) has a very short path length due to many long-distance

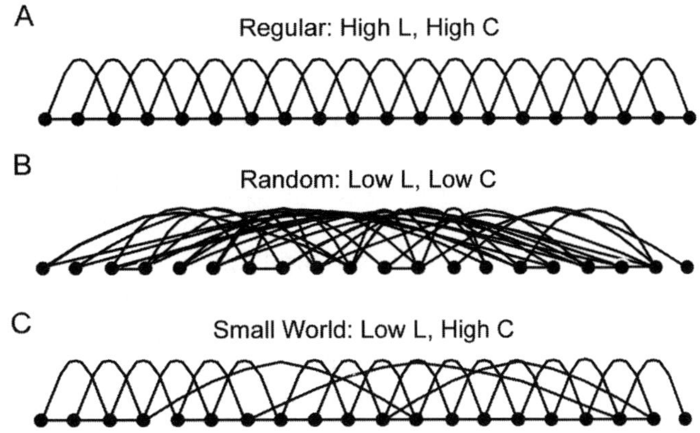

Fig. 10.6. **Schematics of three structural network topologies from graph theory.** (**A**) Regular network topology. The nodes in a regular network are connected to their nearest neighbors, resulting in a high degree of local interconnectedness (high clustering coefficient *C*), but also requiring a large number of steps to reach other nodes in the network from a given starting point (high average path length *L*). (**B**) Random network topology. In a random network, there is no spatial restriction on the connectivity of the individual nodes, resulting in a network with a low average path length *L* but also a low clustering coefficient *C*. (**C**) Small world network topology. Reconnection of even a few of the local connections in a regular network to distal nodes in a random manner results in the emergence of a small world network with a high clustering coefficient *C* but a low average path length *L*. Used with permission from Dyhrfjeld-Johnsen et al., *J Neurophysiol* 2007;97:1566–87.

connections. Local connectivity is minimal as nodes are just as likely to connect to distant nodes as they are to their neighbors. The small world network (**Fig. 10.6C**) contains attributes of both the random and regular graphs. It has high local connectivity combined with ample long-range connections, resulting in a locally and globally well-connected network.

How do the path length and clustering coefficient play a role in understanding the dentate gyrus? Analysis of the structural model of the dentate gyrus in the healthy state, without any mossy fiber sprouting or hilar cell loss, indicated that the dentate gyrus is a small world network. It has a very high clustering coefficient and a path length that is only 2.68, only slightly longer than the path length of the *C. elegans* nervous system, despite having 5,000 times more cells. This finding means that any cell is connected to any of the other one million cells in the dentate network by less than three synapses, and cells are highly connected locally, resulting in fast local computations and the ability to efficiently relay signals to distant parts of the network.

While the analysis of the healthy dentate gyrus provided useful and novel information about its structure, perhaps the most interesting and counterintuitive structural information comes from the analysis of networks undergoing sclerosis (defined here as the concurrent mossy cell + hilar neuron loss and the sprouting of mossy fibers). These networks demonstrate that the small world characteristics of the dentate gyrus are actually *enhanced* up to approximately 80% sclerosis, despite a massive loss of connections (due to the loss of a massive divergence of mossy cell connections onto granule cells as sclerosis progresses and mossy cells die). In fact, the total number of links in the network decreases by 74% while the total number of nodes lost due to hilar cell death is only 4.5% at maximal sclerosis, yet the network becomes *more* locally and globally well connected as the sprouted mossy fibers form local circuits that compensate for the loss of the mossy cell connectivity. This finding predicts that during epileptogenesis, the dentate gyrus may become more readily able to transmit information throughout the network, increasing synchronous firing and leading to a seizure phenotype. It is only at nearly 100% sclerosis that the dentate gyrus network transforms into a more regular network structure, predicting a counterintuitive decrease in hyperexcitability at maximal sclerosis.

5.1. Functional Implications of Structural Alterations

In order to test the predictions made by the structural 1:1 scale model of the dentate, it was necessary to create a functional model network of sufficient size that scaling would not alter the topographical properties of the network. Therefore, the detailed 500 cell dentate gyrus model described previously was expanded to 100 × its size, resulting in a network with over 50,000 cells. Due to the large size of the network, connections between cells were made according to appropriately scaled connection probabilities derived from the 1:1 scale model above. Simulations were then performed at progressive levels of sclerosis (e.g., first in the healthy network at 0%, then at 20% sclerosis, then 405, and so on), and activity in the network was quantified by four different measures: (1) duration of granule cell firing (the time of the last granule cell action potential in the network minus the time of the first granule cell action potential); (2) average number of action potentials per granule cell; (3) time until activity of the most distant granule cells from the stimulation point (i.e., latency to full network activity); and (4) synchrony of granule cell action potentials.

The simulations revealed that network activity very closely parallels structural alterations in the dentate gyrus. The healthy dentate model showed minimal firing in response to a single perforant path stimulation (**Fig. 10.7A**), as in the biological dentate *(31)*. Additionally, as sclerosis progressed, the functional network became increasingly hyperexcitable through approximately 80% sclerosis (**Fig. 10.7B–E**), again in agreement with in vitro measures of epileptiform activity *(32)* and concordant with the enhanced small world features revealed by the structural analysis. Then, at levels of sclerosis exceeding 80%, hyperexcitability of

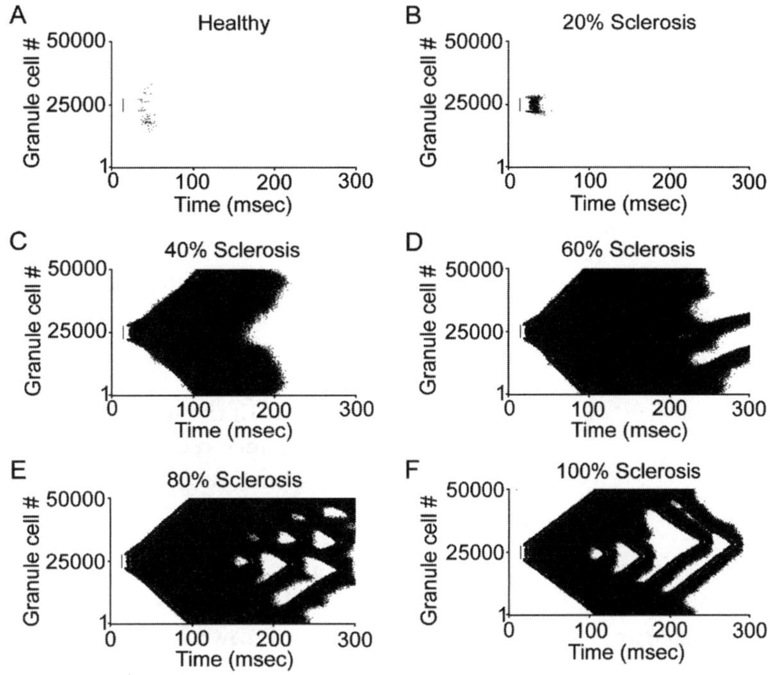

Fig. 10.7. **Effects of the sclerosis-related topological changes on granule cell activity in functional model networks. (*A–F*)** Raster plots of the first 300 ms of granule cell action potential firing in the functional model network (granule cells #1–50,000, plotted on the *y*-axis) at increasing degrees of sclerosis. Network activity was initiated by a single stimulation of the perforant path input to granule cells #22,500–27,499 and to 10 mossy cells and 50 basket cells (distributed in the same area as the stimulated granule cells) at $t = 5$ ms *(9)*. Note that the most pronounced hyperactivity was observed at sub-maximal (80%) sclerosis with a decrease in overall activity at 100% sclerosis. Used with permission from Dyhrfjeld-Johnsen et al., *J Neurophysiol* 2007;97:1566–87.

the functional model network actually decreased (**Fig. 10.7F**), just as predicted by the structural network's transformation from a small world to a more regular network structure. Thus, the functional network followed the same pattern that was seen in the structural network, and the functional effects were solely due to the structural network alterations, as no intrinsic cellular or synaptic properties were altered in the model as sclerosis progressed. Interestingly, network dynamics were also affected by structural changes; a relatively uniform pattern of granule cell activity (from 40 to 60% sclerosis; **Fig. 10.7C,D**) transformed into a pattern with distinct waves of activity (from 80 to 100% sclerosis; **Fig. 10.7E,F**) that could collide and mutually annihilate *(33, 34)*.

Perhaps the most interesting correlate of the findings from these large-scale network simulations is that they are in agreement with experimental observations in both epileptic animals and humans. No studies that have quantified hilar cell loss in animal models of epilepsy ever reported 100% cell loss *(35–43)*. Rather, in surgically removed specimens from pharmacologically intractable human temporal lobe epilepsy patients *(44)*, cell counts showed that only approximately 80% of hilar cells were lost, even in patients with severe sclerosis *(45)*. This finding coincides perfectly with the period of maximal hyperexcitability in the functional model networks, when sclerosis reaches 80%. Intuitively, this correlation between model and biology makes complete sense. Indeed, at levels of sclerosis prior to 80%, there is a high degree of new local circuitry being formed via mossy fiber sprouting. At the same time, the activity in these local circuits is able to spread long distances very quickly via the surviving long-range connections of the mossy cells. Once sclerosis progresses beyond 80%, however, so many of the long-range connections are lost that local foci of activity cannot recruit activity in distant regions and hyperexcitability diminishes.

5.2. Ensuring the Robustness of the Models

Models are not perfect representations of the biological system, and it is therefore extremely important that aspects of the model that are known to differ from reality do not affect the main conclusions. In the dentate network, a number of specific components of the cellular circuitry could not be modeled due to lack of precise data. Therefore, an extensive series of control simulations were performed to test the effects of a wide variety of conditions that were not well constrained by the available experimental data. Structural model controls included

1) variations in cell numbers
2) variations in connectivity estimates
3) inhomogeneous distribution of neuron densities along the septotemporal axis
4) inhomogeneity in connectivity along the transverse axis
5) altered axonal distributions at the septal and temporal poles (the anatomical boundaries of the dentate gyrus)
6) offset degress of sprouting and hilar neuron loss
7) implementation of a bilateral model of the dentate gyrus including commissural projections

All of the controls employed for the structural model were also used as controls for the 50,000 cell functional model. In addition, the functional model had its own set of controls, including

1) double inhibitory synaptic strengths
2) axonal conduction delays
3) spontaneous instead of stimulation-evoked activity

All of the results from these control simulations supported the basic findings of the original models, thus greatly strengthening the primary conclusions.

6. Neuronal Hubs Promote Hyperexcitability in the Dentate Gyrus Model

The modeling results discussed previously showed that after dentate injury, the newly formed connections between GCs are instrumental in creating the hyperexcitable dentate network. However, GC-to-GC connectivity is quite sparse; indeed, the probability of two granule cells connecting, even at maximal levels of sprouting, is only approximately 0.55% (each granule cell contacts 275 other granule cells on average *(26)*, out of a pool of 100,000 postsynaptic target cells within the span of a typical granule cell axonal arbor), and the probability of random formation of a three-neuron "recurrent" circuit where cell A connects to B which connects to C and back to A is miniscule at 4.16×10^{-6}%. Therefore, the specific patterns of GC-to-GC connections that define the microcircuit structure are likely to play a critical role in affecting network excitability.

Might the injury-induced sprouted mossy fibers form non-random patterns of connectivity that can substantially increase the probability of hyperexcitability and seizures? In the previous versions of the model, sprouted mossy fibers contacted other granule cells randomly, constrained only by the Gaussian distribution of their axonal arbors. However, numerous studies have indicated that connectivity in a wide variety of neural systems exhibits highly non-random characteristics. For example, the nervous system of the *C. elegans* worm has local connectivity patterns (network motifs) that are over- or underrepresented compared to what would be present in a random network *(46–48)*. These connectivity patterns have dynamical properties that could influence their abundance, closely tying structure to function *(49)*. Non-random connectivity features such as power-law distributions of connectivity (scale-free topology) *(7, 29, 50)* and non-random distributions of connection strengths *(51)* have been discovered in mammalian cortices as well. Additionally, it has been shown that connection probabilities can demonstrate fine-scale specificity that is dependent both upon neuronal type and the presence or absence of other connections in the network *(52, 53)*.

Morgan and Soltesz *(8)* therefore examined the role of several biologically plausible GC microcircuits on dentate excitability after a moderate insult (resulting in 50% sclerosis) and discovered that a small percentage of highly interconnected granule cells serving as network "hubs" can greatly augment hyperexcitability in the dentate model (**Fig. 10.8A**, diamonds represent hub cells). This change in hyperexcitability was mediated only by a change in the topology of granule cell-to-granule cell connections in the network (i.e., only the granule cell-to-granule cell connections were reconnected such that 5% of all GCs had approximately $5\times$ as

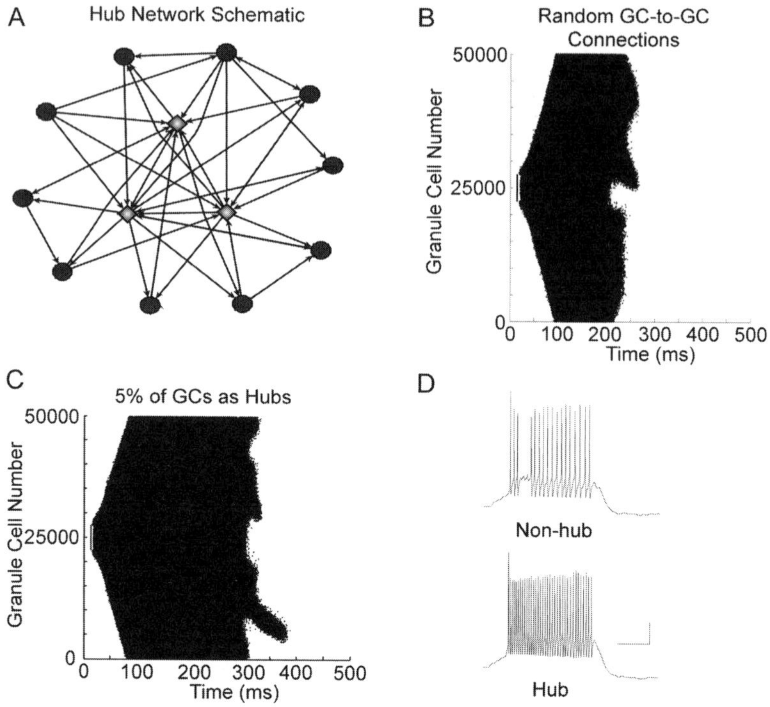

Fig. 10.8. **Granule cell hubs can promote hyperexcitability in the dentate gyrus.** (**A**) Schematic of a hub network. Average granule cells (*black circles*) have relatively few connections compared to the highly interconnected hubs (*gray diamonds*). (**B**) Raster plot of granule cell activity in response to simulated perforant path stimulation of 5,000 granule cells, 10 mossy cells, and 50 basket cells, in the network in which sprouted mossy fibers connected randomly onto other granule cells. Note that this network was implemented at 50% sclerosis. (**C**) Raster plot of granule cell activity in response to the same stimulation as in *B*, but in the network in which 5% of granule cells participated in many more connections than the average granule cell (scale: 20 mV and 100 ms). (**D**) Representative traces of a non-hub (*top*) and hub (*bottom*) granule cell (scale: 20 mV and 100 ms). Used with permission from Morgan and Soltesz, *Proc Natl Acad Sci U S A* 2008;105:6179–84.

many connections as the average granule cell; **Fig. 10.8A**). Importantly, no other connections were altered and the total number of connections in the dentate network remained constant at all times. **Figure 10.8B** displays network activity from the dentate network at 50% sclerosis when granule cell-to-granule cell connections were made randomly. **Figure 10.8C** shows activity of the network in which 5% of granule cells were vastly more interconnected than the average granule cell, creating network hubs (**Fig. 10.8A** – it was necessary for hubs to have both enhanced incoming as well as outgoing connectivity in order to effectively increase network excitability; refer to Fig. 4d in Morgan and Soltesz *(8)*. **Figure 10.8D** shows traces from both an average granule cell and a hub cell, demonstrating the powerful impact that topological changes can exert on cellular activity. Inclusion of hub cells enhanced excitability of the dentate network by approximately 30% without changing the total excitatory drive of the network, demonstrating the potential importance of hub cells in seizures.

The finding that hub cells can promote hyperexcitability in the dentate means little unless there is biological evidence for the model's prediction. Indeed, there are cells in the dentate gyrus that have a very long basal dendrite which receives many times more incoming excitatory synaptic connections than average granule cells. Until recently, however, the main problem with the biological data in support of hubs in the dentate gyrus was that the available data regarding GCs with basal dendrites focused exclusively on the incoming connections to these cells. There were no data suggesting whether GCs with basal dendrites form either more numerous or stronger outgoing connections, and the model's hubs required both an enhancement of incoming and outgoing connectivity for efficacy. In agreement with the model, though, an elegant study of post-status epilepticus GCs recently revealed a previously unseen increase in axonal protrusions (often within the GC layer) among newborn cells that likely corresponds to an increased number of outgoing connections onto other GCs *(54)*. Additionally, over 40% of post-SE newborn GCs retain a basal dendrite in addition to nearly 10% of mature post-SE GCs *(54)*, thereby resulting in the probable existence of GCs with both increased axonal protrusions *and* a basal dendrite (i.e., satisfying the model's requirement for both enhanced incoming as well as outgoing connectivity). While more experimental work is necessary to fully characterize the outgoing connectivity of GCs with basal dendrites, the model's results combined with the described experimental work strongly support the likelihood that neuronal hubs play a hitherto unrecognized, major role in promoting hyperexcitability and seizures.

7. Visualization of Large-Scale Networks

While large-scale, data-driven networks of various brain regions such as the dentate gyrus discussed above are gaining in popularity and usefulness, a significant problem has been the inability to easily visualize the cells and synaptic connections in the network. For many people new to modeling, construction of a network of cells using only text-based programming in a relatively esoteric language such as "hoc" (for the NEURON simulation environment) or GENESIS is tedious and confusing. While these programs have some graphical user interface applications, they are limited in scope and very limited in terms of scaling from single cells or membrane mechanisms to networks. Additionally, it is quite difficult to transport models from one simulation platform to another, so even an experienced modeler can have difficulties when attempting to utilize or visualize networks made in simulation environments they are unfamiliar with. Finally, the vast majority of models that have been constructed to date have been implemented in either one or two dimensions. There is a vast quantity of information that could potentially be lost by neglecting three-dimensional interactions. This loss of information is particularly notable for such issues as topology effects on network activity as described in this chapter and in measurements of spatiotemporal relationships in large active networks. The ability to encode networks in three dimensions is included in simulation programs such as NEURON and GENESIS, but it remains quite difficult to implement, and numerous problems (such as the lack of algorithms to create non-random connectivity structures) still exist even when three-dimensional implementations are successful.

Fortunately, a new software tool has recently been published called neuroConstruct *(10)*. This application is designed to allow interactive, visualized three-dimensional modeling of single cells or networks of varying sizes. Even a model such as the 500 cell dentate gyrus model discussed above can be easily visualized and modified (**Fig. 10.9**). Additionally, neuroConstruct utilizes a simulator-independent model description based on the NeuroML standards *(55, 56)*. This is potentially a dream come true for modelers

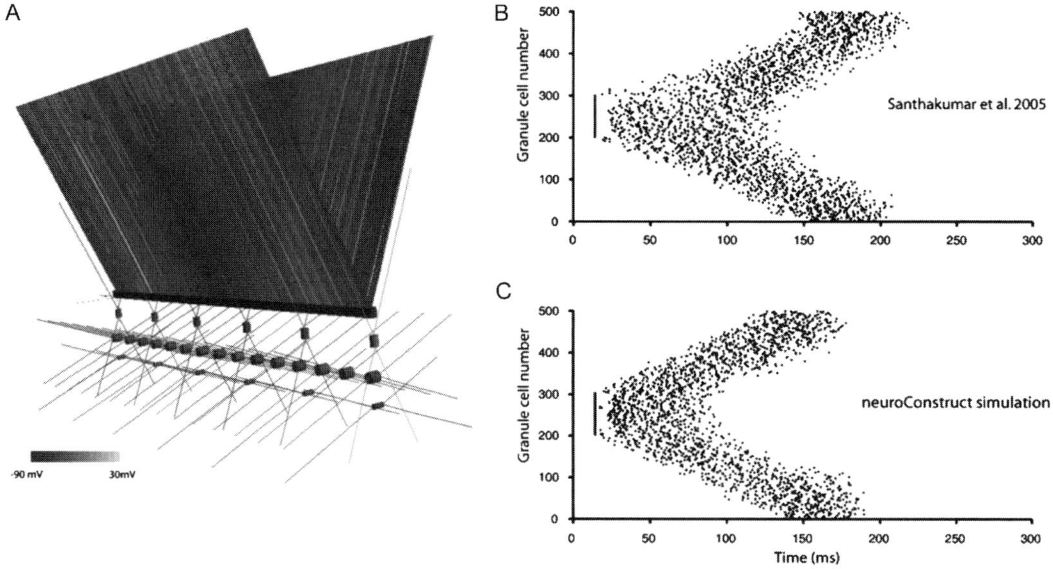

Fig. 10.9. **neuroConstruct implementation of the 500 cell dentate gyrus model. (A)** Replication of the 500 cell dentate gyrus model *(9)* in neuroConstruct *(10)*. The 10,000+ synaptic connections have been removed for clarity. Although shown in grayscale, cell coloring in the actual program gives information about network activity. (**B, C**) Raster plots of dentate gyrus granule cell activity in the original published model and in the neuroConstruct implementation of the network. Used with permission from Gleeson et al., *Neuron* 2007;54:219–35.

as it allows access to models that are coded in simulation environments that are unfamiliar to the user. This program will also be a fantastic tool for people who are new to modeling as the cell and network visualization tools make network construction much more intuitive, interesting, and engaging. Within a few hours of starting up, a naïve user can be well on his way to building a useful, functional network.

8. Conclusions

Computational modeling in epilepsy is a fast-growing field that has tremendous promise in its ability to provide us with invaluable information about seizure generation, propagation, and termination. Epilepsy is an ideal disease for close interaction between experimentalists and computational modelers, and the development of new and exciting hardware and software applications is making modeling accessible to a large population of scientists who would otherwise avoid it. Large-scale modeling such as that discussed in this chapter will greatly enhance our understanding of the pathological processes involved in seizures and will lead to the ability to quickly, efficiently, and cheaply test therapeutic interventions. In the next decade and beyond, the combination of innovative animal and computational models will be an extremely powerful tool for understanding and treating seizure disorders.

Acknowledgments

Funding to IS was provided by NIH grant NS35915 and funding to RM by the UCI MSTP.

References

1. Kalani MY, Vaidehi N, Hall SE, et al. The predicted 3D structure of the human D2 dopamine receptor and the binding site and binding affinities for agonists and antagonists. *Proc Natl Acad Sci U S A* 2004;101:3815–20.
2. Revett K, Ruppin E, Goodall S, Reggia JA. Spreading depression in focal ischemia: a computational study. *J Cereb Blood Flow Metab* 1998;18:998–1007.
3. Soltesz I, Staley KJ. *Computational Neuroscience in Epilepsy.* New York: Elsevier; 2008.
4. Vierling-Claassen D, Siekmeier P, Stufflebeam S, Kopell NJ. Modeling GABA alterations in schizophrenia: A link between impaired inhibition and altered gamma and beta range auditory entrainment. *J Neurophysiol* 2008;99(5):2656–71.
5. Brunel N. Dynamics of networks of randomly connected excitatory and inhibitory spiking neurons. *J Physiol*, Paris 2000;94: 445–63.
6. Brunel N, Wang XJ. What determines the frequency of fast network oscillations with

irregular neural discharges? I. Synaptic dynamics and excitation-inhibition balance. *J Neurophysiol* 2003;90:415–30.

7. Dyhrfjeld-Johnsen J, Santhakumar V, Morgan RJ, Huerta R, Tsimring L, Soltesz I. Topological determinants of epileptogenesis in large-scale structural and functional models of the dentate gyrus derived from experimental data. *J Neurophysiol* 2007;97: 1566–87.
8. Morgan RJ, Soltesz I. Nonrandom connectivity of the epileptic dentate gyrus predicts a major role for neuronal hubs in seizures. *Proc Natl Acad Sci U S A* 2008;105: 6179–84.
9. Santhakumar V, Aradi I, Soltesz I. Role of mossy fiber sprouting and mossy cell loss in hyperexcitability: A network model of the dentate gyrus incorporating cell types and axonal topography. *J Neurophysiol* 2005;93: 437–53.
10. Gleeson P, Steuber V, Silver RA. neuroConstruct: A tool for modeling networks of neurons in 3D space. *Neuron* 2007;54:219–35.
11. Traub RD, Contreras D, Whittington MA. Combined experimental/simulation studies of cellular and network mechanisms of epileptogenesis in vitro and in vivo. *J Clin Neurophysiol* 2005;22:330–42.
12. Traub RD, Wong RK. Cellular mechanism of neuronal synchronization in epilepsy. *Science* 1982;216:745–7.
13. Suffczynski P, Lopes da Silva FH, Parra J, et al. *Dynamics of epileptic phenomena determined from statistics of ictal transitions. IEEE transactions on bio-medical engineering* 2006;53(3):524–32.
14. Wendling F, Bartolomei F, Bellanger JJ, Chauvel P. Epileptic fast activity can be explained by a model of impaired GABAergic dendritic inhibition. *Eur J Neurosci* 2002;15:1499–508.
15. Wendling F, Hernandez A, Bellanger JJ, Chauvel P, Bartolomei F. Interictal to ictal transition in human temporal lobe epilepsy: Insights from a computational model of intracerebral EEG. *J Clin Neurophysiol* 2005;22:343–56.
16. Lytton WW. *From Computer to Brain: Foundations of Computational Neuroscience* New York; 2002.
17. Amaral DG, Witter MP. The three-dimensional organization of the hippocampal formation: A review of anatomical data. *Neuroscience* 1989;31:571–91.
18. Desmond NL, Levy WB. A quantitative anatomical study of the granule cell dendritic fields of the rat dentate gyrus using a novel probabilistic method. *J Comp Neurol* 1982;212: 131–45.
19. Lubke J, Frotscher M, Spruston N. Specialized electrophysiological properties of anatomically identified neurons in the hilar region of the rat fascia dentata. *J Neurophysiol* 1998;79:1518–34.
20. Patton PE, Mcnaughton B. Connection matrix of the hippocampal-formation .1. the dentate gyrus. *Hippocampus* 1995;5: 245–86.
21. Staley KJ, Otis TS, Mody I. Membrane properties of dentate gyrus granule cells: Comparison of sharp microelectrode and whole-cell recordings. *J Neurophysiol* 1992;67:1346–58.
22. Morgan RJ, Santhakumar V, Soltesz I. Modeling the dentate gyrus. *Prog Brain Res* 2007;163:639–58.
23. Ratzliff AH, Santhakumar V, Howard A, Soltesz I. Mossy cells in epilepsy: Rigor mortis or vigor mortis? *Trends Neurosci* 2002;25:140–4.
24. Buckmaster PS, Dudek FE. In vivo intracellular analysis of granule cell axon reorganization in epileptic rats. *J Neurophysiol* 1999;81:712–21.
25. Sutula TP, Hagen J, Pitkanen A. Do epileptic seizures damage the brain? *Curr Opin Neurol* 2003;16:189–95.
26. Buckmaster PS, Zhang GF, Yamawaki R. Axon sprouting in a model of temporal lobe epilepsy creates a predominantly excitatory feedback circuit. *J Neurosci* 2002;22: 6650–8.
27. Ratzliff AH, Howard AL, Santhakumar V, Osapay I, Soltesz I. Rapid deletion of mossy cells does not result in a hyperexcitable dentate gyrus: Implications for epileptogenesis. *J Neurosci* 2004;24:2259–69.
28. Sloviter RS. Hippocampal pathology and pathophysiology in temporal lobe epilepsy. *Neurologia* (Barcelona, Spain) 1996;11(Suppl 4):29– 32.
29. Watts DJ, Strogatz SH. Collective dynamics of 'small-world' networks. *Nature* 1998;393:440–2.
30. Watts DJ. *Small Worlds* Princeton, New Jersey: Princeton UP; 1999.
31. Santhakumar V, Ratzliff AD, Jeng J, Toth Z, Soltesz I. Long-term hyperexcitability in the hippocampus after experimental head trauma. *Ann Neurol* 2001;50:708–17.
32. Rafiq A, Zhang YF, DeLorenzo RJ, Coulter DA. Long-duration self-sustained epileptiform activity in the hippocampal-parahippocampal

slice: A model of status epilepticus. *J Neurophysiol* 1995;74: 2028–42.

33. Netoff TI, Clewley R, Arno S, Keck T, White JA. Epilepsy in small-world networks. *J Neurosci* 2004;24:8075–83.
34. Roxin A, Riecke H, Solla SA. Self-sustained activity in a small-world network of excitable neurons. *Phys Rev Lett* 2004;92:198101.
35. Buckmaster PS, Dudek FE. Network properties of the dentate gyrus in epileptic rats with hilar neuron loss and granule cell axon reorganization. *J Neurophysiol* 1997;77: 2685–96.
36. Buckmaster PS, Jongen-Relo AL. Highly specific neuron loss preserves lateral inhibitory circuits in the dentate gyrus of kainate-induced epileptic rats. *J Neurosci* 1999;19: 9519–29.
37. Cavazos JE, Das I, Sutula TP. Neuronal loss induced in limbic pathways by kindling: evidence for induction of hippocampal sclerosis by repeated brief seizures. *J Neurosci* 1994;14:3106–21.
38. Cavazos JE, Sutula TP. Progressive neuronal loss induced by kindling: a possible mechanism for mossy fiber synaptic reorganization and hippocampal sclerosis. *Brain Res* 1990;527:1–6.
39. Gorter JA, van Vliet EA, Aronica E, Lopes da Silva FH. Progression of spontaneous seizures after status epilepticus is associated with mossy fibre sprouting and extensive bilateral loss of hilar parvalbumin and somatostatin-immunoreactive neurons. *Eur J Neurosci* 2001;13:657–69.
40. Leite JP, Babb TL, Pretorius JK, Kuhlman PA, Yeoman KM, Mathern GW. Neuron loss, mossy fiber sprouting, and interictal spikes after intrahippocampal kainate in developing rats. *Epilepsy Res* 1996;26: 219–31.
41. Mathern GW, Bertram EH, Babb TL, et al. In contrast to kindled seizures, the frequency of spontaneous epilepsy in the limbic status model correlates with greater aberrant fascia dentata excitatory and inhibitory axon sprouting, and increased staining for N-methyl-D-aspartate, AMPA and GABA(A) receptors. *Neuroscience* 1997;77:1003–19.
42. van Vliet EA, Aronica E, Tolner EA, Lopes da Silva FH, Gorter JA. Progression of temporal lobe epilepsy in the rat is associated with immunocytochemical changes in inhibitory interneurons in specific regions of the hippocampal formation. *Exp Neurol* 2004;187:367–79.
43. Zappone CA, Sloviter RS. Translamellar disinhibition in the rat hippocampal dentate gyrus after seizure-induced degeneration of vulnerable hilar neurons. *J Neurosci* 2004;24:853–64.
44. Gabriel S, Njunting M, Pomper JK, et al. Stimulus and potassium-induced epileptiform activity in the human dentate gyrus from patients with and without hippocampal sclerosis. *J Neurosci* 2004;24:10416–30.
45. Blumcke I, Suter B, Behle K, et al. Loss of hilar mossy cells in Ammon's horn sclerosis. *Epilepsia* 2000;41(Suppl 6):S174–80.
46. Milo R, Shen-Orr S, Itzkovitz S, Kashtan N, Chklovskii D, Alon U. Network motifs: Simple building blocks of complex networks. *Science* 2002;298:824–7.
47. Reigl M, Alon U, Chklovskii DB. Search for computational modules in the C. elegans brain. *BMC Biol* 2004;2:25.
48. Sporns O, Kotter R. Motifs in brain networks. *PLoS Biol* 2004;2:e369.
49. Prill RJ, Iglesias PA, Levchenko A. Dynamic properties of network motifs contribute to biological network organization. *PLoS Biol* 2005;3:e343.
50. Barabasi AL, Albert R. Emergence of scaling in random networks. *Science* 1999;286: 509–12.
51. Song S, Sjostrom PJ, Reigl M, Nelson S, Chklovskii DB. Highly nonrandom features of synaptic connectivity in local cortical circuits. *PLoS Biol* 2005;3:e68.
52. Yoshimura Y, Callaway EM. Fine-scale specificity of cortical networks depends on inhibitory cell type and connectivity. *Nat Neurosci* 2005;8:1552–9.
53. Yoshimura Y, Dantzker JL, Callaway EM. Excitatory cortical neurons form fine-scale functional networks. *Nature* 2005;433: 868–73.
54. Walter C, Murphy BL, Pun RY, Spieles-Engemann AL, Danzer SC. Pilocarpine-induced seizures cause selective time-dependent changes to adult-generated hippocampal dentate granule cells. *J Neurosci* 2007;27:7541–52.
55. Crook S, Gleeson P, Howell F, Svitak J, Silver RA. MorphML: Level 1 of the NeuroML standards for neuronal morphology data and model specification. *Neuroinformatics* 2007;5:96–104.
56. Goddard NH, Hucka M, Howell F, Cornelis H, Shankar K, Beeman D. Towards NeuroML: Model description methods for collaborative modelling in neuroscience. *Philos Trans R Soc Lond B Biol Sci* 2001;356:1209–28.
57. Sik A, Penttonen M, Buzsaki G. Interneurons in the hippocampal dentate gyrus: An in

vivo intracellular study. *Eur J Neurosci* 1997;9:573–88.

58. Li XG, Somogyi P, Tepper JM, Buzsaki G. Axonal and dendritic arborization of an intracellularly labeled chandelier cell in the Ca1 region of rat hippocampus. *Exp Brain Res* 1992;90:519–25.
59. Buckmaster PS, Yamawaki R, Zhang GF. Axon arbors and synaptic connections of a vulnerable population of interneurons in the dentate gyrus in vivo. *J Comp Neurol* 2002;445: 360–73.
60. Buckmaster PS, Wenzel HJ, Kunkel DD, Schwartzkroin PA. Axon arbors and synaptic connections of hippocampal mossy cells in the rat in vivo. *J Compar Neurol* 1996;366:270–92.

Chapter 11

Organotypic Hippocampal Slice Cultures as a Model of Limbic Epileptogenesis

Suzanne B. Bausch

Abstract

Organotypic hippocampal slice cultures experience trauma, deafferentation due to cell loss or transection of afferent pathways, and neuronal circuitry rearrangements much like the events that can lead to acquired temporal lobe epilepsy. Organotypic hippocampal slice cultures can be maintained for months in vitro and exhibit a latent period followed by onset of electrographic seizures involving the dentate granule cells, which is a hallmark of epileptogenesis and acquired epilepsy in humans and in vivo animal models. The advantages of organotypic hippocampal slice cultures over in vivo models are that slice cultures exhibit a relatively short latent period and can be treated quickly and easily with a known concentration of reagent without unwanted systemic side effects. They are also more amenable to time-lapse studies and require fewer animals for drug screening and concentration–response analyses. Thus, the in vitro organotypic hippocampal slice culture model is an attractive alternative to in vivo models to begin to elucidate the molecular and cellular mechanisms underlying synaptic rearrangements and epileptogenesis.

Key words: electrophysiology, hippocampus, kainic acid, morphology, mossy fiber, NMDA receptor, seizures, sprouting, temporal lobe epilepsy.

1. Introduction

Epileptogenesis is the process by which a normal brain becomes chronically prone to seizures. Because this process is difficult to investigate in humans, rodent models play an important role in the study of epileptogenesis. The kindling, pilocarpine, and kainate (KA) models are well-established animal models of limbic epileptogenesis. Kindling is a model in which repeated, initially subconvulsive electrical stimuli lead progressively to electrographic and motor seizures and lifelong hyperexcitability *(1)*. The pilocarpine

Scott C. Baraban (ed.), *Animal Models of Epilepsy: Methods and Innovations, vol. 40*

DOI 10.1007/978-1-60327-263-6_11 Springerprotocols.com

(2) and KA *(3–5)* models utilize administration of pilocarpine or KA to generate acute seizures and status epilepticus. Development of chronic spontaneous recurrent seizures occurs after a 'latent period' of weeks to months *(5, 6)*. The 'latent period' in the pilocarpine and KA models and the progressive development of seizures in the kindling model parallel the delay in development of acquired epilepsy following neural insult in humans. Synaptic rearrangements, most notably mossy fiber sprouting, are observed in all three models, similar to temporal lobe epilepsy in humans *(7–11)*. Neuronal loss mimicking aspects of hippocampal sclerosis in humans occur in the KA and pilocarpine models *(2–4, 6, 12, 13)*. Lastly, all three models are chronic, which enables the study of the progression, the mechanisms, and potential therapeutic intervention against epileptogenesis. However, disadvantages of these in vivo models are (1) relatively slow data collection due to the fairly long latent period from initial insult to the appearance of seizures; (2) often complex, inconvenient delivery/dosing of drugs and other reagents to the brain; (3) the need for large numbers of animals for time-lapse studies, drug screening, and concentration-response analyses; and (4) technical difficulties in repeated imaging of fine structures in limbic regions. The purpose of this chapter is to illustrate the utility of the in vitro organotypic hippocampal slice culture model to circumvent these drawbacks for preliminary studies and serve as a relatively simple and rapid first step to study the progression, mechanisms, and therapeutic efficacy of interventions against epileptogenesis.

2. Preparation and Maintenance of Organotypic Hippocampal Slice Cultures

Various methods have been employed over the years to culture brain slices. Today, the two most common methods used to prepare and maintain organotypic hippocampal slice cultures are the roller tube and interface methods.

2.1. Roller Tube Method

Gahwiler pioneered the use of modern day organotypic hippocampal slice cultures using the roller tube technique *(14)*. Using this method, individual 300–400 μm hippocampal slices are embedded in a plasma clot on a glass coverslip. The coverslip is then inserted into a plastic test tube containing 1 mL of medium consisting of 50% Eagle basal medium, 25% Hank's or Earle's balanced salt solution, and 25% heat-inactivated horse serum, supplemented with 6.5 mg/mL glucose (pH 7.4). Test tubes are placed in a roller drum tilted 5° that rotates at approximately 10 revolutions/h to alternate feeding and aeration of the cultures. The entire apparatus holding the roller tubes resides in an incubator maintained at 36±0.5°C in dry air (*see* **Fig. 11.1**, top). Medium is changed weekly. The plasma clot degrades over time in culture. After 2–3 weeks, slice cultures flatten to about 50 μm and form a monolayer. Glia proliferate during the first weeks in culture and eventually cover the slice, but do not compromise viability. Glial proliferation can be reduced with antimitotics. [Methods compiled from *(14–18)*.]

2.2. Interface Method

In 1991, Stoppini and colleagues described the simpler, albeit more expensive, interface method for culturing hippocampal slices *(19)*. Using this method, up to six 300–400 μm thick slices are placed onto a 30 mm semi-porous (0.4 μm pore)

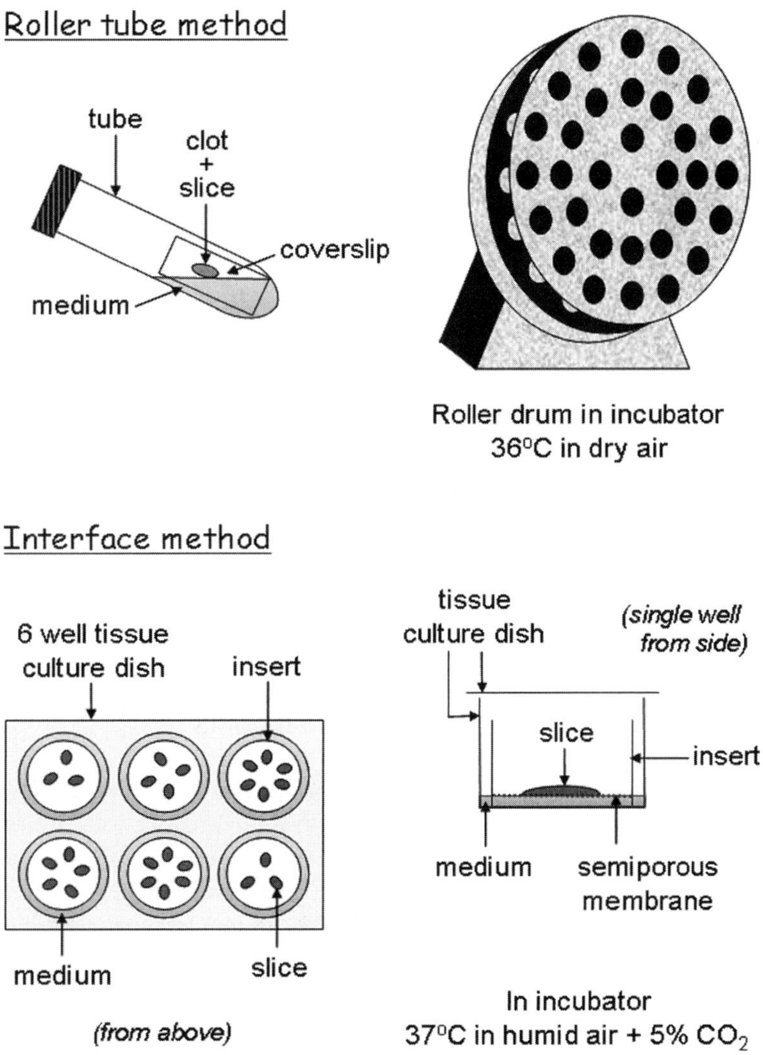

Fig. 11.1. **Schematic illustrates the two most common methods for culturing organotypic hippocampal slices**. *Schematic for roller tube method adapted from (15).*

Millicell-CM membrane insert in a standard 6-well tissue culture plate containing 1 mL of medium consisting of 50% minimal essential medium (MEM), 25% Hank's balanced salt solution, and 25% heat-inactivated horse serum, supplemented with HEPES and 6.5 mg/mL glucose (pH 7.2). The semi-porous membrane serves as an interface between the medium below and the air above. The tissue culture plate is then placed into a humid incubator and maintained at 37±0.5°C under humid room air + 5% CO_2 (*see* **Fig. 11.1**, bottom). No rotation or rocking is required. Medium is changed 2–3 times weekly, depending upon the number of slice cultures on a single insert. The slice cultures become attached to the membrane within a few days and flatten to approximately 150 µm with a thickness of 1–4 layers of neurons after about 1 week. As with roller tube cultures, glia eventually cover the slice cultures. For more detailed methods see Stoppini et al. *(19)*.

Most hippocampal slice cultures are isolated from early postnatal rat or more recently, early postnatal mouse. Modifications for culturing hippocampal slices from older adolescent rodents have been reported and include increased oxygen supply for roller tube cultures *(20)* or reduced temperature and increased K^+ concentrations during the first 7 days in vitro (DIV) for interface cultures *(21)*. In general, the three most important factors for preparation of healthy slice cultures are (1) gentle handling practices; (2) keeping the tissue moist and cold during dissection with frequent application of ice-cold buffer; and (3) minimal time from removal of the brain from the animal to placement of the slice into the incubator. Therefore, while a vibratome is ideal for preparation of acute slices, a McIlwain chopper is faster and preferable for preparation of slice cultures. Slice cultures need no antibiotics but do require proper aeration for survival. Slice cultures will not survive if continually immersed in medium. Under optimal conditions, hippocampal slices can survive for up to several months in culture and are suitable for use in most experimental paradigms designed for acute hippocampal slices with few modifications.

2.3. Comparison Between Methods

The major differences between slice cultures maintained using the roller tube and interface methods are the final thickness of the slice cultures and the presence or absence of the clot. Roller tube slice cultures are thinner, making them preferable for standard imaging and other experiments requiring high optical resolution. Interface slice cultures are thicker and thus maintain more of a three-dimensional structure making them more suitable for electrophysiological recordings in populations of neurons and biochemical studies requiring more tissue. Despite the thickness of interface cultures, individual neurons can be visualized easily using IR-DIC optics for patch clamp recordings or confocal or 2-photon microscopy for live imaging. Since the slice is not embedded in a clot, experiments using interface cultures can be performed within a few hours of dissection (compiled from *16, 19, 22, 23)*.

3. Properties of Organotypic Hippocampal Slice Cultures

Hippocampal slice cultures are similar to acute hippocampal slices in many ways. Both maintain intrinsic hippocampal circuitry and organotypic organization within a relatively thin section. The CA1–3 pyramidal cell and dentate granule cell layers are well defined, though may be slightly more dispersed in hippocampal slice cultures. Principal cells (pyramidal and granule cells) and interneurons retain most of their morphological and electrophysiological features including synaptic plasticity as well as their distinct neurotransmitter and receptor phenotypes. Layer-specific projections in hippocampal slice cultures are similar to those found in slices of adult animals, despite complete denervation from extra-hippocampal afferents *(15, 16, 19, 20, 24–28)*. The generation and regional distribution of newly born neurons resemble those of the adult hippocampus in vivo *(29, 30)*. Microglial morphology, distribution, and interactions are similar to those seen in vivo once the slice culture thins and becomes stable *(31, 32)*. Neuronal and synaptic properties continue to mature in hippocampal slice cultures with a time course that parallels development in vivo *(33–36)*. Lastly, organotypic hippocampal slice cultures prepared from knockout mice retain the same aberrant properties and histopathological features seen in vivo *(37)*.

There are, however, a number of major differences between acute hippocampal slices and organotypic hippocampal slice cultures. The most striking is the extent of synaptic connectivity. Normal synaptic connections in hippocampal slice cultures are estimated to be increased ten times relative to acute slices *(16, 38)*. Dentate granule cell mossy fiber axons in slice cultures exhibit increased collaterals in the granule cell and inner molecular layers and enhanced branching in the hilus *(15, 24, 39, 40)*. The degree of sprouting of mossy fiber collaterals into the granule cell and inner molecular layers (mossy fiber sprouting) displays the same distinct septotemporal gradient as seen in vivo and is dependent upon the integrity of afferents from the entorhinal cortex. More robust sprouting is observed in slice cultures prepared from the ventral compared to dorsal hippocampus and when the slice is cultured without the entorhinal cortex *(41, 42)*. CA1 pyramidal cells also exhibit axonal sprouting in organotypic hippocampal slice cultures. Their total axonal length is increased three to nine times relative to acute slices and axon collaterals can be found in stratum pyramidale and proximal radiatum in addition to their normal distribution in stratum oriens *(22, 43)*. The distribution of CA3 pyramidal cell axons is more widespread in slice cultures than in acute slices and back-projections from CA3 to the dentate gyrus are expanded *(22)*. In addition to making increased numbers of normal synapses, neurons in slice culture also may form aberrant synapses, which may or may not occur in vivo. A reciprocal CA1 pyramidal cell to dentate granule cell synapse, which has not been reported in vivo, has been documented in hippocampal slice cultures *(22, 44)*. Physiological responses recorded in the absence and presence of knife cuts suggest that many of these newly formed synapses are functional *(22, 43, 44)*. Because of the increased synaptic connectivity, especially in layers normally innervated by extrinsic hippocampal afferents, hippocampal slice cultures have been described as a model of hippocampal denervation *(20, 24)*. Since denervation and increased synaptic connectivity are common sequelae of many neurological insults that lead to acquired epilepsy, it is not surprising that another major difference between acute hippocampal slices and organotypic hippocampal slice cultures is their propensity to display epileptiform events and seizures.

4. Organotypic Hippocampal Slice Cultures as a Model of Epileptogenesis

Glutamate-dependent epileptiform activity and seizures have long been observed in the CA1 and CA3 pyramidal cell layers in organotypic hippocampal slice cultures *(45–49)*. More recently such events also have been observed in the normally seizure-resistant dentate granule cells. Dentate granule cells in organotypic hippocampal slices maintained for at least 10 DIV exhibit robust hyperexcitability including hilar-evoked excitatory postsynaptic potentials (EPSPs), multiple hilar-evoked population spikes, epileptiform bursts of EPSPs and population spikes, and very large amplitude hilar-evoked and spontaneous excitatory postsynaptic currents (EPSCs) *(22, 43, 44, 50, 51)*. These responses stand in stark contrast to those observed in acute slices from normal rats under identical recording conditions, but are similar to results reported for acute hippocampal slices isolated from KA-treated epileptic rats or humans with temporal lobe epilepsy *(7, 11, 52–56)*.

More importantly, spontaneous electrographic seizures involving the normally seizure-resistant dentate granule cells also can be induced in hippocampal slice cultures using $GABA_A$ receptor antagonists or low Mg^{2+} recording buffer (**Fig. 11.2**). Seizures are dependent upon glutamatergic synaptic transmission and display both tonic- and clonic-like phases *(22, 44, 51)*. Seizures

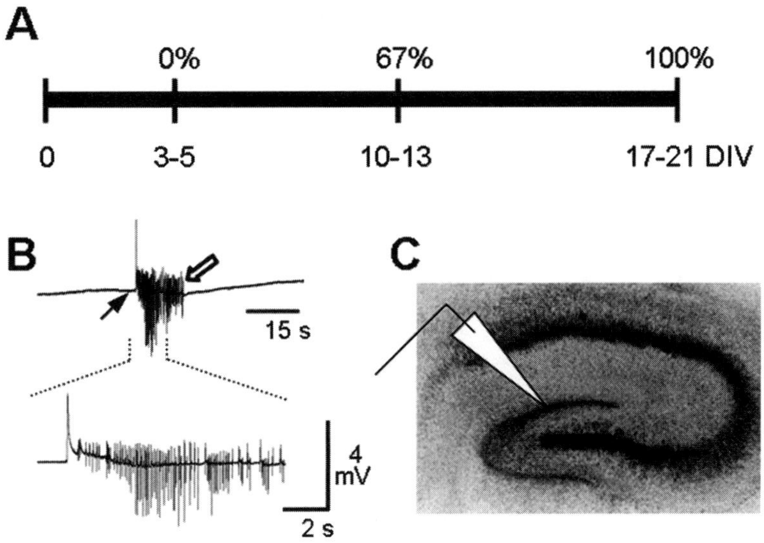

Fig. 11.2. **Organotypic hippocampal slice cultures exhibit a period of epileptogenesis followed by expression of electrographic seizures involving granule cells.** (**A**) Timeline shows the progression in the percentage of organotypic hippocampal slice cultures that exhibit electrographic seizures induced by blockade of inhibition as a function of time in culture. The incidence of seizures at 0 DIV and >17 DIV was significantly different ($p<0.05$, Fisher Exact test). (**B**) Traces show two time scales of a representative electrographic seizure recorded in the granule cell layer of an organotypic hippocampal slice culture following acute blockade of $GABA_A$ receptor-mediated inhibition (bicuculline methiodide, 10 μM). Closed arrow marks the beginning; open arrow indicates the end of the seizure. (**C**) Schematic overlaid on a Toluidine blue-stained organotypic hippocampal slice culture illustrates the location of the extracellular recording electrode used to document electrographic seizures. *Abbreviation:* DIV, days in vitro. *Adapted from (51).*

involving granule cells are not apparent immediately following explantation. Rather, a progressive and significant increase in the proportion of cultures exhibiting electrographic seizures induced via $GABA_A$ receptor blockade occurs as a function of time in culture; 0% of cultures at 3–5 DIV, 67% at 10–13 DIV, and 100% at ≥17 DIV [Fig. 11.2, *(51)*]. This latent period following initial insult and subsequent emergence of seizures is reminiscent of epileptogenesis in vivo and is one major hallmark of acquired epilepsy. The expression of seizures is remarkable given that electrographic seizures involving the dentate granule cells normally cannot be recorded in acute hippocampal slices under the conditions used in these studies. Such an increase in the susceptibility of dentate granule cells to seizures is thought to be a pivotal event in limbic epileptogenesis.

The factors leading to the progressive increase in excitability and seizure expression associated with epileptogenesis in cultured hippocampal slices have not been completely resolved. One possibility is deafferentation-induced expansions in excitatory circuitry, especially since electrographic seizures in slice cultures are dependent upon glutamatergic transmission *(22, 44, 51)*. As already alluded to, organotypic hippocampal slice cultures exhibit an amazing capacity for axonal sprouting and generation of new synapses following injury and deafferentation *(57, 58)*. A number of morphological abnormalities reported in organotypic hippocampal slice cultures also have been described in animal models and humans with temporal lobe epilepsy. In fact, the most extensively studied synaptic rearrangement in the epileptic rat and human hippocampus is mossy fiber sprouting *(7–10, 59–65)*, which is also present to a small degree in long-term hippocampal slice cultures *(40)*. The axonal sprouting and increased synaptic connectivity between CA1 pyramidal cells in organotypic hippocampal slice cultures also has been documented in animal models of temporal lobe epilepsy *(43, 66, 67)*. Another possibility is homeostatic plasticity, which may include upregulation of glutamate receptor-mediated excitatory transmission, downregulation of inhibitory transmission, and alterations in intrinsic excitability *(51, 68, 69–75)* due to loss of excitatory input from cut afferents. Indeed, interictal spikes and/or increased seizure susceptibility have been observed in visually deprived infants and children *(76)*, in rats subjected to hearing deprivation during the early postnatal period *(77)*, and in developing rats exposed chronically to focal blockade of activity with tetrodotoxin (TTX) *(78)*. Like results from in vivo studies, profound exacerbation of electrographic seizures has been documented following chronic blockade of activity with TTX in organotypic hippocampal slice cultures *(51)*. However, this form of plasticity generally occurs within days of reduced excitatory input in vitro rather than the 2–3 weeks required for seizure expression in all organotypic hippocampal slice cultures. This suggests that homeostatic plasticity is unlikely to be the sole factor underlying the delayed development of hyperexcitability and seizures in organotypic hippocampal slice cultures. The third possibility is that re-establishment of excitatory synapses onto neurons that have already been rendered hyperexcitable by deafferentation leads to a dramatic enhancement of neuronal excitability and expression of seizures.

5. Organotypic Hippocampal Slice Cultures as a Model for Drug Screening

The chronic nature of organotypic hippocampal slice cultures, together with the latent period between the insult of preparation and expression of seizures, provides an opportunity to disrupt the process of epileptogenesis. Historically, animal models have played an important role in drug discovery and screening. Acute models are often used to screen for drug efficacy against seizures *(79)*, but prophylactic treatments with drugs effective in these models have not proven effective against epileptogenesis *(80–84)*. The kindling, pilocarpine, or kainate models are established alternative in vivo models for chronic study of limbic epileptogenesis and screening of compounds against complex partial seizures *(1, 5, 6, 13, 79)*. The in vitro organotypic hippocampal slice culture model promises to be a relatively simple and rapid first step to study the effects of drugs on epileptogenesis and offers several advantages over in vivo models. Compared to in vivo models, organotypic hippocampal slice cultures have a relatively short latent period, the diffusion distances for drugs and other reagents to reach their targets are shorter, systemic side effects are absent, reagent delivery and treatment paradigms are simplified, concentration–response studies require fewer animals, time-course studies using the same tissue are possible, and housing requirements are more compact and less expensive in terms of labor and per diem costs. However, like all models, there are also disadvantages. As with all in vitro models, neurons in hippocampal slice cultures are exposed to different extracellular milieus than those in vivo. Tissue used for hippocampal slice cultures is isolated from early postnatal rat brain, prompting some to argue that data from slice cultures may better predict responses in the developing than the mature brain due to differences in axonal guidance cues, receptor distribution and expression, etc. While this is certainly true for tissue isolated from infant rats, hippocampi isolated from P8-11 rats and maintained in culture exhibit many characteristics of mature brain (*see* **Section 3**). In addition, a partial regression to developmental expression patterns is thought to occur in the mature brain following neuronal insults, such as injury, severe hyperactivity, or seizures. Thus, organotypic hippocampal slice cultures hold promise for preliminary studies into drug action following insult to the mature brain. In support of this idea, we showed recently that organotypic hippocampal slice cultures respond to chronic treatment with *N*-methyl-D-aspartate receptor (NMDAR) antagonists in a manner similar to that seen in vivo.

Excessive NMDAR activation is thought to contribute to epileptogenesis because brief blockade of NMDAR activation prior to and/or during brain insult has been shown to inhibit epileptogenesis in some models *(85–91)*. However, most previous studies began brief NMDAR antagonist administration *prior* to initial insult, which may block initiation of pathological processes; therapies in humans must inhibit these cascades once they have begun. Moreover, paradoxical increases in seizure susceptibility and severity have been observed in (1) fully kindled rats after administration of clinically relevant doses of the moderate-affinity uncompetitive NMDAR antagonist, memantine *(90)* and (2) about 40% of epilepsy patients following chronic treatment with the high-affinity competitive NMDAR antagonist, D-CPP-ene ((R)-4-(3-phosphono-2-propenyl)-2-piperazinecarboxylic acid) *(92)*. Similarly, we documented trends toward increased number and duration of recurrent electrographic seizures involving dentate granule cells in organotypic hippocampal slice cultures treated chronically after preparation-induced injury with memantine or the high-affinity competitive NMDAR antagonist, APV, when compared to cultures treated with vehicle *(50, 51)*. This exacerbation of seizures was maintained for at least 48 h after removal of APV (unpublished observations). Since previous studies showed that altered receptor levels elicited by chronic treatment with a combination of NMDAR and AMPAR/KAR antagonists returned to baseline by about 36 h after removal of antagonists *(69)*, these findings suggested that effects of chronic blockade of NMDAR following injury may be longlasting and potentially pro-epileptogenic.

The striking similarities in results from organotypic hippocampal slice cultures and in vivo animal and human epilepsy studies following chronic treatment with moderate-affinity, uncompetitive, and high-affinity, competitive NMDAR antagonists provides preliminary validation for the use of organotypic hippocampal slice cultures for preliminary testing of reagents to inhibit epileptogenesis.

6. Organotypic Hippocampal Slice Cultures to Study Epileptogenesis-Associated Axonal Rearrangements

6.1. Kainic Acid-Induced Axonal Sprouting

Administration of the chemoconvulsants, KA or pilocarpine, to rats causes acute seizures and status epilepticus as well as neuronal loss, mossy fiber sprouting, and CA1 pyramidal cell sprouting that resemble some aspects of hippocampal sclerosis in humans (*see* **Section 1**). Organotypic hippocampal slice cultures treated with pilocarpine also display seizure-like activity followed by mossy fiber sprouting, but little of the neuronal loss usually observed following pilocarpine administration in vivo *(93, 94)*. In contrast, organotypic hippocampal slice cultures treated with KA exhibit seizures, followed by robust mossy fiber sprouting as well as hippocampal sclerosis, which are all strikingly similar to findings from in vivo KA animal models [*(40)*, **Fig. 11.3**]. KA-treated slice cultures also exhibit a significant two-fold increase in total axon length of CA1 pyramidal cells *(43)*, which is similar to that reported for acute slices from KA- or pilocarpine-treated epileptic rats *(66, 67, 95)*. The similar degree of mossy fiber and CA1 pyramidal cell sprouting in vitro and in vivo is remarkable given the increased basal level of mossy fiber and CA1 pyramidal cell

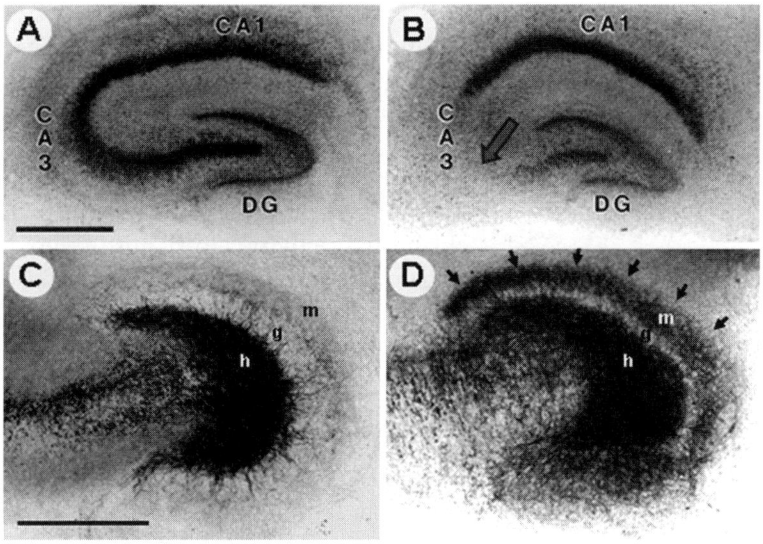

Fig. 11.3. **KA treatment in vitro caused cell loss and mossy fiber sprouting**. (**A, C**) Vehicle- and (**B,D**) KA-treated hippocampal slice cultures were stained with (**A,B**) Toluidine blue or (**C,D**) Timm stain approximately 2 weeks following 48 h treatment with vehicle or KA beginning at 10 DIV. Toluidine blue stain showed (**A**) a normal distribution of pyramidal cells and dentate granule cells in vehicle-treated cultures, but (**B**) a loss of CA3a/b pyramidal cells (arrow) and a slight decrease in granule cells in KA-treated cultures. CA3c pyramidal cells are seen in approximately 50% of KA-treated cultures. Timm stain revealed (**C**) a low level of mossy fiber sprouting into the inner molecular of vehicle-treated cultures. (**D**) Mossy fiber sprouting into the inner molecular layer (arrows) was dramatically increased following KA treatment. *Abbreviations:* DG, dentate gyrus; g, dentate granule cell layer; h, hilus; m, molecular layer. *Scale bar* in A is for A,B; in C is for C,D, 500 μm. *Adapted from (43, 108).*

sprouting in organotypic hippocampal slice cultures *(40, 43, 67, 95)*. Even more remarkable, robust KA-induced mossy fiber sprouting in hippocampal slice cultures occurs in as little as 2 weeks and displays the same strict laminar preference for the inner molecular layer as seen in vivo, despite complete denervation of afferents from the entorhinal cortex in slice cultures *(40)*.

Because sprouting takes place within the confines of the culture, KA-treated organotypic hippocampal slice cultures seemed to be an ideal preparation to investigate the functional consequences of mossy fiber sprouting. Unlike acute slices from in vivo models, newly sprouted axons, which synapse predominantly onto granule cells in vivo *(96)*, would not be severed immediately prior to study. However, despite robust increases in mossy fiber sprouting in KA-compared to vehicle-treated cultures, measures of granule cell hyperexcitability including the incidence of seizures, epileptiform bursts, multiple population spikes, and very large amplitude sEPSCs were similar in KA- and vehicle-treated cultures *(43)*. These findings were surprising because the aberrant recurrent excitatory network between granule cells formed by sprouted mossy fibers is associated with hyperexcitability in acute slices isolated from KA- and pilocarpine-treated epileptic rats *(7, 11, 52, 55, 56)*. The interpretation of these data is complicated by the complex backdrop of extensive synaptic reorganization in organotypic hippocampal slice cultures compared to acute slices. Nonetheless, these findings reinforce the idea that axonal sprouting and synaptic reorganization can contribute to abnormal synchronous activity and raise the possibility that alternate, non-mossy fiber pathways also could play a dominant role. Conversely, one could argue that even the modest degree of mossy fiber sprouting in vehicle-treated organotypic hippocampal slice cultures is sufficient to elicit maximal excitability *(43)*. Regardless of the potential difficulties in studying the functional consequences of a single reorganized pathway in KA-treated slice cultures, the robust axonal reorganization confined to the inner molecular layer and ease with which they can be repeatedly treated with a variety of agents, organotypic hippocampal slice cultures are an ideal preparation for discerning the molecular mechanisms contributing to axonal sprouting and targeting *(97, 98)*.

6.2. Molecular Mechanisms Contributing to Mossy Fiber Sprouting

CD44 is the principal cell surface receptor for the glycoaminoglycan hyaluronan *(99)* and is an important regulator of the hyaluronan-dependant extracellular matrix *(100)* as well as axonal outgrowth, adherence, and guidance *(100–104)*. Both CD44 and hyaluronan are upregulated in the epileptic hippocampus and postulated to play a role in mossy fiber sprouting *(105, 106, 107)*. Interestingly, CD44 upregulation following pilocarpine-induced status epilepticus in mice preceded mossy fiber sprouting, and expression levels of CD44 were inversely associated with those of the axonal growth cone-associated protein, GAP-43 *(107)*. Data from organotypic hippocampal slice cultures also showed an inverse relationship between CD44 expression and KA-induced mossy fiber sprouting *(108)*. In fact, the time course of progressively reduced CD44 expression *(108)* mirrored the temporal progression of increased mossy fiber sprouting following KA treatment of organotypic hippocampal slice cultures *(40)*. Function blocking anti-CD44 antibodies also caused a strong trend toward enhanced mossy fiber sprouting, suggesting that CD44 may play a modest role in limiting mossy fiber sprouting *(108)*. Degradation of hyaluronan with hyaluronidase significantly decreased the width of mossy fiber sprouting in organotypic hippocampal slice cultures, suggesting that hyaluronan may play a permissive role in mossy fiber sprouting *(108)*. Thus, organotypic hippocampal slice cultures were used to confirm that two components of the extracellular matrix, hyaluronan and CD44, are involved in mossy fiber sprouting. However, changes in a myriad of other elements each targeting complementary mechanisms are likely to be responsible for mossy fiber sprouting.

Another factor that has received significant attention in recent years is brain-derived neurotrophic factor (BDNF) and its high-affinity receptor, trk B. BDNF expression and trk B activation are upregulated following seizures, and blockade of increased BDNF/trk B function inhibits the development of epilepsy *(109)*. Whether BDNF affected axonal sprouting or development of new synapses remained unclear due to confounds associated with the in vivo studies *(110–112)*. To circumvent these confounds, a number of investigators used organotypic hippocampal slice cultures combined with increased expression of BDNF following transfection, prolonged hyperactivity, or injury to show that increased expression of BDNF and subsequent activation of trk B receptors did indeed increase axonal sprouting in both pyramidal and granule cells *(113–115*, but *see 116)*.

7. Summary

The studies described in this chapter represent a sample of the approaches and research questions that are highly suited to organotypic hippocampal slice cultures. Some of the other major uses/advantages are as follows. The thin nature and high optical visibility of slice cultures make them superior to acute slices for imaging in a preparation that retains neuronal circuitry [for a recent example see *(117)*]. Because they can be maintained for long periods in vitro, slice cultures also are suitable for long-term live imaging *(23)*. Preparation of slice cultures from fluorescent-protein-expressing mice or co-cultures of preparations from wild-type and fluorescent protein-expressing mice can dramatically simplify and enhance the study of axonal growth and regeneration *(97)* and other structural changes in neurons and glia *(118)*. Preparation of slice cultures from other transgenic or knockout animals is also feasible as aberrant properties and histopathological features are retained in vitro *(37)*. This is particularly advantageous for those transgenic or knockout lines with poor viability. The ease with which slice cultures can be manipulated is important not only for pharmacological and biochemical experiments as described in this chapter but also for molecular biological approaches. Biolistics *(119, 120)*, whole embryo electroporation followed by explantation *(121)*, single-cell electroporation *(122)*, and infection with recombinant viruses *(123, 124)* all have been implemented successfully in slice cultures. Lastly, hippocampal slices can be cultured directly onto multi-electrode arrays for use in both short and long-term electrophysiological recordings *(125–128)*. In summary, organotypic hippocampal slice cultures represent a chronic in vitro model of limbic epileptogenesis and epilepsy that exhibits many of the same electrophysiological and morphological features found in both human and in vivo models. This in vitro model provides a relatively simple system to begin to elucidate the molecular and cellular mechanisms underlying synaptic rearrangements and epileptogenesis and identify potential therapeutic agents to retard epileptogenesis and prevent temporal lobe epilepsy.

Acknowledgments

I thank Natalie White for assistance. Work was funded by the Congressionally Directed Medical Research Program and National Institute of Neurological Disorders and Stroke.

References

1. McNamara JO, Bonhaus DW, Shin C. The kindling model of epilepsy. In: Schwartzkroin PA ed. *Epilepsy: Models, Mechanisms, and Concepts.* New York, NY: Cambridge University Press, 1993:27–47.
2. Turski WA, Cavalheiro EA, Schwarz M,Czuczwar SJ, Kleinrok Z, Turski L. Limbic seizures produced by pilocarpine in rats: Behavioural, electroencephalographic and neuropathological study. *Behav Brain Res* 1983;9:315–335.
3. Nadler JV, Perry BW, Cotman CW. Selective reinnervation of hippocampal area CA1 after destruction of CA3-CA4 afferents with kainic acid. *Brain Res* 1980;182:1–9.
4. Pisa M, Sanberg PR, Corcoran ME, Fibiger HC. Spontaneously recurrent seizures after intracerebral injections of kainic acid in rat: a possible model of human temporal lobe epilepsy. *Brain Res* 1980;200:481–487.
5. Hellier JL, Patrylo PR, Buckmaster PS, Dudek FE. Recurrent spontaneous motor seizures after repeated low-dose systemic treatment with kainate: Assessment of a rat model of temporal lobe epilepsy. *Epilepsy Res* 1998;31:73–84.
6. Cavalheiro EA, Leite JP, Bortolotto ZA,Turski WA, Ikonomidou C, Turski L. Long-term effects of pilocarpine in rats: structural damage of the brain triggers kindling and spontaneous recurrent seizures. *Epilepsia* 1991;32:778–782.
7. Tauck DL, Nadler JV. Evidence of functional mossy fiber sprouting in hippocampal formation of kainic acid-treated rats. *J Neurosci* 1985;5:1016–1022.
8. Sutula T, He X-X, Cavazos J, Scott G. Synaptic reorganization in the hippocampus induced by abnormal functional activity. *Science* 1988;239:1147–1150.
9. Sutula T, Cascino G, Cavazos J, Parada I, Rameriz L. Mossy fiber synaptic reorganization in the epileptic human temporal lobe. *Ann Neurol* 1989;26:321–330.
10. Mello LEAM, Cavalheiro EA, Tan AM, Pretorius JK, Babb TL, Finch DM. Granule cell dispersion in relation to mossy fiber sprouting, hippocampal cell loss, silent period and seizure frequency in the pilocarpine model of temporal lobe epilepsy. In: Engel J Jr, Wasterlain C, Cavalheiro EA, Heinemann U, Avanzini G, ed. *Molecular Neurobiology of Epilepsy, Epilepsy Res* (Suppl 9). BV: Elsevier, 1992:51–60.
11. Wuarin J-P, Dudek FE. Electrographic seizures and new recurrent excitatory circuits in the dentate gyrus of hippocampal slices from kainate-treated epileptic rats. *J Neurosci* 1996;16:4438–4448.
12. Ben-Ari Y. Limbic seizure and brain damage by kainic acid: Mechanisms and relevance to human temporal lobe epilepsy. *Neuroscience* 1985;14:375–403.
13. Turski L, Ikonomidou C, Turski WA, Bortolotto ZA, Cavalheiro EA. Review: Cholinergic mechanisms and epileptogenesis. The seizures induced by pilocarpine: A novel experimental model of intractable epilepsy. *Synapse* 1989;3:154–171.

14. Gähwiler BH. Organotypic monolayer cultures of nervous tissue. *J Neurosci Methods* 1981;4:329–342.
15. Gähwiler BH. Development of the hippocampus in vitro: Cell types, synapses and receptors. *Neuroscience* 1984;11:751–760.
16. Gähwiler BH, Capogna M, Debanne D, McKinney RA, Thompson SM. Organotypic slice cultures: A technique has come of age. *Trends Neurosci* 1997;20:471–477.
17. Del Rio JA, Heimrich B, Soriano E, Schwegler H, Frotscher M. Proliferation and differentiation of glial fibrillary acidic protein-immunoreactive glial cells in organotypic slice cultures of rat hippocampus. *Neuroscience* 1991;43:335–347.
18. Thompson SM, Mason SE. Preparation of organotypic hippocampal slice cultures. In: Celis JE, ed. *Cell Biology: A Laboratory Handbook*, 3rd ed. Amsterdam: Elsevier, 2004.
19. Stoppini L, Buchs P-A, Muller D. A simple method for organotypic cultures of nervous tissue. *J Neurosci Methods* 1991;37: 173–182.
20. Gähwiler BH. Organotypic cultures of neural tissue. *Trends Neurosci* 1988;11:484–489.
21. Xiang Z, Hrabetova S, Moskowitz SI, Casaccia-Bonnefil P, Young SR, Nimmrich VC, Tiedge H, Einheber S, Karnup S, Bianchi R, Bergold PJ. Long-term maintenance of mature hippocampal slices in vitro. *J Neurosci Methods* 2000;98:145–154.
22. Bausch SB, McNamara JO. Synaptic connections from multiple subfields contribute to granule cell hyperexcitability in hippocampal slice cultures. *J Neurophysiol* 2000;84:2918–2932.
23. Gogolla N, Galimberti I, DePaola V, Caroni P. Long-term live imaging of neuronal circuits in organotypic hippocampal slice cultures. *Nat Protoc* 2006;1:1223–1226.
24. Zimmer J, Gahwiler BH. Cellular and connective organization of slice cultures of the rat hippocampus and fascia dentata. *J Comp Neurol* 1984;228:432–446.
25. Caeser M, Aertsen AD. Morphological organization of rat hippocampal slice cultures. *J Comp Neurol* 1991;307: 87–106.
26. Frotscher M, Heimrich B. Formation of layer-specific fiber projections to the hippocampus in vitro. *Proc Natl Acad Sci U S A* 1993;90:10499–10403.
27. Dailey ME, Buchanan J, Bergles DE, Smith SJ. Mossy fiber growth and synaptogenesis in rat hippocampal slices in vitro. *J Neurosci* 1994;14:1060–1078.
28. Frotscher M, Heimrich B, Deller T. Sprouting in the hippocampus is layer-specific. *Trends Neurosci* 1997;20:218–223.
29. Raineteau O, Rietschin L, Gradwohl G, Guillemot F, Gähwiler BH. Neurogenesis in hippocampal slice cultures. *Mol Cell Neurosci* 2004;26:241–250.
30. Kamada M, Li R-Y, Hashimoto M, Okada H, Koyanagi Y, Ishizuka T, Yawo H. Intrinsic and spontaneous neurogenesis in the postnatal slice culture of rat hippocampus. *Eur J Neurosci* 2004;20:2499–2508.
31. Hailer NP, Jarhult JD, Nitsch R. Resting microglial cells in vitro: Analysis of morphology and adhesion molecule expression in organotypic hippocampal slice cultures. *Glia* 1996;18:319–331.
32. Skibo GG, Nikonenko IR, Savchenko VL, Mckanna JA. Microglia in organotypic hippocampal slice culture and effects of hypoxia: Ultrastructure and lipocortin-1 immunoreactivity. *Neuroscience* 2000; 96:427–438.
33. Muller D, Buchs P-A, Stoppini L. Time course of synaptic development in hippocampal organotypic cultures. *Dev Brain Res* 1993;71:93–100.
34. Bahr BA. Long-term hippocampal slices: A model system for investigating synaptic mechanisms and pathologic processes. *J Neurosci Res* 1995;42:294–305.
35. Frotscher M, Zafirov S, Heimrich B. Development of identified neuronal types and of specific synaptic connections in slice cultures of rat hippocampus. *Prog Neurobiol* 1995;45:143–164.
36. De Simoni A, Griesinger CB, Edwards FA. Development of rat CA1 neurones in acute versus organotypic slices: role of experience in synaptic morphology and activity. *J. Physiol* 2003;550:135–147.
37. Wenzel HJ, Tamse CT, Schwartzkroin PA. Dentate development in organotypic hippocampal slice cultures from p35 knockout mice. *Dev Neurosci* 2007;29:99–112.
38. Pavlidis P, Madison DV. Synaptic transmission in pair recordings from CA3 pyramidal cells in organotypic culture. *J Neurophysiol* 1999;81:2787–2797.
39. Frotscher M, Gahwiler BH. Synaptic organization of intracellularly stained CA3 pyramidal neurons in slice cultures of rat hippocampus. *Neuroscience* 1988;24: 541–551.
40. Routbort MJ, Bausch SB, McNamara JO. Seizures, cell death and mossy fiber sprouting in kainic acid-treated organotypic hippocampal cultures. *Neuroscience* 1999;94: 755–765.
41. Coltman BW, Earkey EM, Shahar A, Dudek FE, Ide CF. Factors influencing mossy fiber

collateral sprouting in organotypic slice cultures of neonatal mouse hippocampus. *J Comp Neurol* 1995;362:209–222.

42. Teter B, Harris-White ME, Frautschy SA, Cole GM. Role of apolipoprotein E and estrogen in mossy fiber sprouting in hippocampal slice cultures. *Neuroscience* 1999;91: 1009–1016.
43. Bausch SB, McNamara JO.Contributions of mossy fiber and CA1 pyramidal cell sprouting to dentate granule cell hyperexcitability in kainic acid-treated hippocampal slice cultures. *J Neurophysiol* 2004;92:3582–3595.
44. Gutierrez R, Heinemann U.Synaptic reorganization in explanted cultures of rat hippocampus. *Brain Res* 1999;815:304–316.
45. Fowler J, Bornstein MB, Crain SM. Sustained hyperexcitability elicited by repetitive electrical stimulation of organotypic hippocampal explants. *Brain Res* 1986;378:398–404.
46. McBain CJ, Boden P, Hill RG. The kainate/quisqualate receptor antagonist, CNQX, blocks the fast component of spontaneous epileptiform activity in organotypic cultures of rat hippocampus. *Neurosci Lett* 1988; 93:341–345.
47. McBain CJ, Boden P, Hill RG. Rat hippocampal slices 'in vitro' display spontaneous epileptiform activity following long-term organotypic culture. *J Neurosci Methods* 1989;27:35–49.
48. Malouf AT, Robbins CA, Schwartzkroin PA. Epileptiform activity in hippocampal slice cultures with normal inhibitory drive. *Neurosci Lett* 1990;108:76–80.
49. Scanziani M, Debanne D, Muller M, Gahwiler BH, Thompson SM. Role of excitatory amino acid and $GABA_A$ receptors in the generation of epileptiform activity in disinhibited hippocampal slice cultures. *Neuroscience* 1994;61:823–834.
50. Wang X-M, Bausch SB. Differential effects of distinct classes of*N*-methyl-D-aspartate receptor antagonists on dentate granule cell seizures, neuronal loss and mossy fiber sprouting *in vitro*: suppression by NR2B-selective antagonists. *Neuropharmacology* 2004;47:1008–1020.
51. Bausch SB, He S-J, Petrova Y, Wang X-M, McNamara JO. Plasticity of both excitatory and inhibitory synapses is associated with seizures induced by removal of chronic blockade of activity in cultured hippocampus. *J Neurophysiol* 2006;96:2151–2167.
52. Cronin J, Obenaus A, Houser C, Dudek FE. Electrophysiology in dentate granule cells after kainate-induced synaptic reorganization of the mossy fibers. *Brain Res* 1992;573:305–310.
53. Masukawa LM, Uruno K, Sperling M, O'Connor MJ, Burdette LJ. The functional relationship between antidromically evoked field potential responses of the dentate gyrus and mossy fiber reorganization in temporal lobe epileptic patients. *Brain Res* 1992;579:119–127.
54. Franck JE, Pokorny J, Kunkel DD, Schwartzkroin PA.Physiologic and morphologic characteristics of granule cell circuitry in human epileptic hippocampus. *Epilepsia* 1995;36:543–558.
55. Patrylo PR, Dudek FE. Physiological unmasking of new glutamatergic pathways in the dentate gyrus of hippocampal slices from kainate-induced epileptic rats. *J Neurophysiol* 1998;79:418–429.
56. Okazaki MM, Molnar P, Nadler JV. Recurrent mossy fiber pathway in rat dentate gyrus: Synaptic currents evoked in the presence and absence of seizure-induced growth. *J Neurophysiol* 1999;81:1645–1660.
57. Stoppini L, Buchs PA, Muller D. Lesion-induced neurite sprouting and synapse formation in hippocampal organotypic cultures. *Neuroscience* 1993;57:985–994.
58. McKinney RA, Debanne D, Gähwiler BH, Thompson SM.Lesion-induced axonal sprouting and hyperexcitability in the hippocampus in vitro: implications for the genesis of posttraumatic epilepsy. *Nat Med* 1997;3:990–996.
59. Laurberg S, Zimmer J. Lesion-induced sprouting of hippocampal mossy fiber collaterals to the fascia dentata in developing and adult rats. *J Comp Neurol* 1981;200:433–459.
60. de Lanerolle, NC, Kim, JH, Robbins, RJ, Spencer DD. Hippocampal interneuron loss and plasticity in human temporal lobe epilepsy. *Brain Res* 1989;495:387–395.
61. Houser CR, Miyashiro JE, Swartz BE, Walsh GO, Rich JR, Delgado-Escueta AV. Altered patterns of dynorphin immunoreactivity suggest mossy fiber reorganization in human hippocampal epilepsy. *J Neurosci* 1990;10:267–282.
62. Ribak CE, Peterson GM. Intragranular mossy fibers in rats and gerbils form synapses with the somata and proximal dendrites of basket cells in the dentate gyrus. *Hippocampus* 1991;1:355–364.
63. Represa A, Jorquera I, Le Gal La Salle G, Ben-Ari Y. Epilepsy induced collateral sprouting of hippocampal mossy fibers: Does it induce the development of ectopic synapses with granule cell dendrites? *Hippocampus* 1993;3:257–268.

64. Okazaki MM, Evenson DA, Nadler JV. Hippocampal mossy fiber sprouting and synapse formation after status epilepticus in rats: Visualization after retrograde transport of biocytin. *J Comp Neurol* 1995;352:515–534.
65. Molnar P, Nadler JV. Mossy fiber-granule cell synapses in the normal and epileptic rat dentate gyrus studies with minimal laser photostimulation. *J Neurophysiol* 1999;82:1883–1894.
66. Perez Y, Morin F, Beaulieu C, Lacaille JC. Axonal sprouting of CA1 pyramidal cells in hyperexcitable hippocampal slices of kainate-treated rats. *Eur J Neurosci* 1996;8:736–748.
67. Esclapez M, Hirsch JC, Ben-Ari Y, Bernard C. Newly formed excitatory pathways provide a substrate for hyperexcitability in experimental temporal lobe epilepsy. *J Comp Neurol* 1999;408:449–460.
68. Rao A, Craig AM. Activity regulates the synaptic localization of the NMDA receptor in hippocampal neurons. *Neuron* 1997;19:801–812.
69. O'Brien RJ, Kamboj S, Ehlers MD, Rosen KR, Fischbach GD, Huganir RL. Activity-dependent modulation of synaptic AMPA receptor accumulation. *Neuron* 1998;21:1067–1078.
70. Niesen CE, Ge S. Chronic epilepsy in developing hippocampal neurons: Electrophysiological and morphological features. *Dev Neurosci* 1999;21:328–338.
71. Murthy VN, Schikorski T, Stevens CF, Zhu Y. Inactivity produces increases in neurotransmitter release and synapse size. *Neuron* 2001;32:673–682.
72. Burrone J, O'Byrne M, Murthy VN. Multiple forms of synaptic plasticity triggered by selective suppression of activity in individual neurons. *Nature* 2002;420:414–418.
73. Buckby LE, Jensen TP, Smith PJ, Empson RM. Network stability through homeostatic scaling of excitatory and inhibitory synapses following inactivity in CA3 of rat organotypic hippocampal slice cultures. *Mol Cell Neurosci* 2006;31:805–816.
74. Swanwick CC, Murthy NR, Kapur J. Activity-dependent scaling of GABAergic synapse strength is regulated by brain-derived neurotrophic factor. *Mol Cell Neurosci* 2006;31:481–492.
75. Cai X, Wei D-S, Gallagher SE, Bagal A, Mei Y-A, Kao JPY, Thompson SM, Tang C-M. Hyperexcitability of distal dendrites in hippocampal pyramidal cells after chronic partial deafferentation. *J Neurosci* 2007;27:59–68.
76. Kellaway P. Introduction to plasticity and sensitive periods. In: Kellaway P, Noebels JL ed. *Problems and Concepts in Developmental Neurophysiology*. London: The Johns Hopkins UP, 1989:3–28.
77. Pierson MG, Swann JW. The sensitive period and optimum dosage for induction of audiogenic seizure susceptibility by kanamycin in the Wistar rat. *Hear Res* 1988;32:1–10.
78. Galvan CD, Hrachovy RA, Smith KL, Swann JW. Blockade of neuronal activity during hippocampal development produces a chronic focal epilepsy in the rat. *J Neurosci* 2000;20:2904–2916.
79. White HS, Smith-Yockman M, Srivastava A, Wilcox KS. Therapeutic assays for the identification and characterization of antiepileptic and antiepileptogenic drugs. In: Pitkanen A, Schwartzkroin PA, Moshe SL, ed. *Models of Seizures and Epilepsy*. London:Elsevier,2006:539–549.
80. Penry JK, White BG, Brackett CE.A controlled prospective study of the pharmacologic prophylaxis of posttraumatic epilepsy. *Neurology* 1979;29:600–601.
81. Young B, Rapp RP, Norton JA, Haack D, Tibbs PA, Bean JR. Failure of prophylactically administered phenytoin to prevent late posttraumatic seizures. *J Neurosurg* 1983;58:236–241.
82. Glotzner FL, Haubitz I, Miltner F, Kapp G, Pflughaupt KW. Seizure prevention using carbamazepine following severe brain injuries. *Neurochirurgia* 1983;26:66–79.
83. Temkin NR, Dikmen SS, Wilensky AJ, Keihm J, Chabal S, Winn HR. A randomized double-blind study of phenytoin for the prevention of post-traumatic seizures. *N Engl J Med* 1990;323:497–502.
84. Schierhout G, Roberts I.Prophylactic antiepileptic agents after head injury: A systematic review. *J Neurol Neurosurg Psychiatry* 1998;64:108–112.
85. McNamara JO, Yeh G, Bonhaus DW, Okazaki M, Nadler JV. NMDA receptor plasticity in the kindling model. In: Ben-Ari Y ed. *Excitatory Amino Acids and Neuronal Plasticity*, New York, NY: Plenum Press,1990:451–459.
86. Dingledine R, McBain CJ, McNamara JO. Excitatory amino acid receptors in epilepsy. *Trends Pharmacol Sci* 1990;11:334–338.
87. Sutula T. Reactive changes in epilepsy: Cell death and axon sprouting induced by kindling. *Epilepsy Res* 1991;10:62–70.

88. Meldrum BS.The role of glutamate in epilepsy and other CNS disorders. *Neurology* 1994;44:S14–23.
89. Sutula T, Koch J, Golarai G, Watanabe Y, McNamara JO. NMDA receptor dependence of kindling and mossy fiber sprouting: Evidence that the NMDA receptor regulates patterning of hippocampal circuits in the adult brain. *J Neurosci* 1996;16:7398–7406.
90. Loscher W. Pharmacology of glutamate receptor antagonists in the kindling model of epilepsy. *Prog Neurobiol* 1998;54:721–741.
91. Rice AC, DeLorenzo RJ. NMDA receptor activation during status epilepticus is required for the development of epilepsy. *Brain Res* 1998;82:240–247.
92. Sveinbjornsdottir S, Sander JWAS, Upton D, Thompson PJ, Patsalos PN, Hirt D, Emre M, Lowe D, Duncan JS. The excitatory amino acid antagonist D-CPP-ene (SDZ EAA-494) in patients with epilepsy. *Epilepsy Res* 1993;16:165–174.
93. Poulsen FR, Jahnsen H, Blaabjerg M, Zimmer J. Pilocarpine-induced seizure-like activity with increased BDNF and neuropeptide Y expression in organotypic hippocampal slice cultures. *Brain Res* 2002;950:103–118.
94. Thomas AM, Corona-Morales AA, Ferraguti F, Capogna M. Sprouting of mossy fibers and presynaptic inhibition by group II metabotropic glutamate receptors in pilocarpine-treated rat hippocampal slice cultures. *Neuroscience* 2005;131:303–320.
95. Smith BN, Dudek FE. Short- and long-term changes in CA1 network excitability after kainate treatment in rats. *J Neurophysiol* 2001;85:1–9.
96. Buckmaster PS, Zhang GF, Yamawaki R. Axon sprouting in a model of temporal lobe epilepsy creates a predominantly excitatory feedback circuit. *J Neurosci* 2002;22:6650–6658.
97. Hechler D, Nitsch R, Hendrix S. Green-fluorescent-protein-expressing mice as models for the study of axonal growth and regeneration in vitro. *Brain Res Rev* 2006;52:160–169.
98. Del Turco D, Deller T. Organotypic entorhino-hippocampal slice cultures – A tool to study the molecular and cellular regulation of axonal regeneration and collateral sprouting in vitro. *Methods Mol Biol* 2007;399:55–66.
99. Arrufo A, Stamenkovic I, Melnick M, Underhill SB, Seed B. CD44 is the principal cell surface receptor for hyaluronate. *Cell* 1990;61:1301–1313.
100. Pure E, Cuff CA.A crucial role for CD44 in inflammation. *Trends Mol Med* 2001;7: 213–221.
101. Underhill C.CD-44 – the hyaluronan receptor. *J Cell Sci* 1992;103:293–298.
102. Lin L, Chan SO. Perturbation of CD44 function affects chiasmatic routing of retinal axons in brain slice preparations of the mouse retinofungal pathway. *Eur J Neurosci* 2003;17:2299–2312.
103. Stretavan DW, Feng L, Pure E, Reichardt LF. Embryonic neurons of the developing optic chiasm express L1 and CD44 cell surface molecules with opposing effects on retinal axon outgrowth. *Neuron* 1994;12:957–975.
104. Ponta H, Sherman L, Herrlich PA. CD44: From adhesion molecules to signaling regulators. *Nature Rev Mol Cell Biol* 2003;4:33–45.
105. Perosa SR, Porcionatto MA, Cukiert A, Martins JR, Passerott CC, Amado D, Matas SLA, Nader HB, Cavalheiro EA, Leite JP, Naffah-Mazzacoratti MG. Glycosaminoglycan levels and proteoglycan expression are altered in the hippocampus of patients with mesial temporal lobe epilepsy. *Brain Res Bull* 2002;58:509–516.
106. Perosa SR, Porcionatto MA, Cukiert A, Martins JR, Amado D, Nader HB, Cavalheiro EA, Leite JP, Naffah-Mazzacoratti MG. Extracellular matrix components are altered in the hippocampus, cortex and cerebrospinal fluid of patients with mesial temporal lobe epilepsy. *Epilepsia* 2002;43(Suppl 5):159–161.
107. Borges K, McDermott DL, Dingledine R. Reciprocal changes of CD44 and GAP-43 expression in the dentate gyrus inner molecular layer after status epilepticus in mice. *Exp Neurol* 2004;188:1–10.
108. Bausch SB. Potential roles for hyaluronan and CD44 in kainic acid-induced mossy fiber sprouting in organotypic hippocampal slice cultures. *Neuroscience* 2006;143: 339–350.
109. Binder DK, Croll SD, Gall CM, Scharfman HE.BDNF and epilepsy: Too much of a good thing? *Trends Neurosci* 2001;24:47–53.
110. Kokaia M, Ernfors P, Kokaia Z, Elmer E, Jaenisch R, Lindvall O. Suppressed epileptogenesis in BDNF mutant mice. *Exp Neurol* 1995;133:215–224.
111. Qiao X, Suri C, Knusel B, Noebels JL. Absence of hippocampal mossy fiber sprouting in transgenic mice overexpressing brain-derived neurotrophic factor. *J Neurosci Res* 2001;64:268–276.

112. Shetty AK, Zaman V, Shetty GA. Hippocampal neurotrophin levels in a kainate model of temporal lobe epilepsy: a lack of correlation between brain-derived neurotrophic factor content and progression of aberrant dentate mossy fiber sprouting. *J Neurochem* 2003;87:147–159.
113. Danzer SC, Crooks KR, Lo DC, McNamra JO. Increased expression of brain-derived neurotrophic factor induces formation of basal dendrites and axonal branching in dentate granule cells in hippocampal explant cultures. *J Neurosci* 2002;22:9754–9763.
114. Koyama R, Yamada MK, Fujisawa S, Katoh-Semba R, Matsuki N, Ikegaya Y. Brain-derived neurotrophic factor induces hyperexcitable reentrant circuits in the dentate gyrus. *J. Neurosci* 2004;24:7215–7224.
115. Dinocourt C, Gallagher SE, Thompson SM. Injury-induced axonal sprouting in the hippocampus is initiated by activation of trkB receptors. *Eur J Neurosci* 2006;24:1857–1866.
116. Bender R, Heimrich B, Meyer M, Frotscher M.Hippocampal mossy fiber sprouting is not impaired in brain-derived neurotrophic factor-deficient mice. *Exp Brain Res* 1998;120:399–402.
117. Nestor MW, Mok LP, Tulapurkar ME, Thompson SM. Plasticity of neuron-glial interactions mediated by astrocytic EphARs. *J Neurosci* 2007;27:12817–12828.
118. Noraberg J, Jensen CV, Bonde C, Montero M, Nielsen JV, Jensen NA, Zimmer J. The developmental expression of fluorescent proteins in organotypic hippocampal slice cultures from transgenic mice and its use in the determination of excitotoxic neurodegeneration. *Altern Lab Anim* 2007; 35:61–70.
119. Lo DC, McAllister AK, Katz LC. Neuronal transfection in brain slices using particle-mediated gene transfer. *Neuron* 1994;13:1263–1268.
120. Benediktsson AM, Schachtele SJ, Green SH, Dailey ME. Ballistic labeling and dynamic imaging of astrocytes in organotypic hippocampal slice cultures. *J Neurosci Methods* 2005;141:41–53.
121. Yu JY, Wang TW, Vojtek AB, Parent JM, Turner DL.Use of short hairpin RNA expression vectors to study mammalian neural development. *Methods Enzymol* 2005;392:186–199.
122. Haas K, Sin WC, Javaherian A, Li Z, Cline HT. Single-cell electroporation for gene transfer *in vivo*. *Neuron* 2001;29:583–591.
123. Ehrengruber MU, Lundstrom K, Schweitzer C, Heuss C, SchlesingerS, Gähwiler BH. Recombinant Semliki Forest virus and Sindbis virus efficiently infect neurons in hippocampal slice cultures. *Proc Natl Acad Sci U S A* 1999;96:7041–7046.
124. Gober MD, Laing JM, Thompson SM, Aurelian L. The growth compromised HSV-2 mutant DeltaRR prevents kainic acid-induced apoptosis and loss of function in organotypic hippocampal cultures. *Brain Res* 2006;1119:26–39.
125. Stoppini L, Duport S, Corrèges P. A new extracellular multirecording system for electrophysiological studies: Application to hippocampal organotypic cultures. *J Neurosci Methods* 1997;72:23–33.
126. Karpiak VC, Plenz D. Preparation and maintenance of organotypic cultures for multielectrode array recordings. *Curr Protoc Neurosci* 2002;Chapter 6:Unit 6.15.
127. van Bergen A, Papanikolaou T, Schuker A, Möller A, Schlosshauer B. Long-term stimulation of mouse hippocampal slice culture on microelectrode array. *Brain Res Brain Res Protoc* 2003;11:123–133
128. Hofmann F, Bading H. Long term recordings with microelectrode arrays: Studies of transcription-dependent neuronal plasticity and axonal regeneration. *J Physiol Paris* 2006;99:125–132.

Chapter 12

Seizure Analysis and Detection In Vivo

Javier Echauz, Stephen Wong, and Brian Litt

Abstract

Understanding seizure generation, the transition from interictal to ictal states, and its underlying mechanisms requires continuous electrophysiologic monitoring. Though the duration of monitoring may vary from brief experiments over minutes to months of continuous recording, all recording of this nature requires a similar set of hardware and software tools. This chapter reviews the basic setup and requirements for successful continuous EEG monitoring in vivo and provides a set of computational "tools" that our group has found useful for continuous EEG monitoring in the laboratory. These include methods for seizure detection and basic analysis. References are provided for approaches used by groups involved in this kind of recording in animal models of epilepsy and humans. While not a complete "how to" guide for these procedures, this chapter will provide a good start to groups who are interested in performing this type of recording and data analysis.

Key words: intracranial EEG, seizure detection, signal processing, EEG analysis.

1. Introduction

Continuous EEG recording has become the gold standard for evaluating seizure generation and its underlying mechanisms in animal models. It is also the gold standard in human epilepsy, both for diagnostic purposes and during evaluation for epilepsy surgery. Unfortunately, high-quality human recordings are limited to continuous intracranial monitoring during artificial circumstances, when medications are tapered in hospital over 1–2 week periods to provoke repeated seizures. There has always been concern that this approach may not accurately represent patients' spontaneous seizures and limit proper localization of epileptic networks. Subdural and depth electrode placement in humans, which are vital to mapping epileptic

Scott C. Baraban (ed.), *Animal Models of Epilepsy: Methods and Innovations, vol. 40*

DOI 10.1007/978-1-60327-263-6_12 Springerprotocols.com

networks in all but the most straightforward cases, are limited by clinical necessity and safety concerns. This undersampling curtails our understanding of the temporal and spatial characteristics of physiologic networks that generate ictal events. Continuous electrophysiologic monitoring in animal models of epilepsy provides a powerful tool to approach these challenges. In vivo recording from "higher" animals such as rodents, cats, dogs, and primates allows sampling from systems incrementally approaching the complexity of humans, but with the freedom to record data from structures not approachable in clinical studies where safety is a primary concern. Simpler animal models of epilepsy, such as fruit flies, zebrafish, allow investigators to map more straightforward circuits in their entirety and to study the effect of specific genes on seizure generation. All of these models allow extensive spatial and temporal sampling of network activity in a "more natural" and better controlled state, where seizures have not been precipitated by pharmacologic withdrawal or chemoconvulsant administration. The requirements for prolonged neurophysiologic monitoring are similar in all of these preparations, from humans down to the simplest animal models. The basic principles of chronic recording, storage, and data analysis is similar in all of these systems and is the main subject of this chapter. The chapter below is organized into two major sections: one on the setup for chronic animal recording, analysis, and intervention and the second more technical, focusing on actual analysis algorithms, programming code, formulas, and plots of algorithm output. The latter section is meant to provide a reference and examples of algorithms our group uses to analyze in vivo data both post hoc and in real time.

2. Experimental Setup

2.1. The Mechanics

Seizure detection and analysis begins with a well-designed experimental setup for collecting data. While this may seem obvious, it is commonplace for collaborative teams to spend enormous amounts of time trying to compensate for poor experimental setup and design in post hoc data analysis. For example, our group has spent many hours working on sophisticated methods for removing movement, electrode, and scratch artifact from continuous, in vivo, ambulatory recordings from murine models of epilepsy. These difficulties most commonly arise because impedance in intracranial electrodes and, more commonly, in cables running between brain and amplifier is inordinately high (particularly when commutators are employed). Despite very creative computational methods that can be employed to eliminate artifacts, such as independent component analysis, support vector machines, and other applications, there is no post hoc method that is as effective for getting good signals and results as recording with high signal-to-noise ratio and minimizing sources of artifact. In the long run, time spent up-front, on experimental design, is much more time efficient than post hoc analysis.

Figure 12.1 details a system for continuous, in vivo, rodent modeling that our group has developed over recent years. It is specifically designed with both post hoc and online data analysis in mind. The system is also designed to test strategies for delivering interventions, such as brain stimulation, in real time during experiments. Starting at the bottom right corner of the figure, this particular setup currently houses six rats, each in separate cages. Electrodes of multiple types are implanted stereotaxically in the rodent's brain and connected to a head plug which is anchored to the skull via dental cement. Both macroelectrodes (e.g., dural contacts or bone screws) and arrays of microelectrode wires (usually 40 μm, tungsten) are connected to custom-designed, active MOSFET-based electronics that drive the impedance down in the cable between the mouse head and digital amplifiers, which sit just beyond the commutator. This system allows the animal to freely ambulate in its environment without tangling cables, which are kept safely out of reach. This "head stage" is among the most important elements of a well-designed system to

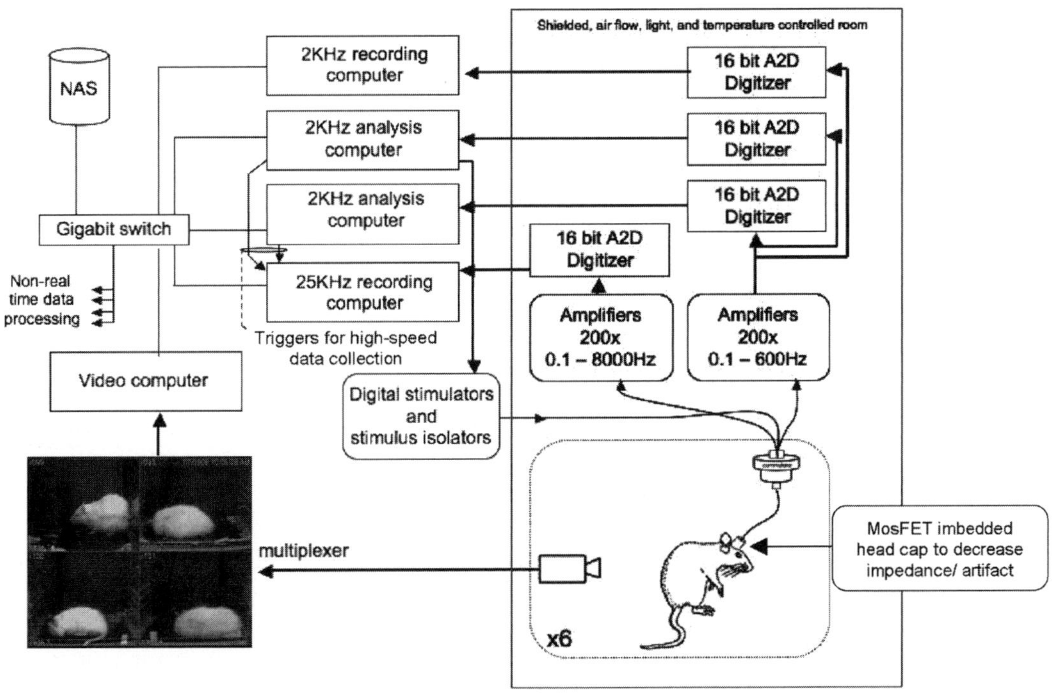

Fig. 12.1. This figure displays a typical experimental setup for continuous intracranial video-EEG monitoring, online analysis, and intervention (e.g., focal brain stimulation). Its components are explained below (courtesy of Jeff Keating, Ph.D., and Marc Dichter, M.D., Ph.D., University of Pennsylvania).

eliminate movement artifact. Of note, we have found it extremely useful to house animals in an electrically shielded, environmentally controlled room, designed not only to filter out environmental electrical noise but also to measure and control variables that might affect electrophysiological signals in unpredictable ways. Light, temperature, day–night cycles, and air flow are all monitored, measured, and controlled so as to give predictable circadian cycles and to minimize interference from the spurious signals outside of the facility, such as elevators, motors, which often plague research buildings.

Amplifiers and analog-to-digital converters are high performance and a minimum of 16 bits resolution, which allows us to record and analyze a broad dynamic range of signals, from single and multiple unit activity to broadly distributed field and evoked potentials. In the diagram above there are two sets of amplifiers, one for standard electrode recordings and the other for broadband recording from microelectrodes and tetrodes to record single units and multiunit activity, as well as much higher frequency oscillations. This aspect is optional and a function of specific details of planned experiments. Next comes analog-to-digital (A to D) converters, each dedicated to a particular electrode set. For some experiments we employ redundant A to D converters for a single experiment, particularly when one data stream is written directly to storage, while a second data stream may be used for real-time online analysis, for example, to control interventions, such as triggering responsive brain stimulation (*see* "connection to digital stimulators and isolators" on diagram). These different A to D converters are housed in separate analysis computers, each with a different purpose. Data flow are controlled by a gigabit Ethernet switch, which controls access to network-attached storage (NAS), in this case consisting of a 10-Terabyte redundant array of inexpensive disks (RAID) that is part of our local recording system, as well as access to larger, off-site backup storage, housed in a central location, accessible for analysis by our distributed research group. Also note that the analysis computers are also connected so that a software "trigger" can be used to activate high speed data collection/storage when a specific "target" event is seen in lower rate sampled data streams. For example, if the investigator chooses to trigger on a particular neurophysiologic event, such as seizure onset, spikes, or a particular frequency oscillation, then the detection software can be used to activate the high-frequency (25 kHz) analysis computer to collect data at higher rates for a predetermined period of time.

Synchronization of all computers is an important requirement of the above system, requiring that a central time clock be attached to the recording network and that each device in the system is set to periodically adjust itself to the central clock. This element is vital not only to real-time operation, particularly when interventions are to be tested, but most importantly to facilitate analyzing data off-line, after recording. Digital video is also recorded in the diagram above. Those requiring great amounts of high-resolution detail from video may choose to record a separate video channel for each cage. The system above multiplexes video data from all cages (four of six cages in the above diagram) into a single data stream that is compressed and recorded. At times our laboratory quantitatively analyzes these digital video signals to aid in seizure detection or for sleep staging/sleep–wake analysis. Commercial software is also available for these purposes which detects digital signatures of movement (e.g., oscillations, such as might be seen during convulsive activity) or long periods in which there is no change in quantitative analysis of video, such as during deep sleep.

Finally, output from detection, prediction, or other types of algorithms can be used to trigger interventions, such as electrical or optical stimulation, focal drug infusion, as "effector" units are connected to analysis computers, "closing" the sensor-therapy loop. It is important to note that the above schematic reflects a multi-purpose experimental setup that has evolved in our laboratory over years. Most laboratories doing in vivo recording may need only some of the features demonstrated above, though all systems will have basic features such as a "head stage" of amplification to reduce noise, amplifiers, A to D converters, network storage, and a central clock.

Figure 12.2 shows different portions of our actually in vivo recording setup as it exists in Marc Dichter's laboratory at the University of Pennsylvania, with five rats being recorded on commutators (**Fig. 12.2A**), multiplexed video from four of these animals on a data collection computer screen (**Fig. 12.2B**), and a bank of analysis computers in the room adjoining the animal lab (**Fig. 12.2C**). **Figure 12.3** depicts a typical long-term recording from the experimental system diagrammed and pictured above, recorded from a rat administered pilocarpine to induce a model of chronic epilepsy. In this figure the raw EEG is replaced by features extracted from the data. Signal energy is computed in hour-long sliding windows and displayed as vertical bar-plot lines on the plot. These sliding window values are "smoothed" in a continuous, running average, which is plotted on the bright green line crossing the figure horizontally. Finally, several different kinds of detections, seizures (red dots) and bursts of complex epileptiform activity (black dots), are plotted at the top of the figure. Daytime periods (12 hours) are marked in the background by white vertical bars and night (12 hours) by gray vertical bars. From these detections and extracted features patterns regarding buildup and decline of energy, epileptiform bursts, and seizure clusters are easily discerned, elements of seizure generation that are impossible to recognize by analyzing the month-long recording by eye alone.

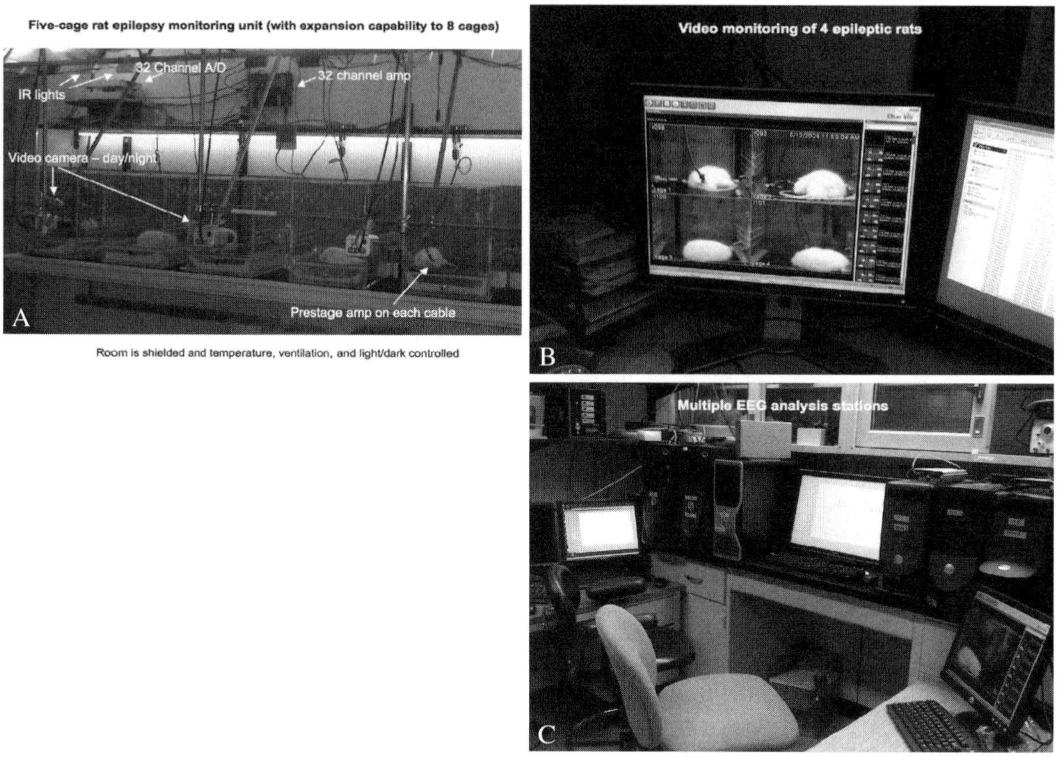

Fig. 12.2. (**A**)–(**C**) demonstrate different portions of our in vivo facility, in which animals are housed in one room (**A**), while analysis and storage computers are kept in an adjacent control room (**B**) and (**C**). (**B**) shows a dedicated screen for watching and analyzing multiplexed video output, while (**C**) shows cascaded and separate analysis stations. It is recommended that wires and computers be housed in appropriately built cabinets, leaving ample workspace for researchers and technicians to accomplish other tasks. The positioning of equipment in (**C**) shows an interim solution prior to construction of appropriate computer housing space. In systems specifically designed from the "ground up," such machines can be rack mounted in a single shielded, cooled, and dust-protected unit for ease of use and maintenance.

3. Analysis of In Vivo Recordings

A core function of this chapter is to present methods we have found useful for analyzing in vivo recordings. There is great controversy in both the animal and human seizure analysis literature about the merits of analyzing data after collection, off-line, as opposed to prospectively, in real time. In this section we explore both approaches, united by the theme of "causality," that is setting up analyses in such a way as to replicate the prospective nature of a data stream arriving as if collected in real time. More specifically, we propose that a discussion of methods for seizure detection in vivo must encompass technical analysis of both: (a) online (physically connected to the living organism), real-time (subject to

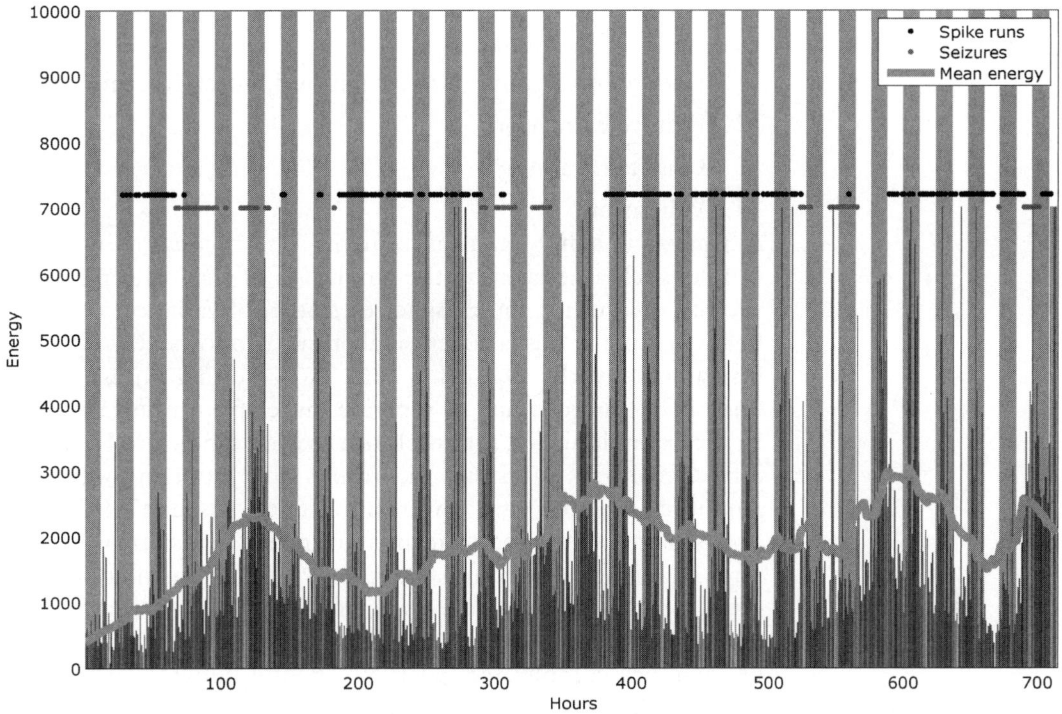

Fig. 12.3. Data recorded from the above experimental setup, with continuous multi-channel intracranial EEG monitoring from a rat with pilocarpine-induced epilepsy. Dark vertical bars denote night time and light daytime. The lines referred to as (**B**) indicate measurements of EEG signal energy in sliding windows of 1 hour, while the lighter horizontal line represents average energy over time (**D**). Discrete black dots (**A**) mark automated detection of bursts of complex epileptiform activity (spike and slow waves) and dots represented as (**C**) seizures. Detections were validated by human readers. The figure displays a cycle of activity leading up to clusters of seizures in which energy appears to build slowly, bursts of complex epileptiform activity occur, signal energy appears to increase, then clusters of seizures occur before signal energy gradually declines and the process starts again. (Courtesy of Marc Dichter, Hal Juul, and Jeff Keating, University of Pennsylvania.)

causality and the Newtonian clock) sensor readings and (b) off-line *recorded* seizure signals/markers as long as causality is enforced. Although technically the latter type of study is ex vivo and in silico (on computers), it can simulate what a passively sensing analyzer would have done in the online, real-time in vivo situation—after all, the signals being processed came from the live setting. Only the analysis is being postponed. We have applied the methods presented not only for causal analysis of digital and video EEG, but also seizure imaging modalities as well, such as recordings from voltage sensitive dyes, fMRI recordings, and evoked potential data streams. This definition necessarily excludes active sensing and interventional/therapy types of studies, because there is no perfect post hoc model to replicate exactly what a living organism would have done once interfered with.

3.1. General Detection Frameworks

A typical framework for quantitative seizure detection from in vivo recordings is the following:

1. Data preprocessing (artifact removal, notch filtering, etc.)
2. Feature extraction
3. Classifier training
4. Validation

A *feature* is simply a measurement. The measurement can be simple, such as measuring the amplitude of a signal, or complex, such as measuring some aspect of statistical variation or chaotic behavior of the signal over small time windows. *Classifiers* are fundamentally the algorithm and the resulting threshold(s) it finds in feature outputs to separate target from non-target data. In the example above, the amplitude feature can be used to process intracranial EEG (IEEG) recordings that contain both high-amplitude seizures and lower amplitude "baseline" EEG data. A classifier is then used to form a boundary that statistically separates seizure epochs from baseline activity. The classifier boundary can be simple logistic regression, for example, to quantitatively define a statistical threshold that separates the two states, or it can be far more complicated, such as a neural network or machine-learning algorithm. Typically, most categorizations of interest are not single-dimensional entities when analyzing EEG but require multiple features to describe or separate them. For example, seizures might present with higher or lower amplitude than background activity, but they are also characterized by specific oscillatory frequency ranges, "spikiness," patterns of spread, etc. In our experience, multiple features extracted from a signal and then "fused" together through a classifier do much better in detecting and defining events of interest, with respect to seizure analysis, than do single features. To state this more formally, in multiple dimensions classifier boundaries correspond to the multi-dimensional realization of thresholds on a single feature and generally exhibit improved performance, provided that data sufficiency conditions are met.

Classifiers can also be grouped according to the types of boundaries that can be created. They range from linear boundaries (regression, perceptrons) to convex boundaries (nearest centroid) to nonlinear, non-convex boundaries (K-nearest neighbors, support vector machines, neural networks). In general, if the features employed in an analytic strategy inadequately separate data into a single continuous region in feature space, classifiers with more flexibility to manipulate boundaries will be necessary for better performance. In the single-dimensional sense, this corresponds to finding the boundaries that separate seizures from other activity that changes frequency, amplitude, or spatial distribution on the EEG, such as movement artifact, scratching behavior, or chewing.

In the case of seizure detection from the electroencephalogram, feature measurements are typically extracted from blocks of single-channel EEG. These features, such as the line-length feature, are designed to efficiently separate seizure from non-seizure data. By extracting multiple features from these EEG blocks, clouds of data will form in the multi-dimensional measurement space formed by these features. Different regions of this feature space will generally contain unique waveform morphologies due to the measurement constraints in these regions. At the single-channel level, therefore, simple linear or convex classifiers will tend to group the EEG by morphology (**Fig. 12.4**). This correspondence is potentially lost with non-convex classifiers, or with the addition of multiple channels of data, with features from each channel in its own dimension. Techniques that deal with this "clustering" of like segments of EEG can be extremely useful for separating out different states of behavior, both normal (such as stages of sleep) from abnormal (e.g., seizures).

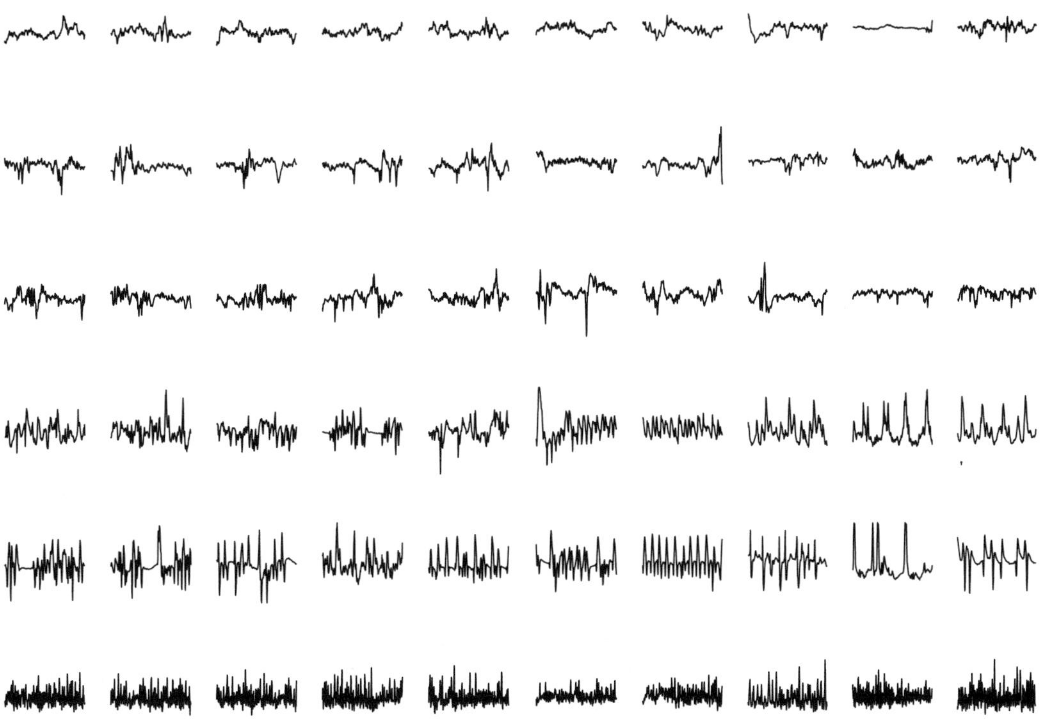

Fig. 12.4. Regions of a multi-dimensional feature space group epochs of single-channel EEG by morphology. Each waveform is a 1 second epoch of IEEG data recorded from epileptic rat hippocampus. Each row corresponds to one of six data-dense centers in a multi-dimensional feature space found by a clustering algorithm. The leftmost column corresponds to the EEG data that produced the feature point nearest to the centroid center. The next four columns are data epochs whose points in feature space that are also nearby the centroids. The final five columns are data epochs randomly chosen from the EEG and classified according to the nearest centroid. Note that excellent morphological agreement can be achieved in this manner.

In a typical supervised classification scheme, labeled seizure and non-seizure feature data marked by experts are given to a classifier for training. A portion of these data, the "hold-out data," are kept hidden from the classifier during training. This is because testing generalization of a classifier's performance on unseen data is paramount and must be optimized with computational feature selection protocols and techniques such as cross-validation. Once classifier boundaries are defined during training using a particular algorithm, the hold-out data set is used to assess the trained classifier's performance in a process called validation, before investigative experiments can move forward. This process, extracting features from data, marking a gold-standard data set, training the classifier, and validating performance with hold-out data are steps we repeat over and over in our laboratory whenever we develop a new technique or apply a technique from our standard repertoire to a new data type. These procedures are a mainstay of EEG analysis and event detection. Finally, a global statistical analysis of results may be required. *It is absolutely essential that careful projections of data requirements and statistical power for experiments be conducted prior to data collection, to be sure that results have real meaning.* Adequate statistical expertise must be on hand for any of the EEG analyses described below. This step is all too frequently overlooked when doing complex in vivo recording experiments.

The performance of a seizure detection scheme is dependent upon many variables, such as artifact, sampling appropriate regions participating in the epileptic network. One clear factor affecting the performance of a classifier is the training data it has received, as a classifier should ideally be trained on a representative sample of all types of data it will encounter. Closely related to this principle is the issue of whether seizure detection should be performed in a patient (or animal)-specific manner (i.e., using training data derived from the specific subject on whom subsequent seizure detection will be performed) or nonspecific manner (i.e., using training data derived from a group of subjects for universal application to others). Even more specifically, it is important to note that the majority of published subject-specific seizure detection studies use seizures from the same complement of electrodes in the same subject. These studies are better described as both patient *and* electrode specific. In general, such protocols enjoy a significant boost in sensitivity and specificity over patient-independent detection schemes *(1–4)*.

3.2. Considerations in In Vivo Recordings

There are two important considerations when performing seizure detection in a real-time, laboratory, or patient setting: (1) computational efficiency and (2) adaptation. As noted above, many algorithms require training prior to implementation. If the amount of data available for training is sufficient such that "the curse of dimensionality" is not an issue (meaning that adding more features does not increase the need for data beyond what is practically possible), additional features for multi-dimensional analysis generally add to both detection sensitivity and specificity. This is provided that what these features measure is not completely covariant with existing features (e.g., additional features must contribute new information). In this setting, computational efficiency becomes limiting. A simple strategy to address this problem is to develop a hierarchical detection framework, whereby candidate seizures are screened by a series of increasingly computationally complex classifiers. The lower level classifiers would ideally have a limited number of efficient features (such as the line-length feature) and would have relatively high-sensitivity, low-specificity characteristics. The higher level classifiers would ideally have the opposite profile, but are engaged only when necessary.

Adaptation is an issue that emerges in all long-term recording scenarios. Signal properties gradually drift as a result of factors not easily controlled in experiments. For example, short-term features change in the day–night cycle, as well as longer periodicities such as menstrual and circaseptan cycles, may influence the baseline EEG as well as seizure characteristics and timing, depending upon which portion of the cycle the subject is in. Electrode gliosis is another potentially confounding factor that may decrease the amplitude of the EEG over time or change quantitative characteristics of extracted features. Continuous in vivo recordings in animal models can span months, enough time to take into account changes in electrical activity due to senescence. In addition, any pharmacologic, electrical, or other interventions may alter the underlying epileptic process, resulting in a change in seizure morphology, timing, and quantitative characteristics. When taken over a long period of time, these long-term nonstationarities may overwhelm algorithm training and shift the detection boundary between seizure and non-seizure in a multi-dimensional feature space. This has the effect of worsening the performance of any static classifier that has been trained to recognize seizures obtained early on during the recording. Though largely undeveloped, autonormalization/change/outlier detection frameworks can potentially address this problem *(5)*. With solid clinical heuristics that describe the typical time course and magnitude of changes due to seizures, it is possible to attain detection performance rivaling classifiers that have been trained by more traditional supervised approaches.

Our anecdotal experience in animal recordings suggests that the seizure signal-to-noise ratio is lower than that of comparable IEEG recordings from human mesial temporal lobe epilepsy (MTLE), though this may greatly vary depending upon recording technique. We have therefore found it absolutely necessary to implement subject-specific seizure detection to achieve our performance requirements for automated EEG analysis. Coupled with the heterogeneity that exists in seizure expression and in electrode placement relative to seizure generators, we have found it challenging to develop automated methods to detect the first seizure for the purpose of collecting a training set. Though we often employ highly sensitive detectors to do data reduction in such situations, such as in long-term epileptogenesis studies, we still find, at present, that some degree of human EEG review is necessary at this time, outside of marking gold-standard events. As above, an outlier detection framework has the potential to recognize the first few seizures, and we have had considerable success with these methods. As the collection of seizures grow, multiple classifiers with different sensitivity/specificity characteristics can be created from the batch of training data and fashioned into a supervised, computationally efficient hierarchical detection framework.

4. Practical Methods: An Overview of Seizure Detection

Long-term seizure detection has progressed from electroencephalographer's 100% visuomanual review and audio transformation to nearly fully automated annotation by computer algorithms, though with variable reliability. The advantages of automated detection are clear: real-time monitoring of patients' seizures, automatic accumulation of diagnostic information, reduction in fatigue, and requirements for human labor. More recently, automated seizure detection has also become an important component of therapeutic neurostimulation devices for epilepsy *(6–10)*.

Detection methods have been categorized into generic vs. patient specific and furthermore into at least three classes of pattern recognition: (1) anthropomimetic/syntactic (mimics human performance using a particular grammar/or rulebase), (2) statistical (classical discriminant analysis), and (3) artificial intelligence (machine learning, neural networks, etc.). All of these methods need to extract quantitative features from EEG. Their differences reside mostly in what features are picked and how they are mapped into detections/decisions. Jean Gotman's mimetic methods in the 1970 s are widely considered seminal in the field. They were originally conceived for spike detection, but evolved to be applicable to seizure detection by quantifying changes in rhythmicity over consecutive half-waves. By anecdotal accounts, the early system could achieve 90% sensitivity, 2.5 false positives per hour, and 24-second detection delays over scalp EEG. In Gotman and Gloor *(11)*, mimetic rules involving thresholds on half-wave amplitude, duration, and sharpness proved to adequately detect a variety of spikes and epileptiform activity. The extension of this method to seizure detection was demonstrated in Gotman *(12)*.

The seizure detection algorithm in Osorio et al. *(13)*, followed up by Osorio et al. *(14)*, demonstrated that a single amplitude-based feature (ratio of short- to long-term median filtered EEG), thresholded with respect to size and duration, can reliably detect seizures and follow their spread across channels in both generic and patient-specific terms. A commercial software package, REVEAL, was recently published and validated *(15)*. The algorithm employs Matching Pursuit to decompose sliding 2-second epochs of scalp EEG into their first two Gabor atoms. Aspects of the atoms' half-waves, such as slope, amplitude, and duration, are subjected to six anthropomimetic rules with emphasis on rhythmicity. Seizure is "perceived" when rhythmic activity has long duration and represents a high percentage of total power in the epoch. A clustering method is used to join a stream of epochs into either "background" or "seizure".

A sampling of the significant studies highlighted in this chapter and a few other methods is summarized in **Table 12.1**. FNR stands for false negative rate (100% minus sensitivity), FPh is the

Table 12.1
Seizure detection performance studies

Detection system	FNR (%)	FPh	Avg. delay	# Features	# Patients	# Szs tested	# Hours tested
Gotman 1982–1990 *(12, 16)*	24	1	10 s	3	49	244	5303
Murro et al. 1991 *(17)*	9	1.5	—	6 (3 2ch)	8	43	—
Olsen et al. 1994 *(28)*	0	0	—	9	2	9	7.5
Qu and Gotman 1997 *(4)*	0	0.02	9.35 s	6	12	47	32
Osorio et al. 1998 *(13)*	0	0	2.1 s	1	16	125	55.5
Gabor et al. 1996 *(18)*	10	0.71	—	6	22	62	528
Gabor 1998 *(19)*	7.2	1.35	—	6	65	181	4554
Echauz et al. 2001 *(31)*	1.5	0.01	5.01 s	3	11	126	1265
Wilson et al. 2004 *(15)*	24	0.11*	—	~4	426	670	1049*
Grewal and Gotman 2005 *(20)*	10.6	0.22	17.1 s	3 × 7bands	19	152	407

*This study measured FPh on 465 additional hours, however, in non-epileptics, so FPh in the target population of this seizure detector remains unknown.

reported number of false positives per hour (note this figure can be misleading, e.g., 1 FP/hour could be a failed system that emits only 1.1 detections per hour), average delay is the latency between expertly marked electrographic onset of seizures on the EEG and the automatic detection (this figure can be 10 seconds or more for scalp than for intracranial EEG), # features is the effective number of scalar quantities extracted from EEG required as input into the detection system, # patients, # seizures, and # hours tested relate to the size of the studies. A dash means not reported. Some of the larger studies serve to validate half-wave, line length, and amplitude-based algorithms for investigational device exemption and regulatory approval of implantable seizure detectors by the US Food and Drug Administration (FDA).

4.1. Need for Real-Time Seizure Detection in Therapeutic and/or Diagnostic Devices

Through implantable devices, automated seizure detection promises to extend patient care from the Emergency Medical Unit and Intensive Care Unit to the long-term out-of-hospital environment. In the basic laboratory, these techniques hold promise to help elucidate mechanisms underlying seizure generation, epileptogenesis, and the spatial and temporal characteristics of epileptic networks. In the case of responsive neurostimulators, seizure detection is a prerequisite function for triggering therapy. The crucial difference between chronic deep brain (i.e., open loop) stimulation and responsive stimulation is that the latter delivers therapy only when presumably needed: during the onset of epileptiform activity (or some other pattern particular to a patient), thus minimizing side effects and resource utilization (e.g., battery life or amount of drug delivered). In seizure "prediction algorithms," ones focused on identifying periods of time with a high probability of seizure onset, seizure detection is important backup, triggering a response in the event that the seizure prediction algorithm fails to alert a subject or deliver effective preventative therapy. Even in the absence of delivering preventative therapy, there is evidence that even seizure warning systems based upon "prediction" might be useful. Block and Fisher report that 50% of TLE patients preserve volitional motor control at the start of seizure, thus in the approximately 10 (range ~2–20) second delay from unequivocal electrographic onset of seizures on EEG to clinical onset, these patients could still remove themselves away from danger given an automated seizure onset indication *(21)*.

In the clinical realm outside of therapeutics, seizure detection could be extremely useful, even if only to report events to persons other than the affected individual. Misinterpretation of seizure-related behavior is common, sometimes with catastrophic consequences. Reports of police brutality, e.g., beatings of seizing individuals because of a lack of understanding of the condition, are not uncommon. Similarly, seizing individuals may be denied help though they are surrounded by people who could render aid, because of fear or lack or knowledge of what constitutes a seizure and how to treat it. A seizure detection device could be of great utility to people who live and function openly in the community for these reasons. Caregivers would want to know that a seizure is in progress, so pager/e-mail alerts as in ICU EEG monitors would be valuable. Automated seizure detection is also crucial during long-term use of an implanted device in order to record the times of the day, durations, locations, and the actual multi-channel EEG epochs during seizures in order to reliably document clinical endpoints through time—a sort of "machine diary." Finally, through a change of software parameters, a seizure detector can be tuned as a hypersensitive and hypospecific predictor, which seems to normalize the background EEG of some patients undergoing neurostimulation responsive to such events, or similarly as a detector of subclinical seizure-like bursts (brief, spatially confined seizures), false starts, chirps, high-frequency epileptiform oscillations, etc., which seem to organize in patient-specific temporal patterns and more generally may be indicative of "proictal" periods when seizure susceptibility/likelihood is higher. In the animal laboratory too, algorithms focused on detecting seizures or abnormal electrophysiology/oscillations associated with epilepsy are extremely useful in elucidating basic mechanisms involving epileptic networks, as well as in developing new algorithms and interventions for translation into humans.

5. Feature Extraction and Pattern Classification

As noted previously, the inputs to decision support systems are features—quantitative or qualitative measures that distill raw data into relevant information. Features can be univariate or multivariate (e.g., single vs. multiple channels of data used as input), scalar-valued vs. vector-valued, single-natured vs. multi-natured (feature coordinate 1 = FFT amplitude in frequency bin 1, feature coordinate 2 = FFT amplitude in frequency bin 2, etc., vs. feature coordinate 1 = energy, feature coordinate 2 = entropy, etc.). For our purposes, a feature is any transformation from a set of numbers to a scalar $\phi:\{R \times R \times \cdots\} \rightarrow R$, for example, from a big matrix containing a page of digitized EEG to one number representing the maximum absolute amplitude in that page. A *feature vector*, element of R^m, can be formed from m scalars regardless of where they came from.

Also, as noted above, in general, the calculation of each feature vector coordinate at a given instant of time requires observation of a "window" of the raw brain signals, opened to a certain extent spatially (multiple channels) and temporally (interval from a point

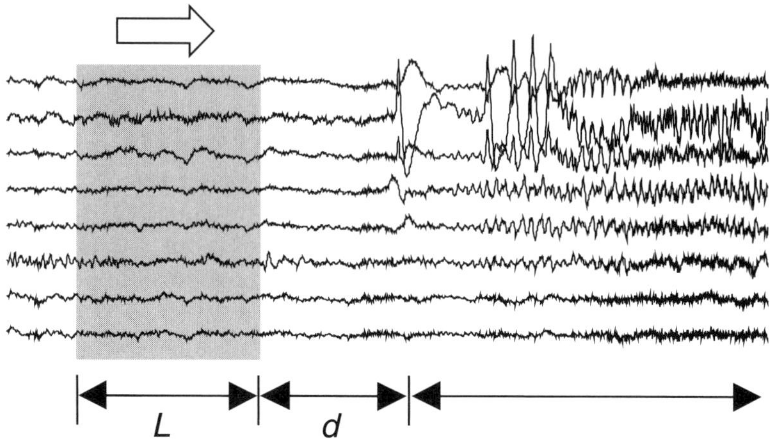

Fig. 12.5. Sliding length-L window for feature extraction. The next frame will advance d samples.

in the past to the present time). For certain features computed recursively (e.g., an IIR filter output), this window is in effect semi-infinitely long but exponentially forgetful of the past. The observation window "slides" across the time axis, typically with a fixed amount of advancement that creates a certain overlap between adjacent windows. Conversion of each data window into a feature vector thus creates a feature time series with built-in compression (going from a big data matrix to a small vector each time, but also skipping a number of time samples in between).

The shaded block in **Fig. 12.5** depicts a sliding *observation window* containing raw multi-channel IEEG. Practical constraints put a limit on the extent and number of channels that can be displayed in this window. For seizure detection, the observation window should have enough temporal resolution to at least capture the seizure onset within the resolution needed for a particular task. For seizure prediction, the period of time from the right edge of this window to the last instant when seizure is still pharmacologically or electrically preventable defines a prediction window or horizon. In addition to the observation and prediction windows, we define a history window as a superset of observation windows. We distinguish multiple levels of features. Instantaneous features are computed from observation windows that are essentially 1.25 seconds or less in duration, in most seizure detection/prediction applications on typical clinical data. Different-sized windows might be used for longer horizon events (e.g., evoked potentials) or shorter scale events (e.g., "microseizures" or unit activity in microelectrodes). Historical features span longer periods and are based on the time *evolution* of instantaneous features. From a geometrical point of view, historical features are features of the trajectories of instantaneous features in feature space—they are "features of the features." This process of nested feature compositions can be repeated ad libitum as long as there is enough buffered data to feed the feature extractors. It is important to note here that some of these exercise, for example, using "features of features," after a time diverge from measures easily pictured in practical applications to more abstract, mathematical constructs that are theoretically based and less easily visualized.

Some of the instantaneous features we have found useful for seizure detection (computer coded and placed in a "feature library" by our research group) include autoregressive coefficients, coherence, cross-covariance, correlation between entropies, energy, energy derivative, entropy, filtered amplitude squared, fractal dimension, fourth power indicator, mean frequency, nonlinear decorrelation lag, nonlinear energy operator, number of zero crossings, Pisarenko harmonic decomposition, power distribution in frequency bands, principal components, principal Lyapunov exponent, real cepstrum, spike (occurrence, amplitude, curvature), third-order spectrum, wavelet subband energy, wavelet compression coefficients, and zero crossings of energy derivatives. In addition, raw EEG signals themselves should be considered as instantaneous features in a special case of zero compression of information.

After features are extracted, one or more stages of processing result in classification output, as noted above. In more detail, the job of a classifier is to categorize its input as belonging to 1 of K classes. Classification is a special case of regression in which the synthesized function is discontinuous and discrete-valued $\hat{f}: \mathrm{R}^n \rightarrow \{\mathrm{C}_1, \mathrm{C}_2, \ldots, \mathrm{C}_K\}$. Detection is a special case of classification in which there are only $K=2$ output classes, seizure or not seizure. The output labels of the classifier can be encoded as one of the integers in $\{1, \ldots, K\}$ or can be spread into a decoder-style, 1-of-K binary vector $[0\ 0\ 1\ 0 \ldots 0]$, where the 1 occurs at the ith position corresponding to the ith class. Two of the K classes can optionally be (1) an outlier class—definitely not one of the recognized classes and (2) a doubt class—where the classifier cannot make a judgment as to which of the classes the event belongs to, so no classification is issued *(22)*. Implementation approaches include k-nearest neighbors (kNN), maximum selector of discriminant functions, decision trees, neural networks, fuzzy production rules, Holland's classifier systems, statistical pattern recognition, syntactic pattern recognition, and support vector machines. These classifiers are mentioned for reference, but a detailed explanation of them and their application is beyond the scope of this chapter.

Classifiers can be created from data using any of the general classes of learning methods: supervised learning, meaning that the classifier is trained, when the different classes of data C_i are known a priori, unsupervised learning, in which the classifier uses intelligent method to train itself, when the number of data classes are not known a priori. In the latter, algorithms often start by clustering input data into groups with similar characteristics, for example, low-amplitude EEG, and then applying the clusters in some decision rule. In reinforcement learning only a high-level "success" vs. "failure" or a delayed performance metric is available. More recently our group and others have worked with classifiers that form a hybrid between unsupervised and supervised learning, one-class novelty detection, where training occurs using only one labeled class of data *(5)*. The actual algorithms used to tune the classifier parameters include genetic algorithms, linear and nonlinear regression using Newton's methods, expectation maximization, and quadratic programming.

Some of the classification methods specifically optimize criteria such as Bayes error (minimum expected probability of error), maximum a posteriori (MAP) probability of data belonging to a particular class given the feature vector, MAP under conditional independence (naive Bayes), maximum likelihood, maximum entropy, minimum error risk, maximum margin. When applied to seizure detection, all of these optimality criteria correspond to different locations on the receiver operating characteristic (ROC) curve, which summarizes the trade-off between the sensitivity and specificity of a detector. The performance of detectors is often framed in terms of mean detection latency, confusion matrix, sensitivity (recall), specificity, misclassification rate, positive predictive value (precision), negative predictive value, number of false-positive detections per hour, extremal margin, and others.

During training and measuring performance it is crucial to understand that if a model is too complex, it can memorize peculiarities of data sparsity or irrelevant details about noise. If the model is too simple, it cannot even approximate noise-free training data. Thus, the right balance of model complexity and straightforwardness is needed to achieve the best possible generalization performance. Several approaches exist to attempt to minimize the expected prediction error over future input data as opposed to the overoptimistic (on average) apparent error over the training set. These include split sample testing (dividing data into training vs. hold-out test groups); group method of data handling (GMDH); cross-validation to estimate nearly unbiased (though potentially highly variable in small samples) prediction error *(23)*; jackknife, bootstrap to estimate bias, variance, confidence intervals, *p*-values; Vapnik–Chervonenkis theory leading to support vector machines (SVMs) with a built-in tight upper bound on future misclassification rate *(24)*; and regularization formulas that approximate bias correction from a single data sample, including Akaike's final prediction error (FPE), an information criterion (AIC), Rissanen's minimum description length (MDL), Barron's predicted square error (PSE), Moody's generalized prediction error (GPE), and others.

The latter regularization methods facilitate model selection (e.g., "size" determination) with a variable structure (e.g., growing number of parameters) by adding a complexity term to the apparent error rate. This is similar to ridge regression in statistics, weight decay in artificial neural networks, and the fact that most weights are set to zero during quadratic programming in SVMs. Another method to address generalization with a variable size structure is pruning, where instead of growing, a model starts with large structure and then weights or connections are eliminated gradually over time. An example is optimal brain damage algorithms (OBD) in neural networks *(25)*. Other methods address generalization with a fixed structure, including early stopping where training occurs until performance on a "validation" or checking data set starts degrading, and jittering, where noise is artificially injected to the input data.

6. Epileptiform EEG Analysis Tools

In the sections that follow, several specific tools and models useful for application to detecting seizures or oscillations of interest in the epileptiform EEG are presented. Their descriptions below, and following examples, are of a more technical nature than the text above and are aimed at an audience who might use these sections as a reference and actually implement these algorithms/models for event detection.

6.1. Chirps as Models of Epileptiform Events

Chirps are important spectrographic signatures of seizure onsets *(26)* and can serve as models of epileptiform events to be detected. A general chirp is a function $x(t) = A(t)\cos[\varphi(t)]$ or its complex-valued version $A(t)\exp[j\varphi(t)]$. The *local* amplitude, phase, and frequency are $A(t)$, $\varphi(t)$, and $d\varphi(t)/dt$, respectively. The *instantaneous* amplitude, phase, and frequency are $|z(t)|$, $\angle z(t)$, $d\angle z(t)/dt$, respectively, where $z(t)$ is the analytic signal of $x(t)$

$$z(t) = x(t) + jH\{x\}(t),$$

and $H\{x\}(t)$ is the Hilbert transform (HT; also a function of time)

$$H\{x\}(t) = x(t)^* 1/(\pi t),$$

where $\star$ denotes convolution, or

$$\mathrm{real}(\mathrm{IFFT}\{-j^* \mathrm{sgn}([1 : \mathrm{N}/2 \; -\mathrm{N}/2 : -1])^* \, \mathrm{FFT(x)}\})$$

for the power-of-2 length discrete-time version. When $A(t)$ varies slowly compared to the oscillations $\cos[\varphi(t)]$, the local and instantaneous counterparts are approximately the same. For example, one type of AM demodulator is $A(t) \approx \sqrt{x^2(t) + H^2\{x\}(t)} = |z(t)|$. In the case of linear frequency modulation, $\phi(t) = 2\pi f_o t + \pi\alpha t^2$ (instantaneous $= 2\pi f_o + \alpha t$), another parameter of interest is the chirp rate α. Here, we are interested in the more general changes in frequency $\varphi'(t)$ (rad/s, or $\varphi'(t)/2\pi$ in Hz).

Methods for detecting chirps based on cross-ambiguity function, Radon, chirplet, Hough continuous-time wavelet, and other transforms involve computation of multiple inner products (correlations) for which no fast algorithms are known. Faster methods that can track the time-varying frequency component of a chirp include spectrogram peak (FFT), Hilbert transform, wavelet packets, and line length.

6.2. Line Length

A computationally efficient signal feature that is sensitive to the chirp-like signatures usually seen at seizure onset is line length. Kircher and Raskin reported that "line length," the sum of absolute differences between adjacent sample points,

$$LL[n] = \sum_{n'=n-L+1}^{n} |x[n'] - x[n'-1]|,$$

where L is the length of the sliding window and n is the current time index, is a good measure of respiration suppression for use in computerized lie detection *(27)*. In image processing, this feature is known as 1-D total variation, or L_1 norm of a gradient. A group at The Johns Hopkins University disclosed line length as one of the features for automatic seizure detection *(28)*. The JHU Applied Physics Lab group also evaluated about 10,000 features for different signal analysis applications and line length made it to their PolyScore lie detection system (which is identical to their EEG patent except that the input into this system is electrodermal). A variant of line length was implemented as Euclidean distances in Katz fractal dimension and can be seen upsampled in Higuchi's method *(29)*. Line length (LL) can be used in a universal seizure detector algorithm suitable for implantable medical devices *(30–31)* and is currently implemented in mixed-logic hardware and software in NeuroPace's Responsive Neurostimulator System RNS-300®.

Line length linearly tracks both amplitude and frequency, as in simultaneous AM and FM demodulation. Consider a sinusoid $x(t) = A\cos[2\pi ft]$, with frequency of oscillation f slow enough for the digitally sampled waveform to still resemble the sinusoid (up to $\sim F_s/8$, or $4\times$ oversampling). The LL of this sinusoid over an observation window of length L is approximately the LL of one half-wave (sum of vertical segments from one extremum to the next $= 2A$) times the number of half-waves that fit in the length-L window. Equivalently, that is $2A$ times *twice* the number of full-waves that fit in the length-L window (f such waves fit in F_s samples), so $LL \approx (2A)(2)(f\,L/F_s) = (4L/F_s)Af$. Thus, the LL of an oversampled sinusoid is proportional to its amplitude and frequency as claimed (and obeys homogeneity but not additivity). From this, the estimate of frequency in Hz is

$$f_{\mathrm{LL}}[n] = \frac{F_s LL}{4AL}.$$

Note direct dependence on A, so for this estimate to work on a chirp, the amplitude has to vary very slowly compared to the oscillation.

6.3. Application Examples

In the following section we illustrate some of the techniques discussed thus far by applying audio transformation, line length, classical spectrogram, wavelet packet spectrum, joint sign periodogram event characterization transform (JSPECT) "inverted Vs" *(32)*, and filtering around an 8 Hz center frequency, to a 6 minute and 46 seconds long EEG data segment containing five annotated "beta buzz" chirps.

Figure 12.6 shows 10 seconds of the 6^+-minute segment at the start of one of the EEG chirps. **Figure 12.7** shows channel RT3 overlayed and stretched against the chirp tickmarks such that they correspond in time. When the signal is played back at 10 KHz, it falls well within the human hearing range. Since the data were originally sampled at 200 Hz, it only takes a few seconds to hear

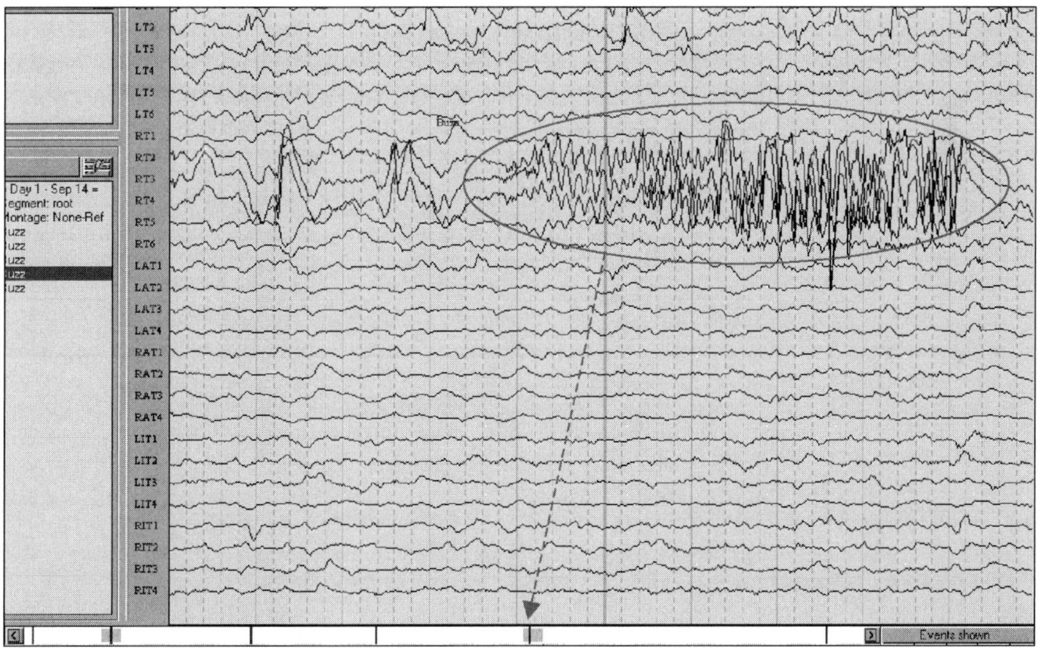

Fig. 12.6. Ten-second page displaying one of five EEG chirps in the 6.77 minute data segment. The leftmost tickmark at the bottom indicates beginning of the file. The next five tickmarks indicate the time location of the chirps relative to the data segment.

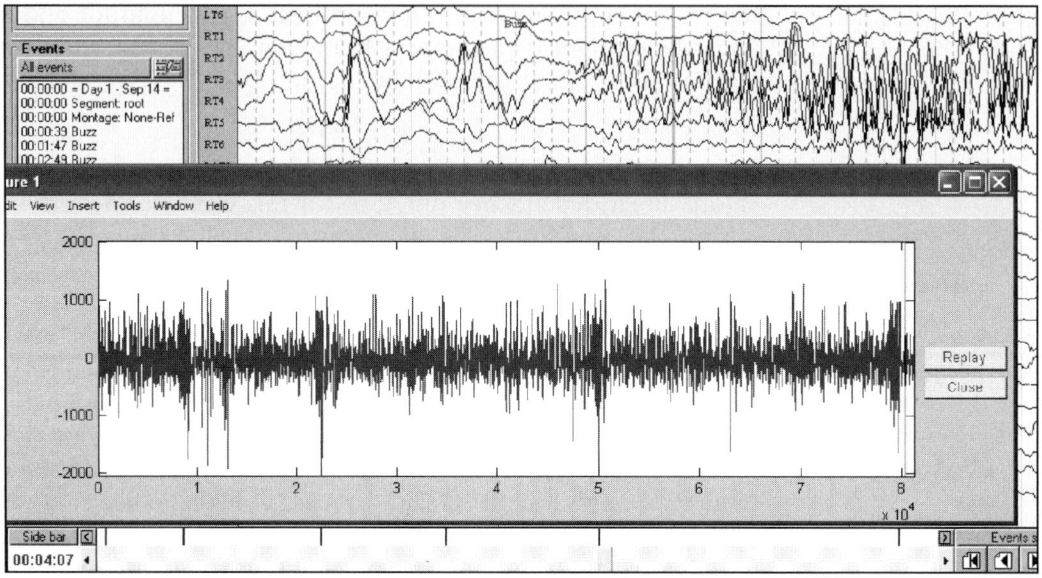

Fig. 12.7. Channel RT3 is overlayed and stretched against the chirp tickmarks such that they correspond in time in preparation for audio transformation.

the entire 6^+-minute segment. The signal sounds like the crackle and pop of an old vinyl record, with five small scratches at the time of the markings. There is not enough time to hear frequency "evolution" in these bursts.

Note that an extreme downward excursion of the signal in **Fig. 12.7** can identify four of the five chirps, but is an equivocal feature producing two false hits and a miss. A better targeted feature is line length, as shown in **Fig. 12.8**. Clearly, this universal feature is already capable of highlighting the magnitude and location of the five chirps. Because implementation is a

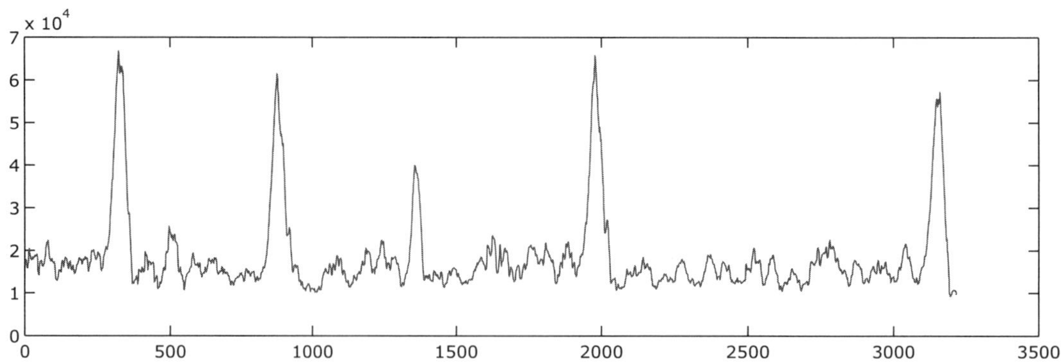

Fig. 12.8. Sliding 4-second line length of the RT3 signal updated every 0.125 seconds. Time axis in number of line-length samples.

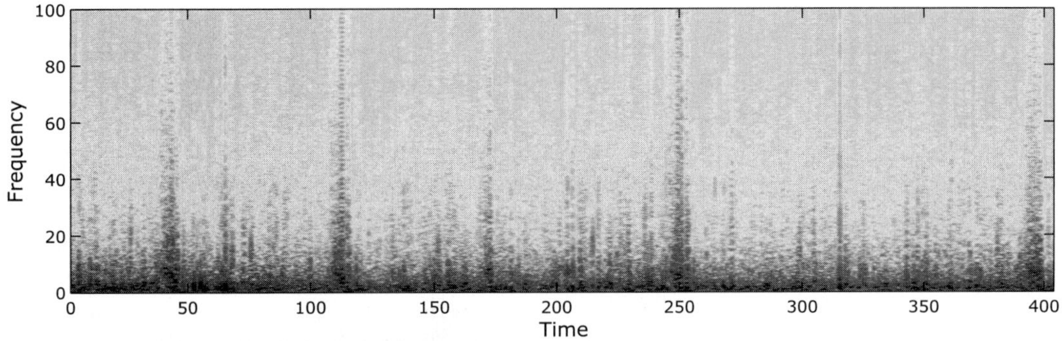

Fig. 12.9. Classical spectrogram using a 1024-point FFT for each slide of a 4-second window, displaced every 1/8 of a second. This figure shows the classical spectrogram (sliding log|*FFT*|) of the data segment using an equivalent 4-second window and 0.125-second displacement parameters. The five thickest columns above correspond to the chirps marked as noteworthy. Compared to line length, this spectrogram takes seven times longer to compute.

simulation of real time with causality enforced, a 2-second minimum delay is expected between onset of a chirp and line length rising above background.

Figure 12.9 shows the classical spectrogram (sliding log|*FFT*|) of the data segment using an equivalent 4-second window and 0.125-second displacement parameters. The five thickest columns above correspond to the chirps marked as noteworthy. Compared to line length, this spectrogram takes seven times longer to compute.

A similar "spectrogram" can be obtained in one shot instead of gradually as a window slides by computing the wavelet packet (WP) spectrum of the data segment. **Figure 12.10** shows the result of using a level-5 WP tree with discrete Meyer wavelet (FIR filter length 62). The image matrix is formed by stacking the rows of WP coefficients in natural frequency ordering, each row representing one of 32 nodes, and keeping only the central $\lceil 81{,}200/32 \rceil = 2{,}532$ columns most closely corresponding to the original time axis. The five broadband chirps are still visible. Of

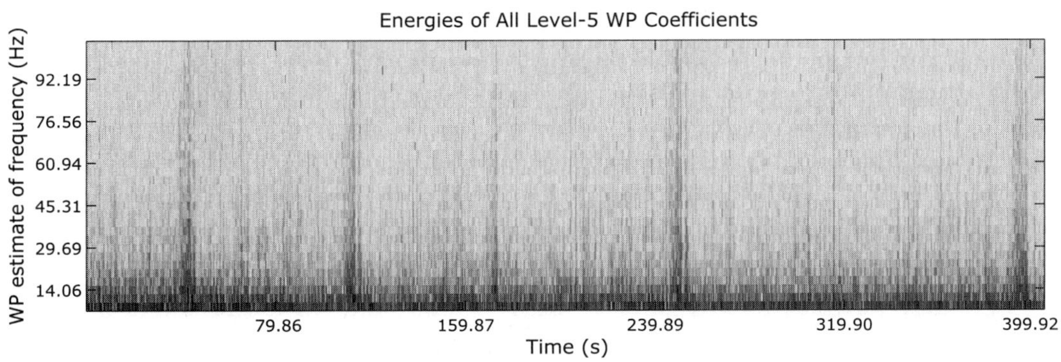

Fig. 12.10. One-shot wavelet packet spectrum of the entire 6.77-minute data segment using level 5 tree with discrete Meyer wavelet.

note, this image is computed five times faster than the spectrogram shown previously, but there is a causality trade-off since the whole signal needs to be buffered before one can see the image.

Figure 12.11 shows yet another transform, joint sign periodogram event characterization transform (JSPECT) *(32)*, used to detect chirps on the IEEG. This transform is basically obtained as the spectrogram of hardlimited pre-whitened EEG of the data segment using the same parameters. This makes almost everything uniform in the image due to the noise immunity of the signum function, except for some frequency variation. For EEG chirps, this typically looks like faint "inverted Vs," for chirps that commonly modulate up and down in frequency, a common characteristic of seizure-like events. In this case, JSPECT shows an inverted V around time 200 seconds between the third and fourth chirps, which was not evident before.

Figure 12.12 shows the effect on line length of filtering the EEG channel around a center frequency of 8 Hz ± 2 Hz prior to feature extraction. In agreement with JSPECT, the epileptiform formation in the middle of the record is now highlighted too.

In the following section we illustrate a seizure prediction technique by applying audio transformation, line length, and prediction with trending and feature composition (features of features), to a 1 hour 21 minutes and 23 seconds (1.35 hours) long EEG data segment containing 18 chirps (including the 5 analyzed before) leading to seizure. We compose the moving

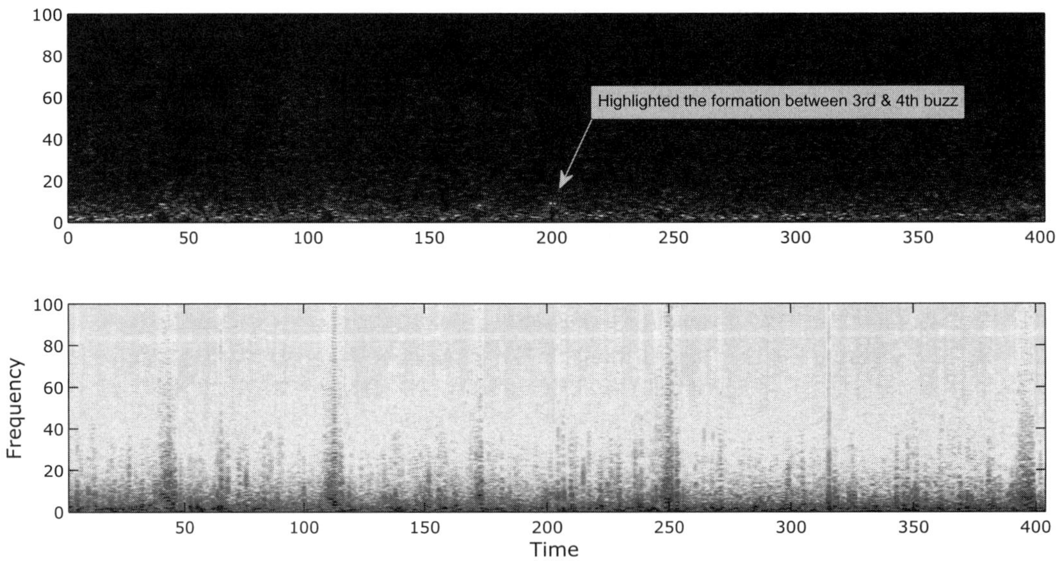

Fig. 12.11. Top: JSPECT transform of data segment. Magnitude squared ()○2 instead of log() was used in order to enhance changes against background. Bottom: Classical spectrogram added for reference to event times only.

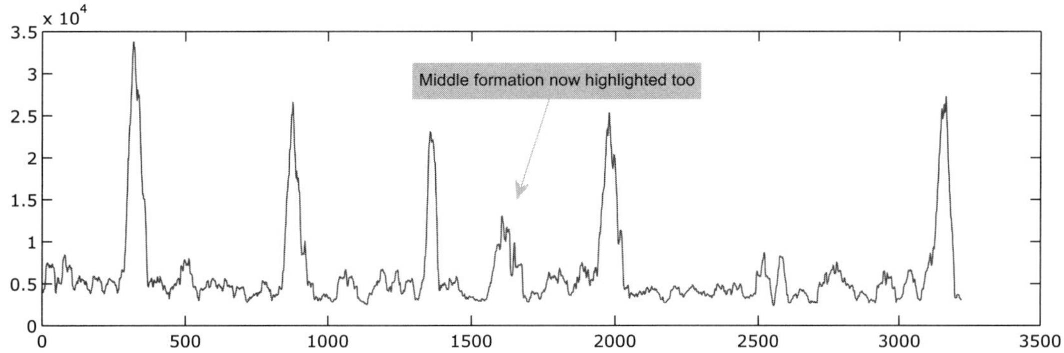

Fig. 12.12. Line length after narrow zero-phase filtering around the 8 Hz center frequency of the chirps.

average of line length, the moving variance of line length, and their ratio as the predictor of a particular seizure pattern in the illustrated patient.

Figure 12.13 shows channel RT3 of the 1.35-hour-long data segment. The 12-bit amplitude resolution and gain combination results in clipping at integer –2048 during seizure. When played back at 10 KHz, the 18 prechirps are confirmed by ear (2 softer than the rest). The seizure sounds like a turntable stylus sweeping across a vinyl record. The typical up-then-down frequency evolution during seizure is readily apparent. The postictal period sounds like a warble.

Figure 12.14 shows the 4-second line length updated every 0.125 seconds for the entire record. As expected, line length picks out seizure out of the background quite distinctly and quantifies the typical postictal depression. It also suggests the chirp events are "false starts" or even small seizures themselves.

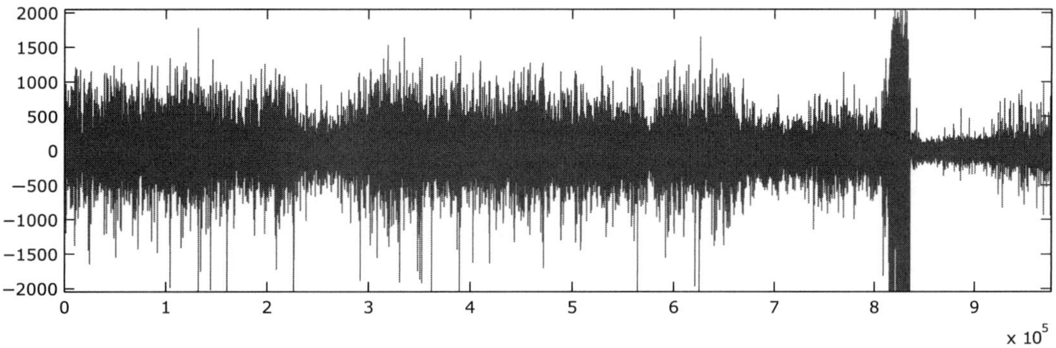

Fig. 12.13. Channel RT3 from an intrahippocampal depth electrode in a 1.35-hour-long EEG data segment leading to temporal lobe seizure onset.

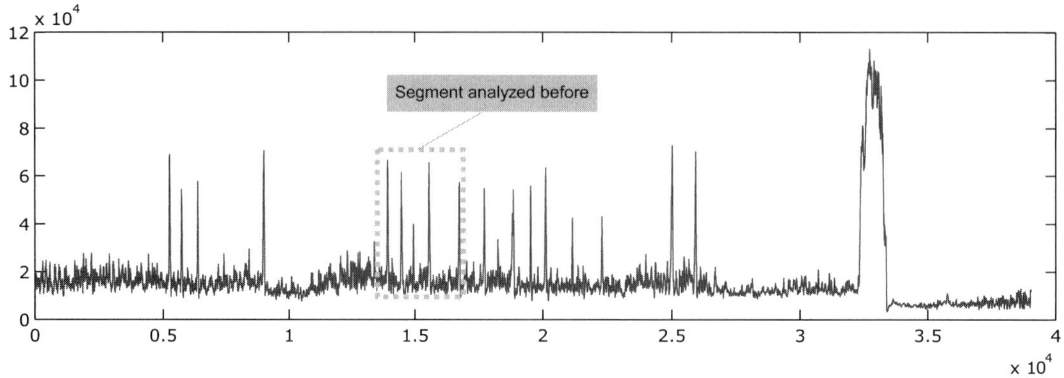

Fig. 12.14. Sliding 4-second line length of the RT3 signal over the 1.35-hour-long record updated every 0.125 seconds, with temporal relationship to segment analyzed before. Time axis in number of line-length samples.

6.4. Seizure Prediction with Trending and Feature Composition

A sliding 4 second window leads to decision making based on very recent, time-localized information (trained to make comparisons with respect to a distant past, but of microtemporal behavior). This time scale is naturally suited for detection of paroxysmal events. However, it does not necessarily capture bigger dynamical patterns such as infra-, circa-, and ultradian rhythms or any historical transitional behavior over minutes, hours, days, months, etc. Trending refers to smoothing local features over bigger time scales to determine any tendencies. More generally, feature composition refers to features of features, which in turn can be composed further to any number of levels as features of {...features of {features of features}...}.

Figure 12.15 shows the 5-minute trend (moving average) of line length updated every 15 seconds. Trending introduces delay in the sense that the bigger the scale the more sluggish the reaction to a local event. Despite this, upslopes in **Fig. 12.15** seem to

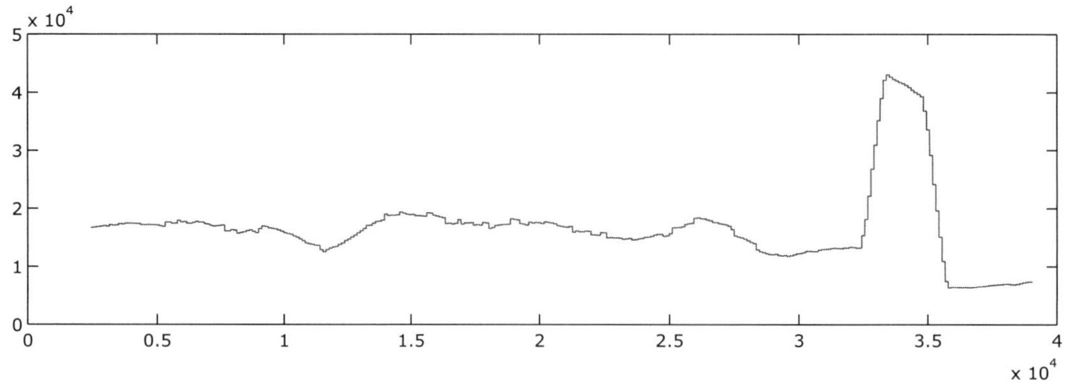

Fig. 12.15. Five-minute trend of line length updated every 15 seconds over the 1.35-hour segment. A right-alignment convention is used to strictly enforce causality of the computation.

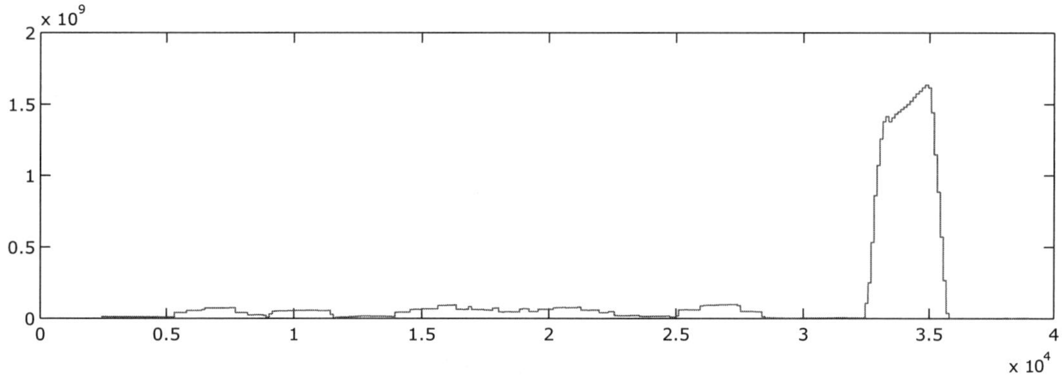

Fig. 12.16. Five-minute moving variance of line length updated every 15 seconds over the same 1.35-hour segment, in lockstep with the moving average of Fig. **12.15**.

correspond to a state more "proictal" than not. Instead of exploiting this observation, we will add information from a second composed feature for a more robust indication.

Figure 12.16 shows the 5-minute moving variance of line length updated every 15 seconds over the same data segment. Examination of **Figs. 12.15 and 12.16** reveals that the pointwise *ratio* of these quantities is a better warning feature than either feature individually. **Figure 12.17** shows the ratio of moving average of line length to moving variance of line length in units comparable to EEG. An approximately 8-minute long warning emerges as a period of high average line length relative to "too quiet" variation in line length.

The perceptive reader will note that a single example of a warning pattern may be spurious and may not generalize to future seizures. Indeed, the subsequent seizure hours later (not shown) possesses a nearly identical predictive pattern for the ratio in **Fig. 12.17**, but this is not true for all seizures in this patient.

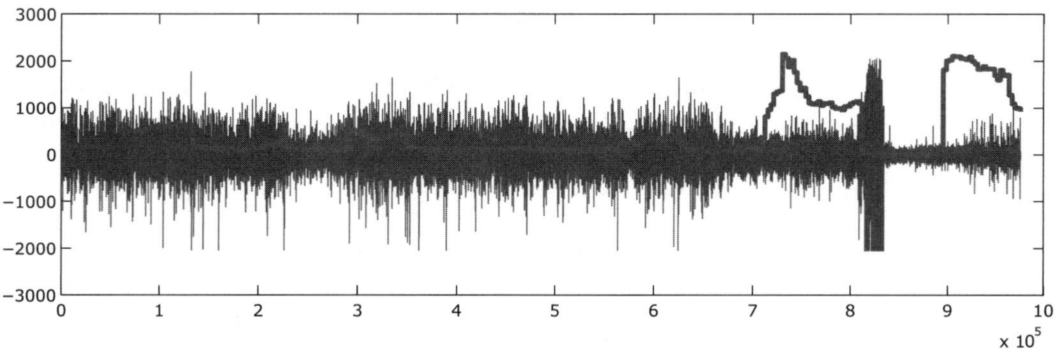

Fig. 12.17. Ratio of moving average to moving variance of line length (foreground trace) overlaid in units comparable to EEG (background trace). Zero-order hold is used to expand from feature time to EEG time samples.

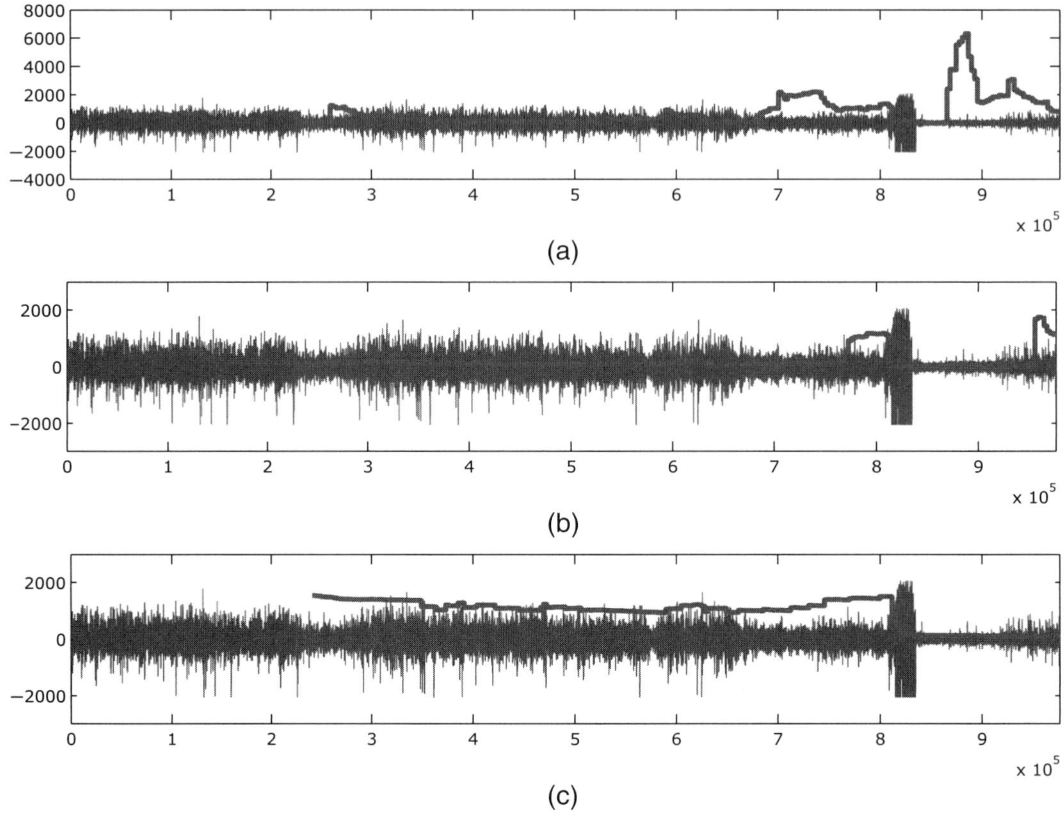

Fig. 12.18. Effect of trend time scale on the duration and differentiability of the ratio warning signal. Sliding window has length (a) $L = 1200$ (2.5 minutes), (b) $L = 4800$ (10 minutes), (c) $L = 9600$ (20 minutes; amplitude of red trace was magnified $10\times$ with respect to the others in order to see shape of signal).

This underlines the importance of a constructive approach to seizure prediction where an overall system is built from complementary pieces of evidence.

We now investigate the duration of the warning period with respect to trends calculated over other time scales. With $L = 1200$ (2.5-minute sliding window), warning persists around ~9 minutes (**Fig. 12.18A**). With $L = 4800$ (10-minute sliding window), warning becomes ~4 minutes (**Fig. 12.18B**). Finally, with $L = 9600$ (20-minute sliding window), the warning washes out to just an upslope, and a "baseline" cannot be measured anymore (**Fig. 12.18C**).

6.5. Discussion and Summary

The above derivations and application examples, with all their complexity, are not meant to obfuscate the area of event detection during in vivo recording. Rather, they are meant to give practical examples of methods we have to date found useful in performing these tasks in chronic in vivo human and animal IEEG recordings. We have purposefully not tried to gloss over details or challenges in the implementation of these methods. Rather, we have tried to honestly portray their complexity, utility, and relative early success in reducing data and elucidating physiological events related to seizures and their generation. We have presented them in the hope that the examples will enable readers to experiment with implementing these approaches in their own laboratories. The relatively unpolished nature of these presentations are also meant to accurately portray that this area is not one that has already been exhaustively mined for research. To the contrary, these approaches are relatively novel in many of their applications, and their use is revealing new aspects of epileptiform electrophysiology that has eluded other simpler or more straightforward methods in the past.

It is our impression that these sorts of tools and perhaps more sophisticated ones will be necessary to explain the complex, multi-scale, distributed cellular dynamics that are responsible for creating seizures. Already our group and others are finding new, cortical column-scale phenomena being called "microseizures" that reveal a new dynamic structure in the chain of events leading to seizures, as measured by arrays of microelectrodes in humans (personal communication and joint unpublished research in submission with Gregory Worrell, M.D., Ph.D., Mayo Clinic). It was not possible to really appreciate these phenomena before engineering tools like those demonstrated above were turned to analyze multi-Terabyte streams of continuous 32 kHz-sampled intracranial EEG. It is our hope that as these types of quantitative methods are more commonly applied to biological systems, and particularly in neuroscience, they will become less mysterious to those involved in translational research. It is our firm belief that there is a real need for applying more engineering methods to systems neuroscience, in the basic laboratory, in the clinic world, and most importantly in helping to translate basic investigations into new discoveries and therapies that will help us better understand and treat human disease. Work in this area is rapidly accelerating and catching on among research groups worldwide, suggesting that others agree with these ideas.

References

1. Wilson SB, "Algorithm architectures for patient dependent seizure detection," *Clinical Neurophysiology*, 117(6):1204–16, 2006.
2. Wilson SB, "A neural network method for automated and incremental learning applied to patient-dependent seizure detection," *Clinical Neurophysiology*, 116:1785–95, 2005.
3. Shoeb A, Edwards H, Connolly J, Bourgeois B, Treves ST, Guttag J. "Patient-specific seizure onset detection," *Epilepsy and Behavior*, 5:483–98, 2004.
4. Qu H and Gotman J, "A patient-specific algorithm for the detection of seizure onset in long-term EEG monitoring: Possible use as a warning device," *IEEE Transactions of Biomedical Engineering*, 44(2):115–22, 1997.
5. Gardner A, Krieger A, Vachtsevanos G, and Litt B, "One class novelty detection for seizure analysis from intracranial EEG," *Journal of Machine Learning Research*, 7:1025–44, 2006.

6. Osorio I, Frei MG, Manly BF, Sunderam S, Bhavaraju NC, and Wilkinson SB, "An introduction to contingent (closed-loop) brain electrical stimulation for seizure blockage, to ultra-short-term clinical trials, and to multidimensional statistical analysis of therapeutic efficacy," *Journal of Clinical Neurophysiology*, 18(6):533–44, 2001.
7. Murro AM, Park YD, Greene D, Smith JR, Ray P, King DW, Loring DW, and Lee KH, "Closed loop neurostimulation in patient with intractable epilepsy," in *Proc. American Clinical Neurophysiology Society Annual Meeting*, New Orleans, LA, Sept. 20–21, 2002.
8. Bergey G, Britton J, Cascino G, Choi H, Karceski S, Kossoff E, Meador K, Murro A, Park Y, Ritzl E, Smith J, Spencer D, Spencer S, Ray P, Greenwood J, and Greene D, "Implementation of an external responsive neurostimulator system (eRNS) in patients with intractable epilepsy undergoing intracranial seizure monitoring," *Epilepsia*, 43(suppl 7):191, 2002.
9. Kossoff EH, Ritzl EK, Politsky JM, Murro AM, Smith JR, Duckrow RB, Spencer DD, and Bergey GK, "Effect of an external responsive neurostimulator on seizures and electrographic discharges during subdural electrode monitoring," *Epilepsia* 45(12):1560–7, 2004.
10. Worrell G, Wharen R, Goodman R, Bergey G, Murro A, Bergen D, Smith M, Vossler D, and Morrell M, "Safety and evidence for efficacy of an implantable responsive neurostimulator (RNS®) for the treatment of medically intractable partial onset epilepsy in adults," *Epilepsia*, 46(suppl 8):226, 2005.
11. Gotman J and Gloor P. "Automatic recognition and quantification of interictal epileptic activity in the human scalp EEG," *Electroencephalography and Clinical Neurophysiology*, 41(5):513–29, 1976.
12. Gotman J, "Automatic seizure detection: Improvements and evaluation," *Electroencephalography and Clinical Neurophysiology*, 76(4):317–24, 1990.
13. Osorio I, Frei M, and Wilkinson S. "Real-time automated detection and quantitative analysis of seizures and short-term prediction of clinical onset," *Epilepsia*, 39(6):615–27, 1998.
14. Osorio I, Frei MG, Giftakis J, Peters T, Ingram J, Turnbull M, Herzog M, Rise MT, Schaffner S, Wennberg RA, Walczak TS, Risinger MW, and Ajmone-Marsan C. "Performance reassessment of a real-time seizure-detection algorithm on long ECoG series," *Epilepsia*, 43(12):1522–35, 2002.
15. Wilson SB, Scheuerb ML, Emerson RG, Gabor AJ, "Seizure detection: Evaluation of the reveal algorithm," *Clinical Neurophysiology*, 115:2280–91, 2004.
16. Gotman J. "Automatic recognition of epileptic seizures in the EEG," *Electroencephalography and Clinical Neurophysiology*, Nov;54(5):530–40, 1982.
17. Murro AM, King DW, Smith JR, Gallagher BB, Flanigin HF, and Meador K. "Computerized seizure detection of complex partial seizures," *Electroencephalography and _Clinical Neurophysiology*, Oct;79(4): 330–3, 1991.
18. Gabor A, Leach R, and Dowla F. "Automated seizure detection using a self-organizing neural network," *Electroencephalography and Clinical Neurophysiology*, 99(3): 257–66, 1996.
19. Gabor AJ. "Seizure detection using a self-organizing neural network: validation and comparison with other detection strategies," *Electroencephalography and Clinical Neurophysiology*, Jul;107(1):27–32, 1998.
20. Grewal S and Gotman J. "An automatic warning system for epileptic seizures recorded on intracerebral EEGs," *Clinical Neurophysiology*, Oct;116(10):2460–72, 2005.
21. Block A and Fisher R, "Can patients perform volitional motor acts at the start of a seizure?," *Journal of Clinical Neurophysiology*, 16(2):141–5, 1999.
22. Ripley RD, *Pattern Recognition and Neural Networks.* Cambridge (U.K) University Press, 1996.
23. Efron B and Tibshirani R *An Introduction to the Bootstrap*, Boca Raton, FL: CRC Press, 1993.
24. Cherkassky V and Mulier F, *Learning from Data: Concepts, Theory, and Methods.* NY: Wiley, 1998.
25. Le Cun Y, Denker JS, and Solla SA, "Optimal brain damage," in D. Touretzky (ed.) *Advances in Neural Information Processing Systems 2*, San Mateo, CA Morgan Kaufmann, 598–605, 1990.
26. Schiff SJ, Colella D, Jacyna GM, Hughes E, Creekmore JW, Marshall A, Bozek-Kuzmicki M, Benke G, Gaillard WD, Conry J, Weinstein SR, "Brain chirps: Spectrographic signatures of epileptic seizures," *Clinical Neurophysioly*, 111(6):953–8, 2000.
27. Kircher J and Raskin D, "Human versus computerized evaluations of polygraph

data in a laboratory setting," *Journal of Applied Psychology*, 73:291–302, 1988.

28. Olsen D, Lesser R, Harris J, Webber R, and Cristion J, "Automatic detection of seizures using electroencephalographic signals," U.S. Patent 5,311,876, 1994.
29. Esteller R, *Detection of Seizure Onset in Epileptic Patients from Intracranial EEG Signals*, Ph.D. dissertation, Georgia Institute of Technology, 2000.
30. Esteller R, Echauz J, Tcheng T, Litt B, Pless B, "Line length: An efficient feature for seizure onset detection," in *Proc. 23rd Annual International Conference of the IEEE Engineering in Medicine and Biology Society*, Istanbul, Turkey, Oct. 25–28, 2:1707–10, 2001.
31. Echauz J, Esteller R, Tcheng T, Pless B, Gibb B, Kishawi E, Litt B, "Long-term validation of detection algorithms suitable for an implantable device," *Epilepsia*, 42(Suppl 7):35–6, 2001.
32. Niederhauser J, Esteller R, Echauz J, Vachtsevanos G, and Litt B, "Detection of seizure precursors from depth-EEG using a sign periodogram transform," *IEEE Transactions on Biomedical Engineering*, 50(4): 449–58, 2003.

Chapter 13

Viral Vector Gene Therapy for Epilepsy

Stacey B. Foti, Shelley J. Russek, Amy R. Brooks-Kayal, and Thomas J. McCown

Abstract

Theoretically, gene therapy offers an attractive alternative for the treatment of focal epilepsies, and recently, studies have established the basic viability of anti-seizure gene therapy by employing a number of diverse approaches. Using recombinant adeno-associated virus (AAV) vectors, significant seizure suppression has been obtained *in vivo* by interrupting NMDA receptor function or by altering GABA receptor composition. Similarly, engineering cells to release GABA or adenosine has been shown to exert significant anti-seizure actions. In addition, studies have reported seizure suppression using in vivo, viral vector-mediated expression of GDNF, ICP10PK, or clostridial light chain. At present though, studies with the neuroactive peptides, galanin and neuropeptide Y (NPY), have progressed the furthest. AAV vector-mediated expression and constitutive secretion of galanin essentially prevents electrographic and behavioral seizure activity induced by peripheral kainic acid administration, prevents seizure-induced cell damage, and attenuates the seizure threshold in previously kindled animals. For NPY, AAV vector-mediated expression of prepro-neuropeptide Y delays seizure onset and increases kindling seizure thresholds, while expression and constitutive secretion of NPY (13-36) attenuates kainic acid-induced seizures. These results have proven quite promising, but several key issues remain, including the adequate transduction of chronic seizure-exposed tissue. In light of the present advances, however, gene therapeutic approaches will provide an effective treatment alternative for intractable, focal seizures.

Key words: Epilepsy, gene therapy, adeno-associated virus, viral vectors, NMDA, GABA, galanin, neuropeptide Y.

1. Introduction

About 2.1 million people in the United States have epilepsy *(1)*, and with a worldwide prevalence of approximately 1% *(2)*, epilepsy represents one of the most common neurological diseases. Not only can epilepsy be debilitating for patients, it also exacts a significant economic toll. In the United States the estimated annual

Scott C. Baraban (ed.), *Animal Models of Epilepsy: Methods and Innovations, vol. 40*

DOI 10.1007/978-1-60327-263-6_13 Springerprotocols.com

economic cost of epilepsy is 12.5 billion dollars, a figure that includes both the direct cost of treatment and the indirect costs such as lost productivity and wages *(3)*. Unfortunately, even with the availability of new anti-epileptic drugs (AEDs), the number of drug-resistant epilepsies has not decreased *(4)*. In fact, 30–40% of all patients with epilepsy have medically intractable seizures, and only half of these patients are candidates for surgery *(5–8)*. Of those patients that undergo temporal lobe surgery, 20% still have seizures post-op *(9)*, and surgical complications along with post-operative memory and language deficits can occur *(8)*. For all these reasons, improving epilepsy treatment remains an important priority.

Theoretically, a gene therapy approach to seizure suppression is an attractive alternative treatment option for focal epilepsy, the most common adult form *(10)*. Focal epilepsy arises from a circumscribed, hyperexcitable network of neurons whose synchronous electrical discharge creates seizure activity. To date, most gene therapy studies have employed viral vectors based upon adenovirus (Ad), adeno-associated virus (AAV), human immunodeficiency virus (HIV), or herpes simplex virus (HSV). In all cases, these viral vectors have proven capable of transducing post-mitotic neurons which provide direct access to the cells responsible for generating and propagating the seizure activity. Furthermore, these viral vectors have proven capable of influencing brain function on a regional basis, so viral vector gene therapy should have the ability to influence sites of seizure genesis. However, even with the promising results obtained using viral vectors (*see* **Table 13.1**),

Table 13.1
Potential Therapeutic Targets that Influence Seizure Behavior

Target	Placement	Delivery vector	Experimental model	Functional effect	References
Adenosine	Intracerebroventricular	Grafting of fibroblasts or myoblasts	Hippocampal kindling	Suppression of generalized seizures	Huber et al. *(49)*, Guttinger et al. *(61)*
Anti-NMDA	Collicular cortex	AAV-CMV vector	Focal electrical stimulation	Increased seizure threshold	Haberman et al. *(17)*
		AAV-TET-Off vector		Decreased seizure threshold	Haberman et al. *(17)*
ASPA	Intracerebroventricular	Ad-CB vector	Spontaneous seizures in SER	Transiently reduced incidence of tonic seizures (2 weeks)	Seki et al. *(62)*
Cholecystokinin	Intracerebroventricular	Lipofectin and plasmid	Audiogenic seizure-prone rats	Transient seizure inhibition (1 week)	Zhang et al. *(63)*

(continued)

Table 13.1 (continued)

Target	Placement	Delivery vector	Experimental model	Functional effect	References
GABAAR	Hippocampus	AAV-GABAR4	Pilocarpine-induced status epilepticus	Transiently reduced spontaneous seizures (4 weeks)	Raol et al. *(27)*
GAD65	Anterior SN	Grafting of neuronal or glial cell lines	Entorhinal kindling	Delayed rate of kindling	Thompson et al. *(33)*
	Posterior SN	Grafting of neuronal or glial cell lines		Faster rate of kindling	Thompson et al. *(33)*
	Anterior SN	Grafting neuronal cell line	Pilocarpine-induced status epilepticus	Reduced spontaneous seizures	Thompson and Suchomelova *(36)*
	Piriform cortex	Grafting neuronal cell line	Amygdala kindling	Increased threshold to seizures	Gernert et al. *(34)*
GAD67	SN pars reticulata	Grafting of mixed neuronal and glial cell line	Kainic acid i.p.	Delayed seizure behavior	Castillo et al. *(35)*
Galanin	Collicular cortex	AAV-TET-Off vector + FIB secretory sequence		Increased seizure threshold	Haberman et al. *(50)*
	Hippocampus	AAV-TET-Off vector + FIB secretory sequence	Kainic acid i.p.	Neuroprotection	Haberman et al. *(50)*
		AAV-NSE vector + WPRE	Kainic acid i.h.	Reduced number of seizures	Lin et al. *(52)*
	Piriform cortex	AAV-CB vector + FIB secretory sequence	Kainic acid i.p.	Seizure inhibition	McCown *(51)*

(continued)

Table 13.1(continued)

Target	Placement	Delivery vector	Experimental model	Functional effect	References
			Focal electrical stimulation	Increased seizure threshold	McCown *(51)*
GDNF	Hippocampus	AAV-CB vector + WPRE	Hippocampal kindling	Decreased number of generalized seizures, increased seizure threshold	Kanter-Schlifke et al. *(42)*
	Hippocampus	Ad-CMV vector	Kainic acid i.p.	Delayed seizure behaviors, neuroprotective	Yoo et al. *(41)*
ICP10PK	Intranasal vaccine	HSV-2ΔRR −ICP10 vector	Kainic acid i.p.	Reduced seizure behaviors, neuroprotection	Laing et al. *(44)*
Clostridial toxin light chain	Motor cortex	Ad-CMV vector	Focal penicillin	Reduced electrographic and behavioral seizures	Yang et al. *(47)*
NPY	Hippocampus	AAV-NSE vector + WPRE	Kainic acid i.h.	Delayed seizure behavior	Richichi et al. *(53)*
			Hippocampal kindling	Increased seizure threshold and delayed rate of kindling	Richichi et al. *(53)*
	Piriform cortex	AAV-CB vector + FIB secretory sequence	Kainic acid i.p.	Reduced and delayed seizure behaviors	Foti et al. *(54)*
NPY 13-36	Piriform cortex	AAV-CB vector + FIB secretory sequence	Kainic acid i.p.	Reduced and delayed seizure behaviors	Foti et al. *(54)*

Abbreviations: AAV: adeno-associated virus; ASPA: aspartoacylase; Ad: adenoviral; CB: cytomegalovirus enhancer, chicken beta-actin promoter; CMV: cytomegalovirus; GAD: glutamic acid decarboxylase; GDNF: glial cell line-derived neurotrophic factor; ICP10PK: ICP 10 protein kinase; i.h.: intrahippocampal; i.p.: intraperitoneal; NMDA: *N*-methyl-D-aspartate; NPY: neuropeptide Y; NSE: neuron- specific enolase; SER: spontaneously epileptic rats; SN: substantia nigra; TET-Off: tetracycline-off regulatable promoter; WPRE: woodchuck hepatitis virus post-translational regulatory element.

a number of factors must be considered when contemplating a gene therapy for the treatment of epilepsy.

1.1. Identifying Potential Therapeutic Targets

In general, epilepsy arises from a loss of inhibition, an enhanced state of excitation, or a combination of both processes, so given this wide range of influences, a wealth of potential therapeutic targets exists. On the most basic level, altering postsynaptic receptor or membrane-bound ion channel-mediated function clearly has been shown to attenuate seizure activity. For example, AEDs can exert their actions by *(1)* increasing GABAergic inhibition, *(2)* decreasing excitatory glutamatergic excitation, *(3)* decreasing voltage-dependant calcium release, or *(4)* decreasing voltage-dependant Na^+conductance *(11)*. Because Ad, AAV, HIV-1, and HSV vectors exhibit a clear neurotropism, any of these targets are readily accessible by these viral vectors. In addition, basic research has identified a number of other modulatory factors that likewise can attenuate seizure activity *in vivo*. Prominent among this class are the neuropeptides, such as galanin and neuropeptide Y *(12–14)*. Although this extensive list of potential targets would seem to offer many opportunities, the major difficulty is establishing a strategy that is both effective and non-toxic.

1.2. Excitatory Amino Acid Receptors

Excitatory amino acid receptors comprise an obvious target for anti-epileptic gene therapy, and certainly many studies have shown that *N*-methyl-D-aspartic acid (NMDA) receptors influence seizure sensitivity. Because a fully functional receptor requires the NMDA receptor 1 protein (NMDAR1), removal of this subunit protein should significantly reduce NMDAR1-mediated excitation *(15, 16)*. Haberman et al. *(17)* showed that AAV vector-mediated delivery of an antisense RNA specific to the NMDAR1 could indeed reduce NMDA receptor function *in vivo* and NMDAR1 protein *in vivo*. Using a focal seizure model, CMV promoter-driven expression of the NMDAR1 antisense significantly decreased the seizure susceptibility, proving the principle that viral vector-derived antisense could influence native receptor function in vivo. However, when expression of the same antisense construct was driven by a tetracycline-regulated minimal CMV promoter *(18)*, the seizure susceptibility actually increased. The fact that NMDA receptor-mediated excitation can drive endogenous GABA inhibition provided a tenable explanation for these diametrically opposed results. In the case of the CMV promoter, excitatory principal neurons likely comprised the preponderance of transduced cells, thus removal of NMDA receptor excitation blunted the seizure susceptibility. In the case of the regulated minimal CMV promoter, the preponderance of gene expression probably occurred in inhibitory GABAergic interneurons, so removal of the endogenous excitatory drive to these interneurons would cause a state of increased seizure sensitivity. The likelihood of such an occurrence was validated, when equal amounts of each AAV vector were administered together using different reporter genes. Some neurons supported expression from both promoters, but significant portions exhibited transduction from only one or the other of the promoters. Thus, a slight change in only the promoter resulted in a dramatic shift in the final outcome of the gene expression. These surprising findings greatly complicate the targeting of neurotransmitter receptor proteins or ion channels for the treatment of epilepsy. Certainly, it is possible that the viral vector tropism and/or promoter tropism could differ between experimental animals and humans. Thus, without some detailed *a priori* knowledge of the transduction pattern in humans, any manipulation of neurotransmitter receptor proteins or ion channels could lead either to the desired reduction of seizure susceptibility or a paradoxical heightening of seizure susceptibility.

1.3. Inhibitory GABA Receptors

GABA is the major inhibitory neurotransmitter in the adult brain. Most fast synaptic inhibition in the mature brain is mediated by $GABA_A$ receptors (GABRs), pentameric anion-selective channels composed of multiple subunit subtypes ($\alpha 1-6$, $\beta 1-3$, $\gamma 1-3$, δ, ϵ, π, θ, and $\rho 1-3$). GABAergic signaling has long been hypothesized to play an important role in the genesis of epilepsy, and many of the oldest and most commonly used anticonvulsants such as benzodiazepines and barbiturates act, at least in part, through augmentation of GABR activity. Mutations in several different GABR subunits have been identified in families with generalized epilepsies *(19)*. In addition, several laboratories have identified changes in GABR function and subunit composition in human temporal lobe epilepsy (TLE) and in animal models of TLE *(20–23)*. Studies in humans with temporal lobe epilepsy (TLE) and in adult rodent models of TLE have found reduced expression of GABR $\alpha 1$ subunit gene and increased expression of GABR $\alpha 4$ subunit gene in the dentate gyrus (DG) *(22–25)*. These findings suggest that diminished $\alpha 1$ levels in dentate granule neurons (DGN) may contribute to epileptogenesis and/or that elevated $\alpha 1$ levels may be protective. Certainly, reduction of GABR $\alpha 1$ subunit protein using an AAV that expressed a GABR $\alpha 1$ antisense RNA significantly increased focal seizure susceptibility *(26)*.

To directly test the hypothesis that the expression of higher $\alpha 1$ subunit levels inhibits development of epilepsy after SE, an AAV vector serotype 5 was designed to express a bicistronic RNA that codes for both the GABR $\alpha 1$ subunit as well as the reporter, enhanced yellow fluorescence protein (eYFP) *(27)*. Expression of this RNA was placed under control of the GABR $\alpha 4$ core promoter region, because it was previously shown to be markedly activated in DG following SE *(28)*. AAV vectors containing either the $\alpha 1$/eYFP-fused cDNA (AAV-$\alpha 1$) or the eYFP reporter only (AAV-eYFP) were injected into DG of adult rats, and SE was induced

2 weeks later by intraperitoneal injection of pilocarpine (385 mg/kg). Rats injected with AAV-α1 showed threefold higher levels of α1 subunits in DG by 2 weeks after SE compared to the control groups. AAV-α1 injection resulted in a threefold increase in the mean time to the first spontaneous seizure following SE, and only 39% of AAV-α1 injected rats were observed to develop spontaneous seizures in the first 4 weeks after SE, as compared to 100% of rats receiving sham injections. Because all groups of rats experienced similar SE after pilocarpine injection, these findings provide the first direct evidence that increasing the levels of a single GABR subunit in DG can inhibit the development of spontaneous seizures after SE .

Another unique aspect of this particular study involves the use of a condition-specific promoter which when upregulated during a disease process drives expression of a therapeutic gene. In this case, the upregulation of the GABR α4 promoter drove expression of the α1 transgene which enhanced α1 expression in DG. Unfortunately, this vector-derived expression lasted for only the first 2 weeks after SE. To differentiate whether the decline in α1 subunit levels by 4 weeks after SE was due to loss of recombinant or endogenous α1 subunits, mRNA levels for eYFP (produced from the vector containing the fused α1/eYFP cDNAs) were assayed in AAV-injected rats sacrificed at 1 or 4 weeks after SE. With presence of eYFP mRNA as a marker for co-presence of α1, recombinant α1 mRNA levels were found to decline over sixfold in AAV-injected rats between 1 and 4 weeks after SE. This finding suggests that either the GABRA4 promoter in the AAV vector was transcriptionally downregulated or that the recombinant α1/eYFP mRNAs were degraded. A number of studies have demonstrated robust, long-term gene expression in the hippocampus after AAV transduction *(29, 30)*, but this stable expression is promoter dependent. For example, McCown et al. *(31)* showed that when gene expression was driven by a CMV promoter, the initial level of AAV-mediated gene expression in the hippocampus declined over a period of 4 weeks whereas some other areas of brain exhibited no decline in gene expression. Further studies will be required to define the precise mechanism responsible for the proposed decrease in activity of the recombinant GABRA4 promoter over time and to rule out whether a change in mRNA stability may also play a role in declining levels of α1/eYFP transcript.

In addition to the effects on seizures, AAV-mediated elevation of α1 expression was associated with a behavioral phenotype in a fraction of the AAV-α1-injected rats. Most of the AAV-α1-injected rats behaved similar to sham-injected rats, but 30% exhibited abnormal behaviors including excessive sedation, anorexia, and weight loss that persisted for days to weeks following SE. This effect was not seen after SE in any of the other groups (including the AAV-eYFP control), suggesting that the effect likely resulted from elevated $\alpha 1$ levels rather than the effects of AAV injection or SE. Such effects are not surprising given that $\alpha 1$-containing receptors are the major form of GABRs mediating sedation in the nervous system *(32)*.

Genetic engineering approaches have also been used to augment other aspects of GABAergic neurotransmission. Specifically, two groups have transplanted immortalized neurons genetically engineered to stably express the GABA synthesizing enzyme glutamic acid decarboxylase (GAD65 or GAD67) into different brain regions and examined the effects on induced and spontaneous seizures. Transplant of these conditionally immortalized neurons engineered to produce GABA into either substantia nigra (SN) or piriform cortex of rats has been found to inhibit kindling *(33, 34)* and reduce the number and severity of kainic acid-induced seizures *(35)*. Transplantation of GABA-producing neurons into SN also has been shown to reduce spontaneous seizures in a TLE model *(36)*. About 45–65 days after lithium-pilocarpine-induced status epilepticus, substantia nigra injection of GABA-producing neurons under the control of tetracycline reduced the number of spontaneous seizures and epileptiform discharges at 7–8 days after transplantation. No seizure suppression was found in animals that received control cells or that received the GABA-producing cells plus doxycycline to inhibit GABA production *(36)*. Similarly, transplantation of GABA-producing cells into dentate gyrus raised GABA tissue concentrations, increased the afterdischarge threshold, shortened the afterdischarge duration, and prolonged the latency to expression of the first stage 5 seizure in a hippocampal kindling model *(37)*. Although the engineered cells show evidence of integration, this cell transplantation approach remains hampered by the limited long-term survival of the transplanted cells *(36)*.

Taken together, it is clear that by controlling GABAergic signaling either through discrete regulation of its receptor isoforms or through accessibility to its neurotransmitter GABA, it may be possible in the future to treat a multitude of diseases that stem from an imbalance between excitatory and inhibitory neurotransmission. The challenge remains to develop better vehicles and approaches for the selective delivery of gene products over time in a therapeutically useful manner.

2. Diverse Unrelated Targets

A number of diverse, unrelated targets also have emerged from basic research that focused upon neuroprotection, not epilepsy. Glial cell line-derived neurotrophic factor (GDNF) exerts a potent neuroprotective action in the CNS, especially on dopamine-containing neurons *(38, 39)*, and when infused intraventricularly, GDNF attenuates kainic acid-induced seizures *(40)*. Using viral vectors, Yoo et al. *(41)* subsequently showed that Ad-mediated GDNF expression in the hippocampus attenuated behavioral seizures and cell death induced by peripheral kainic acid administration. Similarly, Kanter-Schlifke et al. *(42)* used an AAV vector to overexpress GDNF in the hippocampus and found that this overexpression reduced the seizure severity in both hippocampal kindling and kindled self-sustained status epilepticus. Thus, viral vector-mediated expression of GDNF can influence seizure sensitivity. Another anti-apoptotic agent was identified by Perkins et al. *(43)*, who reported that the herpes simplex virus type 2 R1 protein kinase (ICP10 PK) exerted an anti-apoptotic action in hippocampal cultures. Based upon these findings, Laing et al. *(44)* showed that nasal inoculation with a herpes simplex virus type 2 that contains the anti-apoptotic gene ICP10PK prevented both behavioral seizures and neuronal death elicited by peripheral kainic acid administration. Thus, it appears that at least in these two instances, gene products with neuroprotective actions also can significantly influence seizure sensitivity.

A second group of therapeutic targets have emerged from diverse modulators of synaptic excitation. The protein homer 1a is induced by excessive excitation and is thought not only to modulate synaptic plasticity but also to reduce postsynaptic calcium function *(45)*. Using this information, Klugman et al. *(46)* found that AAV-mediated overexpression of Homer 1a attenuated electrographic seizure activity in a hippocampal status epilepticus model. Another means to reduce excitation would be to inhibit the efficacy of synaptic transmission. Clostridial toxin light chain digests the vesicle docking protein, synaptobrevin, which results in impaired synaptic transmission, so Yang et al. *(47)* used Ad vectors to express this protein in rat cortex. Ad-mediated expression of this protein attenuated both electrographic and behavioral seizures elicited by subsequent penicillin injection into the cortex. Finally, a number of studies have established that adenosine potently inhibits neuronal activity *(48)*, so Huber et al. *(49)* engineered fibroblasts to secrete adenosine. When these cells were implanted into the ventricles of rats, both behavioral and electrographic seizures were attenuated in electrically kindled rats. Although these studies established the anti-seizure potency of adenosine *in vivo*, the effect waned significantly 24 days post-transplantation. In total, these studies illustrate the wide range of targets by which various gene products can reduce neuronal excitation by *in vivo* viral gene therapy.

2.1. Neuroactive Peptides

Like other gene therapy studies, previous pharmacological findings with neuroactive peptides have provided a strong rationale for application in anti-epileptic gene therapy. Of the many neuropeptides that can modulate neuronal excitability, galanin and neuropeptide Y (NPY) have been most thoroughly characterized. For example, galanin and NPY both attenuate seizure activity in vivo *(12–14)*. Key to any neuropeptide gene therapy, though, is the consideration of endogenous neuropeptide function. Most neuroactive peptides are expressed initially as prepro-peptides, a form that presumably contains the appropriate information for trafficking to the pre-synaptic terminal and packaging into vesicles. Thus, if a viral vector expresses a prepro-neuropeptide sequence, one must assume that the transduced cells indeed contain the appropriate pathways for trafficking and release. Whether determined by the vector or the promoter, the actual tropism of viral vectors is unknown in the human CNS, so as demonstrated for NMDA receptor gene therapy, successful neuropeptide gene therapy might require transduction and gene expression in the appropriate neuronal population. Given these potential pitfalls, Haberman et al. *(50)* devised a novel approach to peptide gene therapy focusing upon galanin. The laminar protein, fibronectin, occurs in most cell types and normally is secreted in a constitutive manner. This constitutive secretion is determined by the secretory signal sequence for fibronectin. Using AAV vectors Haberman et al. *(50)* combined the secretory signal sequence for the laminar protein fibronectin with the coding sequence for the active galanin neuropeptide (AAV-FIB-GAL). By attaching this secretory signal sequence to the coding sequence for the active galanin peptide, it was reasoned that the viral vector-mediated expression would lead to the constitutive secretion of active galanin. Thus, if the appropriate receptors exist in the area of transduction, different patterns of neuronal transduction should not adversely influence the outcome. In fact these investigators showed that AAV-FIB-GAL supported constitutive secretion of galanin in vitro, and in vivo sufficient galanin was secreted to significantly attenuate focal seizure activity and prevent kainic acid-induced neuronal cell death in the hippocampus. If galanin was expressed in the absence of the FIB, or GFP was constitutively secreted, no effects were found on focal seizure sensitivity or cell death in the hippocampus. Using this gene therapy approach, subsequent studies found that bilateral infusion of AAV-FIB-GAL into the piriform cortex prevented both electrographic and behavioral seizure activity induced by peripheral kainic acid administration and significantly elevated the seizure initiation threshold in previously kindled rats (*see* **Fig. 13.1**) *(51)*. In contrast, Lin et al. *(52)* used AAV vectors to express human galanin in the hippocampus. Even though these studies achieved much higher levels of vector-derived gene expression in comparison to Haberman et al. *(50)*, the effects upon local

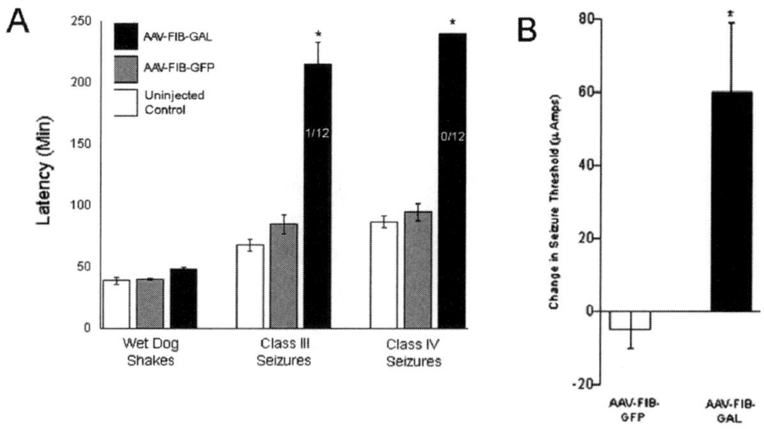

Fig. 13.1. **The effects of AAV-FIB-GFP and AAV-FIB-GAL vectors on the expression of limbic seizure behaviors.** As seen in **(A)** 7 days after infusion of AAV-FIB-GFP (N = 8) into the rat piriform cortex, kainic acid administration (10 mg/kg, i.p.) elicited wet dog shakes, class III and class IV seizure activity with the same latency as in untreated controls (t-test, $p>0.1$). Seven days after infusion of AAV-FIB-GAL (N = 12) into the piriform cortex, kainic acid administration elicited wet dog shake behaviors with the same latency as in untreated control animals. This result would be expected, because the vectors were infused into the piriform cortex while wet dog shakes arise from the ventral hippocampus. In marked contrast to the AAV-FIB-GFP group, however, the only seizure activity in this AAV-FIB-GAL group was a single, brief class III seizure exhibited by only 1 of the 12 rats over the 240 minute post-kainic acid observation period. None of the rats exhibited any class IV seizure behaviors over the entire 240 minute post-kainic acid observation period (*t-test, $p<0.01$) (Values = mean $\pm$ SEM). **(B)** shows the effects of AAV-FIB-GAL (N = 9) or AAV-FIB-GFP (N = 5) infusion on the threshold for seizure genesis in rats that previously had been electrically kindled to class V seizure activity. AAV-FIB-GFP infusion did not alter the stimulation threshold for seizure genesis, but AAV-FIB-GAL infusion significantly elevated the amount of current necessary to elicit seizure activity (paired t-test, $p<0.02$) (Values = mean $\pm$ SEM). Figure adapted from McCown *(51)* with permission from Nature Publishing Group.

infusion of kainic acid were restricted to a decrease in the number of seizures, not the seizure latency. Furthermore, no neuronal protection was evident. When these investigators further explored the effects of this approach, they found that galanin overexpression did not alter the course of electrical hippocampal kindling or modulate short-term plasticity of mossy fiber CA3 synapses. Thus, just because high levels of a peptide are expressed *in vivo*, such expression does not necessarily lead to an increase in the peptide's in vivo actions.

Neuropeptide Y (NPY), like galanin, exhibits anti-seizure activity in vivo *(12, 13)*, and a recent set of studies by Richichi et al. *(53)* have shown that AAV-mediated expression of prepro-NPY exerts a range of anti-seizure effects. Expression of a human prepro-NPY gene prevented the appearance of status epilepticus after intracerebroventricular administration of kainic acid. Also, hippocampal expression of prepro-NPY increased the electrical seizure threshold in the hippocampus and significantly retarded the rate of kindling epileptogenesis. Another important facet of these studies involved the exploitation of different AAV serotypes. A hybrid AAV1/2 serotype transduced a substantially wider range of neurons, when compared to the AAV2 serotype, and this increased transduction translated to a greater effect of prepro-NPY expression. Thus, these studies clearly established the anti-epileptic potential for vector-derived prepro-NPY expression in vivo.

A subsequent study by Foti et al. *(54)* utilized the constitutive secretion approach to demonstrate not only the efficacy of vector-derived NPY, but also the efficacy of the NPY receptor-specific fragment NPY (13-36). Although some question remains as to which of the NPY receptors (Y1–Y5) mediates the anti-seizure activity *(55, 56)*, several studies have suggested a critical role for the Y2 receptor in mediating anti-epileptic actions *(57–59)*. Because acute intracerebral delivery of the Y2 receptor preferring agonist NPY (13-36) reduces seizure susceptibility following systemic kainic acid administration *(60)*, Foti et al. *(54)* used an AAV vector to express and constitutively secrete NPY or NPY (13-36) in the rat piriform cortex. Similar to previous results with galanin *(51)*, expression of either NPY or NPY (13-36) significantly suppressed limbic seizures elicited by subsequent peripheral kainic acid administration. Thus, these studies showed that the constitutive secretory approach can effectively express and secrete receptor-specific peptide fragments in vivo.

3. Critical Considerations and Future Directions

The wide range of gene therapy studies amply illustrate the potential of viral vector gene therapy as an effective treatment modality for epilepsy, but even with these positive findings, there are a number of considerations that will inevitably impact both the efficacy and the safety of this therapeutic approach. Although many viral vectors exhibit a neuronal tropism *in vivo*, one must always take into account how specific patterns of transduction might alter the outcome of therapeutic gene expression. Viral vector tropism aside, as shown previously *(17)* different promoters can lead to differential gene expression even within the context of the same viral vector. Because such a change can result in an unwanted outcome, one first must consider the need to transduce a specific cellular population and the consequences if such specificity is not achieved. One obvious solution involves the use of cell-specific promoters, such as that used by Raol et al. *(27)* where expression would only be achieved in those cells that contain

GABR $\alpha 4$ promoter activity. However, as also seen by Raol et al. *(27)* endogenous promoters may be susceptible to promoter silencing. Second, the most likely clinical population for gene therapy encompasses those individuals with intractable temporal lobe epilepsy who are surgical candidates. In such a case, these individuals will have extensive seizure histories and inevitably varying degrees of hippocampal sclerosis. At present, most viral vector studies have examined the ability to transduce normal tissue. Thus, a second consideration is whether the viral vectors will support sufficient neuronal transduction in epileptic tissue with its associated pathology. Finally, it is incumbent to demonstrate that the proposed gene therapy can prevent spontaneous seizure activity. Thus, to best model an effective gene therapy, viral vector transduction should be initiated in animals that exhibit spontaneous seizures. As studies focus upon these aspects, it is likely that an effective antiepileptic gene therapy will emerge.

References

1. Hirtz D, Thurman DJ, Gwinn-Hardy K, Mohamed M, Chaudhuri AR, Zalutsky R. "How common are the "common" neurologic disorders?" *Neurology* 2007;68: 326–37.
2. Hauser WA, Hesdorffer DC. *Epilepsy frequency, causes, and consequences.* New York, Demos Publications, 1990.
3. Shafer PO, Begley C. The human and economic burden of epilepsy. *Epilepsy Behav* 2000;1:91–92.
4. Loscher W, Leppik IE. Critical re-evaluation of previous preclinical strategies for the discovery and the development of new antiepileptic drugs. *Epilepsy Res* 2002;50:17–20.
5. Shafer SQ, Hauser WA, Annegers JF, Klass DW. EEG and other early predictors of epilepsy remission: A community study. *Epilepsia* 1988;29:590–600.
6. Engel JJr, Levesque MF, Shields WD. Surgical treatment of the epilepsies: Presurgical evaluation. *Clin Neurosurg* 1992;38: 514–34.
7. Sander JW. Some aspects of prognosis in the epilepsies: A review. *Epilepsia* 1993;34: 1007–16.
8. Siegel AM. Presurgical evaluation and surgical treatment of medically refractory epilepsy. *Neurosurg Rev* 2004;27:1–18.
9. Spencer DD, Spencer SS, Mattson RH, Williamson PD, Novelly RA. Access to the posterior medial temporal lobe structures in the surgical treatment of temporal lobe epilepsy. *Neurosurgery* 1984;15:667–71.
10. Semah F, Picot MC, Adan C, Broglin D, Arzimanoglou A, Bazin B, Cavalcanti D, Baulac M. Is the underlying cause of epilepsy a major prognostic factor for recurrence? *Neurology* 1998;51:1256–62.
11. Ure JA, Perassolo M. Update on the pathophysiology of the epilepsies. *J Neurol. Sci* 2000;177:1–17.
12. Woldbye DPD, Larsen PJ, Mikkelsen JD, Klemp K, Madsen TM, Bolwig TG. Powerful inhibition of kainic acid seizures by neuropeptide Y via Y5-like receptors. *Nat Med* 1997;3:761–4.
13. Baraban SC, Hollopeter G, Erickson JC, Schwartzkroin PA, Palmiter RD. Knockout mice reveal a critical role for neuropeptide Y. *J Neurosci* 1997;17:8927–36.
14. Mazarati AM, Lie H, Soomets U, Sankar R, Shin D, Katsumori H, Langel U, Wasterlain CG. Galanin modulation of seizures and seizure modulation of hippocampal galanin in animal models of status epilepticus. *J Neurosci* 1998;18:10070–7.
15. Monyer H, Sprengel R, Schoepfer R, Herb A, Higuchi M, Lomeli H, Burnashev N, Sakmann B, Seeburg PH. Heteromeric NMDA receptors: Molecular and functional distinction of subtypes. *Science* 1992;256:1217–21.
16. Hollmann M, Heinemann S. Cloned glutamate receptors. *Ann Rev Neurosci* 1994;7: 31–108.
17. Haberman RP, Criswell HE, Snowdy S, Ming Z, Breese GR, Samulski RJ, McCown

TJ. Therapeutic liabilities of *in vivo* viral vector tropism: Adeno-associated virus (AAV) vectors, NMDAR 1 antisense and focal seizure sensitivity. *Mol Ther* 2002;6:495–500.

18. Haberman RP, McCown TJ, Samulski RJ. Inducible long-term gene expression in brain with adeno-associated virus gene transfer. *Gene Ther* 1998;5:1604–11.
19. Macdonald RL, Gallagher MJ, Feng HJ, Kang J. GABA(A) receptor epilepsy mutations. *Biochem Pharmacol* 2004;68:1497–506.
20. Buhl E, Otis T, Mody I. Zinc-induced collapse of augmented inhibition by GABA in a temporal lobe epilepsy model. *Science* 1996;271:369–73.
21 Gibbs JW 3rd, Shumate MD, Coulter DA. Differential epilepsy-associated alterations in postsynaptic $GABA_A$ receptor function in dentate granule cells and CA1 neurons. *J Neurophysiol* 1997;77:1924–38.
22. Brooks-Kayal AR, Shumate MD, Jin H, Rikhter TY, Coulter DA. Selective changes in single cell $GABA_A$ receptor subunit expression and function in temporal lobe epilepsy. *Nat Med* 1998;4:1166–72.
23. Brooks-Kayal AR, Shumate MD, Jin H, Lin DD, Rikhter TY, Holloway KL, Coulter DA. Human neuronal γ-aminobutyric $acid_A$ receptors: Coordinated subunit mRNA expression and functional correlates in individual dentate granule cells. *J Neurosci* 1999;19:8312–18.
24. Peng Z, Huang CS, Stell BM, Mody I, Houser CR. Altered expression of the delta subunit of the GABAA receptor in a mouse model of temporal lobe epilepsy. *J Neurosci* 2004;24:10167–75.
25. Zhang N, Wei W, Mody I, Houser CR. Altered localization of GABA(A) receptor subunits on dentate granule cell dendrites influences tonic and phasic inhibition in a mouse model of epilepsy. *J Neurosci* 2007;27:7520–31.
26. Xiao X, McCown TJ, Li J, Breese GR, Morrow AL, Samulski RJ. Adeno-associated virus (AAV) vector antisense gene transfer in vivo decreases $GABA_A$ $\alpha 1$ containing receptors and increases collicular seizure sensitivity. *Brain Res* 1997:576:76–83.
27. Raol YH, Lund IV, Bandyopadhyay S, Zhang G, Roberts DS, Wolfe JH, Russek SJ, Brooks-Kayal AR. Enhancing GABA(A) receptor $\alpha 1$ subunit levels in hippocampal dentate gyrus inhibits epilepsy development in an animal model of temporal lobe epilepsy. *J Neurosci*2006;26:11342–6.
28. Roberts D, Raol YSH, Budrick E, Lund I, Passini M, Wolfe J, Brooks-Kayal AR, Russek SJ. Egr3 stimulation of GABRA4 promoter activity as a mechanism for seizure-induced up-regulation of $GABA_A$ receptor alpha4 subunit expression. *Proc Natl Acad Sci U S A* 2005;102:11894–9.
29. Klein RL, Meyer EM, Peel AL, Zolotukhin S, Meyers C, Muzyczka N, King MA. Neuron-specific transduction in the rat septohippocampal or nigrostriatal pathway by recombinant adeno-associated virus vectors. *Exp Neurol* 1998;150:183–94.
30. Burger C, Gorbatyuk OS, Velardo MJ, Peden CS, Williams P, Zolotukhin S, Reir PJ, Mandel RJ, Muzyczka N. Recombinant AAV viral vectors pseudotyped with viral capsids from serotypes 1, 2 and 5 display differential efficiency and cell tropism after delivery to different regions of the central nervous system. *Mol Ther* 2004;10:302–17.
31. McCown TJ, Xiao X, Li J, Breese GR, Samulski RJ. Differential and persistent expression patterns of CNS gene transfer by an adeno-associated virus (AAV) vector. *Brain Res* 1996;713:99–107.
32. Mohler H, Fritschy JM, Rudolph U. A new benzodiazepine pharmacology. *J Pharmacol Exp Ther* 2002;300:2–8.
33. Thompson KW, Anantharam V, Sehrstock S, Bongarzone E, Campagnoni A, Tobin AJ. Conditionally immortalized cell lines, engineered to produce and release GABA, modulate the development of behavioral seizures. *Exp Neurol* 2000;161:481–9.
34. Gernert M, Thompson KW, Loscher W, Tobin AJ. Genetically engineered GABA-producing cells demonstrate anticonvulsant effects and long-term transgene expression when transplanted into the central piriform cortex of rats. *Exp Neurol* 2002;176: 183–92.
35. Castillo CG, Mendoza S, Freed WJ, Giordano M. Intranigral transplants of immortalized GABAergic cells decrease the expression of kainic acid-induced seizures in the rat. *Behav Brain Res* 2006;171:109–15.
36. Thompson KW, Suchomelova LM. Transplants of cells engineered to produce GABA suppress spontaneous seizures. *Epilepsia* 2004;45:4–12.
37. Thompson KW. Genetically engineered cells with regulatable GABA production can affect afterdischarges and behavioral seizures after transplantation into the dentate gyrus. *Neurosci* 2005;133:1029–37.

38. Choi-Lundberg DL, Lin Q, Chang YN, Hay CM, Mohajeri H, Davidson BL, Bohn MC. Dopaminergic neurons protected from degeneration by GDNF gene therapy. *Science* 1997;275:838–41.
39. Mandel RJ, Spratt SK, Snyder RO, Leff SE. Midbrain injection of recombinant adeno-associated virus encoding rat glial cell line-derived neurotrophic factor protects nigral neurons in a progressive 6-hydroxydopamine-induced degeneration model of Parkinson's disease in rats. *Proc Nat Acad Sci U S A* 1997;94:14083–8.
40. Martin D, Miller G, Rosendahl M, Russell DA. Potent inhibitory effects of glial derived neurotrophic factor against kainic acid mediated seizures in the rat. *Brain Res* 1995;683:172–8.
41. Yoo YM, Lee CJ, Lee U, Kim YJ. Neuroprotection of adenoviral-vector-mediated GDNF expression against kainic-acid-induced excitotoxicity in the rat hippocampus. *Exp Neurol*206;200:407–17.
42. Kanter-Schlifke I, Georgievska B, Kirik D, Kokais M. Seizure suppression by GDNF gene therapy in animal models of epilepsy. *Mol Ther* 2007;15:1106–3,2007.
43. Perkins D, Pereira EF, Aurelian L, The herpes simplex virus type 2 R1 protein kinase (ICP10 PK) functions as a dominant regulator of apoptosis in hippocampal neurons involving activation of the ERK survival pathway and upregulation of the antiapoptotic protein Bag-1. *J Virol* 2003;77: 1292–305.
44. Laing JM, Gober MD, Golembewski EK, Thompson SM, Gyure KA, Yarowsky PJ, Aurelian L. Intranasal administration of the growth-compromised HSV-2 vector DeltaRR prevents kainate-induced seizure and neuronal loss in rats and mice. *Mol Ther* 2006;13:870–81.
45. Ango R, Prezeau L, Muller T, Tu JC, Xiao B, Worley PF, Pin JP, Bockaert J, Fagni L. Agonist-independent activation of metabotropic glutamate receptors by the intracellular protein Homer. *Nature* 2001;411:962–5.
46. Klugman M, Symes CW, Leichtlein CB, Klaussner BK, Dunning J, Fong D, Young D, During MJ. AAV-mediated hippocampal expression of short and long Homer 1 proteins differentially affect cognition and seizure activity in adult rats. *Mol Cell Neurosci* 2005;28:347–60.
47. Yang J, Teng Q, Fererici T, Najm I, Chabardes S, Moffitt M, Alexopoulos A, Riley J, Boulis N. Viral clostridial light chain gene-based control of penicillin-induced neocortical seizures. *Mol Ther* 2007;15: 542–51.
48. Dunwiddie RV, Masino SA. The role and regulation of adenosine in the central nervous system. *Annu Rev Neurosci* 2001;24:31–55.
49. Huber A, Padrun V, Deglon N, Aebischer P, Mohler H, Boison D. Grafts of adenosine-releasing cells suppress seizures in kindling epilepsy. *Proc Nat Acad Sci U S A* 2001; 98:7611–16.
50. Haberman RP, Samulski RJ, McCown TJ. Attenuation of seizures and neuronal death by adeno-associated virus (AAV) vector galanin expression and secretion. *Nat Med* 2003;9:1076–80.
51. McCown T J. Adeno-associated virus-mediated expression and constitutive secretion of galanin suppresses limbic seizure activity in vivo. *Mol Ther* 2006;14: 63–8.
52. Lin ED, Richichi C, Young D, Baer K, Vezzani A, During MJ. Recombinant AAV-mediated expression of galanin in rat hippocampus suppresses seizure development. *J Neurosci* 2003;18:2087–2003.
53. Richichi C, Lin EJ, Stefanin D, Colella D, Ravizza T, Grignaschi G, Veglianese P, Sperk G, During MJ, Vezzani A. Anticonvulsant and antiepileptogenic effects mediated by adeno-associated virus vector neuropeptide Y expression in the rat hippocampus. *J Neurosci* 2004;24:3051–9.
54. Foti S, Haberman RP, Samulski RJ, McCown TJ. Adeno-associated virus-mediated expression and constitutive secretion of NPY or NPY13-36 suppresses seizure activity *in vivo*. *Gene Ther* 2007;14: 1534–6.
55. Marsh DJ, Baraban SC, Hollopeter G, Palmiter RD. Role of the Y5 neuropeptide Y receptor in limbic seizures. *Proc Nat Acad Sci U S A*. 1999;96:13518–23.
56. Lin EJ, Young D, Baer K. Herzog H, During MJ. Differential actions of NPY on seizure modulation via Y1 and Y2 receptors: Evidence from receptor knockout mice. *Epilepsia* 2006;47:773–80.
57. Colmers WF, Klapstein GJ, Fournier A, St-Pierre S, Treherne KA. Presynaptic inhibition by neuropeptide Y in rat hippocampal slice in vitro is mediated by a Y2 receptor. *Br J Pharmacol* 1991;102:41–4.
58. Greber S, Schwarzer C, Sperk G. Neuropeptide Y inhibits potassium-stimulated glutamate release through Y2 receptors in rat hippocampal slices in vitro. *Br J Pharmacol* 1994;113:737–40.

59. El Bahh B., Balosso S., Hamilton T., Herzog H., Beck-Sickinger A.G., Sperk G, Gehlert D.R., Vezzani A., Colmers W.F. The anti-epileptic actions of neuropeptide Y in the hippocampus are mediated by Y and not Y receptors. *Eur J Neurosci* 2005;22:1417–30.
60. Vezzani A, Moneta D, Mule F, Ravizza T, Gobbi M, French-Mullen J. Plastic changes in neuropeptide Y receptor subtypes in experimental models of limbic seizures. *Epilepsia* 2000;(Suppl.6):S115–21.
61. Guttinger M, Fedele D, Koch P, Padrun V, Pralong WF, Brustle O, Boison D. Suppression of kindled seizures by paracrine adenosine release from stem cell-derived brain implants. *Epilepsia* 2005;46:1162–9.
62. Seki T, Matsubayashi H, Amano T, Kitada K, Serkawa T, Sasa M, Sakai N. Adenoviral gene transfer of aspartoacylase ameliorates tonic convulsions of spontaneously epileptic rats. *Neurochem Int* 2004;45:171–8.
63. Zhang LX, Li XL, Smith MA, Post RM, Han JS. Lipofectin-facilitated transfer of cholecystokinin gene corrects behavioral abnormalities of rats with audiogenic seizures. *Neurosci* 1997;77:15–22.

Chapter 14

Neural Stem Cells in Experimental Mesial Temporal Lobe Epilepsy

Michelle M. Kron and Jack M. Parent

Abstract

Neurogenesis persists in the adult mammalian hippocampal dentate gyrus and is influenced by epileptogenic insults. Studies of rodent mesial temporal lobe epilepsy (mTLE) models indicate that status epilepticus acutely increases dentate granule cell (DGC) neurogenesis, but in chronic stages neurogenesis may decrease. The functional implications of altered neurogenesis in either stage of mTLE are poorly understood. Accumulating evidence suggests, however, that altered neurogenesis contributes to several well-characterized cellular abnormalities seen in human and experimental mTLE. These abnormalities include mossy fiber sprouting, DGC layer dispersion, aberrant migration of DGC progenitors, and the appearance of DGCs in ectopic locations or with hilar basal dendrites. The mechanisms underlying these structural abnormalities are beginning to be characterized, and ongoing work aims to define their role in epileptogenesis or associated cognitive dysfunction. Other research on neural stem or progenitor cells suggest that cell grafts may offer a strategy to repair epileptogenic damage or to deliver anticonvulsant treatments. Thus, both manipulating endogenous neural progenitors and neural stem cell transplantation offer potential therapeutic strategies for mTLE. This chapter describes ongoing investigations of experimental mTLE aimed at elucidating the roles of endogenous neural stem cells in acquired epileptogenesis, as well as the potential use of neural stem cell grafts to treat epilepsy.

Key words: seizure, epilepsy, neurogenesis, stem cell, dentate granule cell, hippocampus, mossy fiber sprouting, neuronal migration.

1. Introduction

Neural stem cells (NSCs) persist in the adult mammalian brain, and both embryonic and adult NSCs can be expanded in vitro as a source of transplantable cells for repair *(1)*. Although the reparative potential of endogenous or graftable NSCs is obvious, less intuitive is the idea that endogenous neural progenitors contribute to disease

Scott C. Baraban (ed.), *Animal Models of Epilepsy: Methods and Innovations, vol. 40*

DOI 10.1007/978-1-60327-263-6_14 Springerprotocols.com

pathophysiology. In terms of epilepsy, NSC research in the field has focused on the latter, the influence of endogenous NSCs, based on findings of aberrant hippocampal neurogenesis in rodent epilepsy models. Controversy exists as to the net effects of injury-induced neurogenesis in terms of contributing to epileptogenesis or associated morbidities versus compensating by restoring inhibition to hyperexcitable networks. Only recently have NSC transplantation studies been undertaken to examine their potential as an antiepileptogenic or anticonvulsant therapy. In this chapter, we will discuss recent work examining the influence of endogenous and transplanted NSCs in animal models of mesial temporal lobe epilepsy (mTLE). Consideration will be given to alterations in endogenous neurogenesis during epileptogenesis and the use of grafted NSCs to restore inhibition or as vehicles to deliver anticonvulsant agents.

2. Adult Mammalian Hippocampal Neurogenesis

Neurogenesis occurs throughout adulthood in specific regions of the mammalian brain, including the forebrain subventricular zone and the hippocampal dentate gyrus *(2)*. Within the mammalian hippocampal formation, dentate granule cells (DGCs) are a unique cell population. The majority are generated in early postnatal life, and new DGCs are produced at a lower rate throughout adulthood and into senescence *(3–7)*. The human dentate gyrus generates neurons into at least the 7th decade of life *(8, 9)*. New neurons make up about 6% of the granule cell layer in adult rats *(10)* and the adult-born DGCs acquire characteristics of mature DGCs *(11, 12)*. Bromodeoxyuridine (BrdU) mitotic labeling of DGCs generated during adulthood show that the cells that integrate into circuits and survive to maturity are very stable and may permanently replace DGCs born during development *(13)*. Importantly, adult-generated DGCs develop electrical activity with functional synaptic transmission *(11)* and successfully integrate into the hippocampal network *(14–16)*. These properly integrated cells are thought to play an important role in certain types of learning and memory *(17–21)*.

3. Mesial Temporal Lobe Epilepsy

Approximately 50 million people worldwide suffer from epilepsy. Forty percent of those have mTLE, making it the most common form of adult epilepsy *(22)*. Furthermore, about 35% of patients with mTLE have chronic seizures that are pharmacoresistant *(23)*, making the development of new therapeutics crucial. In addition to the recurrent seizures that define mTLE, the syndrome has associated chronic comorbidities that include problems with both learning and memory *(24, 25)* and depression *(26, 27)*. mTLE often begins with an initial precipitating event such as a complicated febrile seizure, followed by a latent period and ultimately the development of epilepsy. This progression of events is mirrored in a variety of rodent models. Chemoconvulsant models of epilepsy begin with a "precipitating event" of status epilepticus (SE), also followed by a latent period during which structural and functional epileptogenic changes occur. In animal models of mTLE, hippocampal pathways show structural plasticity that mirrors the changes seen in humans with mTLE, recapitulating much of the human pathology *(28–30)*.

4. Altered Neurogenesis in mTLE

Mesial temporal lobe epilepsy disrupts normal neurogenesis and induces a complex set of pathophysiological changes. Seizure-induced injury appears to influence both pre-existing and adult-born DGCs in the epileptogenic hippocampal formation. Studies of adult rodent models of mTLE suggest that prolonged seizures potently stimulate DGC neurogenesis in the short term. Dentate gyrus cell proliferation increases 5- to 10-fold after a latent period of several days and this leads to increased numbers of neuroblasts *(31–33)*. In addition to stimulating neurogenesis, the epileptic dentate gyrus is associated with numerous cellular abnormalities including mossy fiber sprouting (MFS), DGCs with persistent hilar basal dendrites (HBDs), DGC layer dispersion, and ectopically located DGCs. The role of adult-born neurons in these abnormalities will be considered in the following sections.

4.1. Mossy Fiber Sprouting in mTLE

In the normal brain, granule cells receive excitatory input from both the perforant path and the mossy cells of the dentate hilus. Granule cell axons, or mossy fibers, innervate hilar interneurons and CA3 pyramidal cells. Prolonged seizures have been shown to kill hilar mossy cells, perhaps serving as a trigger for DGCs to develop mossy fiber collaterals that synapse with other granule cells or with interneurons (reviewed in *(34)*). In the epileptic human dentate gyrus, Timm staining, dynorphin immunoreactivity, and biocytin fills reveal mossy fiber sprouting into the supragranular inner molecular layer *(35–37)*. Mossy fiber sprouting (MFS) is thought to begin during the second week after SE and peak around 2–3 months *(38, 39)*.

Initial evidence from chemoconvulsant mTLE models suggested that newborn DGCs contribute to aberrant mossy fiber remodeling *(31)*. Subsequent work, however, indicated that MFS occurs in older cells, evidenced by the fact that killing new cells via irradiation or pretreatment with cyclohexamide does not decrease MFS *(31, 40)*. Furthermore, retroviral studies indicate that adult-generated DGCs born at least 4 weeks before SE, but not 1 week before or after, contribute to sprouting *(31, 41)*. Whether fully mature or only late developing DGCs contribute to mossy fiber sprouting is unknown.

4.2. Ectopic Dentate Granule Cells in mTLE

A second prominent cellular abnormality seen after experimental SE is the appearance of ectopic DGCs. Normally, the vast majority of newborn neurons in the subgranular zone migrate into the granule cell layer. BrdU labeling and Prox-1 immunohistochemistry used to identify newborn granule cells, however, demonstrate that many progenitors migrate aberrantly to the hilus and molecular layer after prolonged seizures and differentiate into ectopic DGCs in the rat *(36, 42)* (**Fig. 14.1A-D**). Hilar and molecular layer-ectopic DGCs have been observed by a number of laboratories using different mTLE models *(42–47)*. A similar pathological abnormality is seen in human mTLE *(36, 42)* (**Fig. 14.1E,F**). Interestingly, ectopic cells are thought to still receive proper synaptic input from the perforant path despite their abnormal locations *(48)*.

The exact mechanism through which cells migrate ectopically remains unclear. Evidence suggests that abnormal chain migration is involved, as neuroblast marker expression indicates the delayed appearance of atypical chain-like progenitor cell formations extending into the hilus and molecular layer *(36, 42, 49)*. One potential mechanism underlying this abnormality involves decreased expression of Reelin, a signaling molecule normally expressed in interneuron subsets typically lost in human and experimental mTLE. Prolonged seizures decrease Reelin immunoreactivity in the adult rat dentate gyrus, and blockade of Reelin signals increases chain migration *(50)*. This suggests that Reelin modulates DGC progenitor migration to maintain normal DGC integration, and loss of Reelin expression in epilepsy likely contributes to ectopic chain migration of newborn DGCs. Decreased Reelin expression is unlikely to be the sole defect responsible for seizure-induced aberrant cell migration, however, as other factors implicated in regulating DGC progenitor migration are altered after status epilepticus *(51)*.

4.3. Hilar Basal Dendrites in mTLE

Under normal conditions, the dendrites of most rodent DGCs arise from the apical pole of the cell body and extend into the molecular layer, while the axons extend from the hilar (basal) pole and terminate in the stratum lucidum of CA3 *(52)*. During early

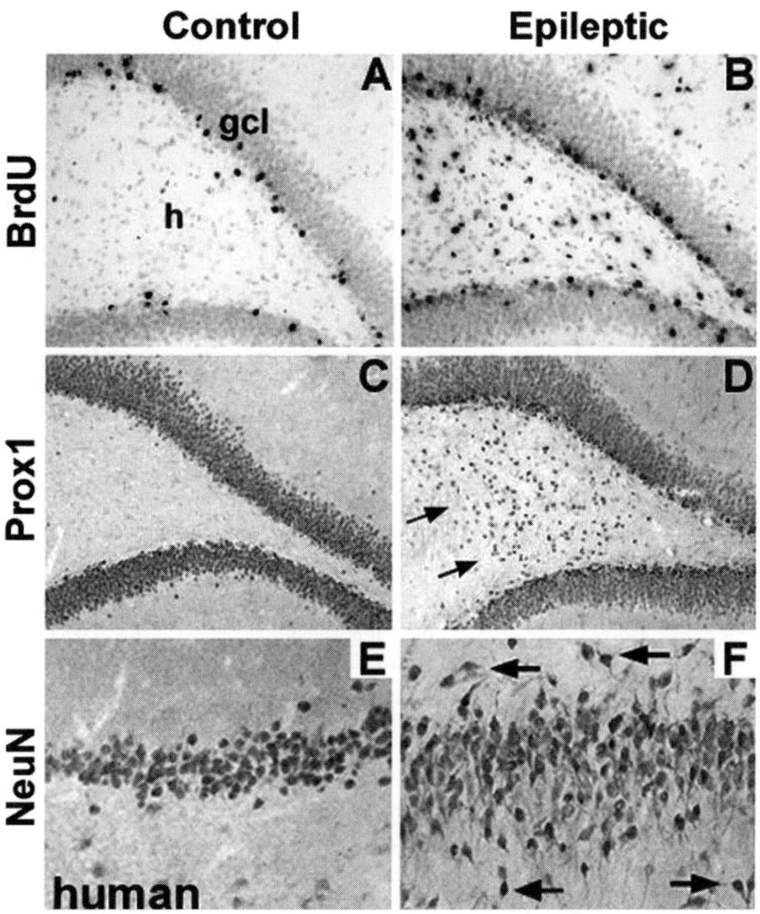

Fig. 14.1 **Hilar ectopic granule cells in experimental and human mTLE.** Bromodeoxyuridine (BrdU; **A, B**) and Prox1 (**B, C**)-immunolabeled sections through the adult rat dentate gyrus show marked increases in recently born (BrdU+) cells and those that express the DGC-specific marker Prox1 in the hilus (h) of animals 35 days after pilocarpine-induced status epilepticus (**B, D**) compared to controls (**A, C**). BrdU was given 7 days after seizure induction. (**E, F**) Surgically resected hippocampal tissue from a patient with intractable mTLE immunostained for NeuN also shows hilar and molecular layer (ml) ectopic granule-like neurons (arrows in **F**) not seen in an autopsy control (**E**). gcl, granule cell layer. Scale bar: 50 μm for (**A, B**); 150 μm for (**C, D**); 30 μm for (**E, F**). Modified with permission from Parent et al. *(42)*.

postnatal ages, hilar basal dendrites (HBDs) transiently appear on normal granule cells *(53)*, but they do not persist on mature DGCs *(54–56)*. In rats experiencing SE, however, a large portion of newborn DGCs generated around the time of the prolonged seizures exhibit HBDs that persist for many months *(45–47)* (**Fig. 14.2**). Using Thy1-GFP mice, Walter et al. demonstrated that almost 50% of immature granule cells exposed to pilocarpine-induced SE exhibited HBDs, and recently born cells were even more severely impacted than older but still immature cells

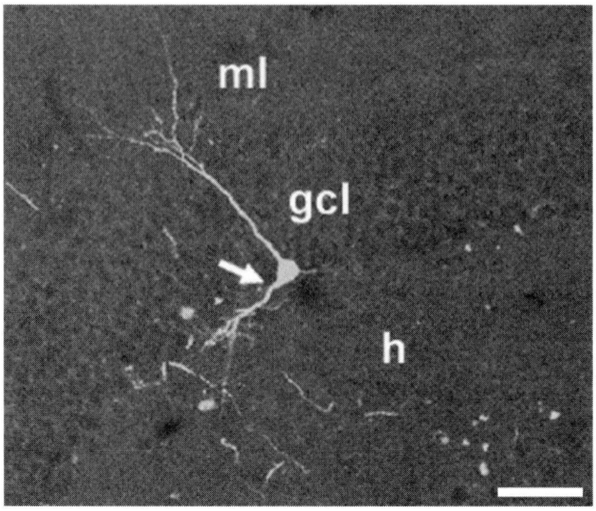

Fig. 14.2. **GFP-labeled, adult-born DGC at the granule cell layer (gcl)/hilar (h) border.** A retroviral-GFP reporter was injected into the dentate gyrus of this adult rat 2 days before status epilepticus was induced with pilocarpine, and the animal was killed 21 days after status epilepticus. An apical dendrite extends into the molecular layer (ml), and a prominent basal dendrite is evident (arrow) that runs parallel to the gcl. Scale bar: 50 μm.

(58). Many of these basal dendrites have numerous spines, suggesting they are postsynaptic to axon terminals *(54, 56)*. The molecular mechanisms responsible for the extension of HBDs remain unclear, but may involve an alteration in glial scaffolding *(47)*.

HBDs can form on granule cells as early as 1 week following SE, consistent with the hypothesis that HBDs on granule cells may be generated from seizure-induced, de novo granule cells *(59)*. However, alternative explanations that some or all HBDs arise from older dentate granule cells cannot be excluded at present.

5. Functional Implications of Seizure-Induced Neurogenesis

The functional implications of seizure-induced neurogenesis are poorly understood. Several studies demonstrate that ablation of neurogenesis after SE ameliorates epileptic symptoms *(60, 61)*. For example, the ablation of new neurons generated in response to seizures by cytosine-b-D-arabinofuranoside (AraC) administration reduces the frequency and duration of spontaneous seizures *(60)*. Furthermore, post-seizure cognitive impairment is reduced when seizure-induced neurogenesis is inhibited with valproic acid *(62)*. Although these studies suggest that neurons born after seizure may contribute to epileptogenesis and long-term learning and memory impairments seen in patients with mTLE, the implications of specific morphological abnormalities remain unclear.

The role of ectopic cells in the post-seizure environment is also receiving increasing attention, with accumulating data suggesting that ectopic cells are hyperexcitable. Ultrastructural studies indicate that hilar ectopic DGCs are postsynaptic to mossy fibers and have less inhibitory input onto their somata and proximal dendrites than properly located granule cells *(45)*. These morphological alterations are consistent with electrophysiological findings showing that hilar ectopic granule cells from epileptic rats are much more excitable than DGCs in the layer as they burst fire in synchrony with spontaneous, rhythmic bursts of CA3 pyramidal neurons *(46)*. Also consistent with these data is the finding that ablating neurogenesis after SE attenuates subsequent epileptogenesis, with a reduction of the frequency and severity of spontaneous recurrent seizures *(60, 61)*. However, one caveat is that antimitotic treatment caused knockdown of all dividing cell types *(60, 61)*. Because both interventions have other non-specific effects, it is difficult to interpret and draw definitive conclusions from these data.

Whereas past findings led to hypotheses that seizure-induced neurogenesis contributes to epilepsy *(63)*, a more recent study suggests that post-seizure neurogenesis under some circumstances may restore inhibition and potentially play a compensatory role. Using retroviral reporter labeling of adult-born cells coupled with electrophysiological recordings, Jakubs and colleagues (2006) demonstrated that new granule cells generated after electrically induced SE exhibit functional connectivity consistent with reduced excitability, reflected by reduced mean frequency of excitatory postsynaptic currents *(64)*. These data point out the possibility that new cells born after epileptogenic insult, at least those that migrate normally into the DGC layer, may in fact decrease excitability and, therefore, be beneficial to the epileptic brain.

The functional role of HBDs has been less extensively examined, but several studies suggest that increased numbers of DGCs in epileptic rats extend basal dendrites into the hilus, providing a potential route for recurrent excitation *(54, 56, 65, 66)*. Electron microscopy demonstrates that HBDs form asymmetric synapses and are innervated by mossy fibers, further supporting recurrent excitatory circuit formation *(55)*. These findings, therefore, raise the possibility that HBD formation is pro-epileptogenic *(45, 60)*.

Although neurogenesis immediately following SE increases, controversy exists about the long-term impact of SE on neurogenesis. Some data suggest that chronic mTLE is associated with a decrease in neurogenesis *(44)*. Supporting this hypothesis is evidence of fewer newborn neurons in the dentate gyrus of children that had experienced frequent seizures *(67)*. Furthermore, adults with mTLE also appear to have decreased neurogenesis, as evidenced by a decrease in mRNA for hippocampal doublecortin, a neuroblast marker, as well as the absence of immunoreactivity for Ki67, a marker of actively cycling cells *(68)*. Hypothesized reasons for this possible decrease include exhaustion of the progenitor pool, loss of needed growth/trophic factors, or altered cellular interactions in the stem cell niche.

Much additional work is needed to clarify the role of altered DGC neurogenesis in mTLE. A better understanding of the functional effects of the aberrant integration of neurons generated during epileptogenesis will require the more selective ablation of adult-born neurons. Even more useful for examining the functional importance of seizure-induced neurogenesis would be strategies to prevent only aberrant neurogenesis and maintain potentially compensatory neuronal birth. Moreover, if DGC neurogenesis is involved in learning and memory, the lack of neurogenesis may explain the disabling memory disturbances seen in mTLE. More comprehensive behavioral testing in experimental mTLE models will be necessary to sort out how aberrant or depleted neurogenesis impacts cognitive function. A last consideration is the relevance, if any, of aberrant or compensatory neurogenesis to human mTLE. As described above, evidence of impaired neurogenesis from human mTLE surgical specimens exists but is not strong, and elucidating the pathophysiological effects of neurogenesis will be critical for devising stem cell-specific antiepileptogenic therapies.

6. Exogenous Stem Cells in Epilepsy

The lack of antiepileptogenic therapies has led to the study of stem cells as a potential tool for restoring the hippocampus to a more normal state after epileptogenic insult. Given that mTLE is associated with substantial neuronal cell death and hyperexcitable network function, potential roles for exogenous stem cell therapy in the treatment of mTLE include restoring lost neurons or increasing inhibitory connections. Studies are only now starting to address these concepts, and much work is needed to identify the best cellular sources for therapeutic grafts.

One cell source, neural progenitors from E19 hippocampus, has the ability to give rise to both hippocampal pyramidal-like neurons and interneurons and may be a good source of cells for grafting into the hippocampus in mTLE *(69)*. Early studies using supplemented grafts have shown some promise. Grafts supplemented with brain-derived neurotrophic factor, neurotrophin-3 and a caspase inhibitor, or fibroblast growth factor and a caspase inhibitor display increased long-term survival and reverse some of the abnormal synaptic reorganization in the injured hippocampus such as MFS in the dentate supraganular layer *(70)*. Furthermore, hippocampal fetal cell grafts survive robustly in the chronically epileptic rodent hippocampus and reduce the frequency of chronic recurrent seizures *(71)*.

The epileptic brain represents a hyperexcitable environment with decreased numbers of interneurons. For example, in situ hybridization or immunohistochemistry for GAD65, parvalbumin, or somatostatin demonstrates loss of inhibitory interneurons in the rodent hippocampus after chemoconvulsant-induced SE *(72, 73)*. Furthermore, the frequency of GABA(A) receptor-mediated mIPSCs is reduced after prolonged seizures *(74)*. Despite this reported decrease in GABA-mediated inhibition in the epileptic brain, GABA-mimetic therapies for controlling seizures in pharmacoresistant mTLE are generally ineffective *(75)*, likely due to their non-specific actions. Because of this, a more targeted approach to increasing GABA in the brain, such as stem cell grafting, may have greater efficacy for the long-term control of epilepsy. Thus, replacing lost hippocampal interneurons may be even more important than regenerating pyramidal cells for treating mTLE.

One important caveat for transplanting neural stem or progenitor cells to restore GABAergic interneurons is the need for them to integrate appropriately; obviously, if they were to aberrantly inhibit other interneurons, the epilepsy could worsen considerably. Although this may seem like an insurmountable obstacle, one recent study using grafts of embryonic medial ganglionic eminence (MGE) cells suggests that the cues to establish appropriate functional inhibitory synapses persist in the postnatal brain *(76)*. These cells have been demonstrated to disperse, migrate long distances in the adult brain, and mature into neurons with morphologies resembling GABA-expressing interneurons *(77)*. Grafted MGE-derived neurons appear to differentiate into GABA-producing inhibitory neurons and fully differentiate into functionally integrated interneurons within 1 month after transplantation. After integration, the grafted cells significantly increase GABA-mediated synaptic inhibition in regions containing the transplanted MGE cells. Consistent with an enhancement of GABAergic tone, the investigators found a significant increase in the frequency of spontaneous IPSCs. Furthermore, the transplantation of MGE cells did not appear to alter synaptic excitation in regions of the brain containing transplanted cells. From these experiments the authors conclude that grafted MGE-derived cells function as inhibitory interneurons. These cells receive excitatory input from local pyramidal neurons and were shown to functionally integrate into cortical and hippocampal synaptic circuitry of the host brain in such a manner as to modify the overall level of inhibition *(76)*.

Fetal CA3 grafts can also restore GAD interneuron numbers through graft axon reinnervation of the host. In studies using this approach, GAD interneuron density was analyzed in the adult rat hippocampus 6 months after kainic acid administration followed by grafting of fetal-mixed hippocampal, CA3, or CA1 cells into the CA3 region. In dentate and CA1 regions of the lesioned hippocampus receiving grafts, GAD interneuron density significantly increased. Thus, grafts containing CA3 cells restored CA3 lesion-induced depletions in hippocampal GAD interneurons. This specific graft-mediated effect is beneficial because reactivation of interneurons could ameliorate both loss of functional inhibition and hyperexcitability in CA3-lesioned hippocampus *(73)*.

Although many of the data seem promising, one theoretical problem with the grafting of GABAergic cells mentioned above is the potential for grafted cells to integrate in an undesirable manner. For instance, if grafted GABAergic cells were to synapse onto pre-existing GABAergic interneurons, they may inhibit these inhibitory neurons and possibly increase overall excitation. A second potential problem is the idea that a critical period may exist during neuronal differentiation in which developing neurons are vulnerable to being recruited into epileptogenic neuronal circuits *(58)*. It is not clear if grafted neural progenitors, or specific sources of progenitors, are less vulnerable to being recruited into such circuits than endogenous adult neural progenitors or if they lack the plasticity to be influenced by the post-seizure environment.

Alternative stem cell therapies for epilepsy may also include grafts that secrete other substances, and recent work involving adenosine-secreting grafts shows promise.

Recipients of adenosine-releasing neural progenitors displayed sustained protection from developing generalized seizures. In contrast, recipients of control cells failed to display a significant effect on kindling development. The therapeutic effect of adenosine-secreting grafts was demonstrated to be due to graft-mediated adenosine release, as blocking adenosine A1 receptors resulted in the ability to provoke seizures. Furthermore, histological analysis revealed intrahippocampal graft-derived cells expressing mature neuronal markers. These data may indicate a potential antiepileptogenic function of stem cell-mediated adenosine delivery, although more research must be done *(78)*.

7. Conclusions

Our understanding of neural stem cells (NSCs) has dramatically increased over the past decade, leading some to propose stem cell therapies for various brain disorders. Future NSC therapies for epilepsy may include those that alter endogenous neurogenesis as well as NSC transplantation.

In epilepsy, the stimulation of endogenous NSCs may have opposing consequences for epileptogenesis after acute insults depending upon whether the newborn neurons integrate normally or aberrantly. In the chronic stages, depletion of NSCs may be pathologic. Therefore, manipulation of both short-term neurogenesis coupled with chronic stimulation of endogenous neurogenesis may be required. Further understanding of how DGC neurogenesis is regulated and the functional effects of seizure-induced neurogenesis is needed before therapeutic manipulation of endogenous neurogenesis to treat epilepsy or its comorbidities is realized.

NSC transplantation may also offer useful antiepileptogenic or anti-ictogenic therapies via several different mechanisms, including principal cell replacement, restoring interneuron function, or as vehicles to focally deliver anticonvulsant agents. Further research into these grafts is needed to answer questions about proper timing of the introduction of graft cells as well as how these cells migrate, differentiate into various neuronal subtypes, and integrate into the pre-existing circuitry. Nevertheless, both endogenous and exogenous NSCs offer great potential to treat numerous brain disorders, including mTLE. Rigorous studies in animal models must be conducted, however, to ascertain their precise roles. Only then can we determine the most beneficial ways to optimize NSC therapies.

References

1. Reynolds BA, Weiss S. Generation of neurons and astrocytes from isolated cells of the adult mammalian central nervous system. *Science* 1992;255 (5052):1707–10.
2. Taupin P, Gage FH. Adult neurogenesis and neural stem cells of the central nervous system in mammals. *J Neurosci Res* 2002; 69(6):745–9.
3. Schlessinger A, Cowan WM, Gottlieb D. An autoradiographic study of the time of origin and the pattern of granule cell migration in the dentate gyrus of the rat. *J Comp Neurol* 1975;159(2):149–75.
4. Kuhn HG, Dickinson-Anson H, Gage FH. Neurogenesis in the dentate gyrus of the adult rat: age-related decrease of neuronal progenitor proliferation. *J Neurosci* 1996; 16(6):2027–33.
5. Altman J, Das GD. Post-natal origin of microneurones in the rat brain. *Nature* 1965;207(5000):953–6.
6. Kaplan MS, Hinds JW. Neurogenesis in the adult rat: electron microscopic analysis of light radioautographs. *Science* 1977;197 (4308):1092–4.
7. Cameron HA, Woolley CS, McEwen BS, Gould E. Differentiation of newly born neurons and glia in the dentate gyrus of the adult rat. *Neuroscience* 1993;56(2): 337–44.
8. Eriksson PS, Perfilieva E, Bjork-Eriksson T, et al. Neurogenesis in the adult human hippocampus. *Nat Med* 1998;4(11):1313–7.
9. Kempermann G, Jessberger S, Steiner B, Kronenberg G. Milestones of neuronal development in the adult hippocampus. *Trends Neurosci* 2004;27(8):447–52.
10. Cameron HA, McKay RD. Adult neurogenesis produces a large pool of new granule cells in the dentate gyrus. *J Comp Neurol* 2001;435(4):406–17.
11. Song HJ, Stevens CF, Gage FH. Neural stem cells from adult hippocampus develop essential properties of functional CNS neurons. *Nat Neurosci* 2002;5(5):438–45.
12. Ge S, Goh EL, Sailor KA, Kitabatake Y, Ming GL, Song H. GABA regulates synaptic integration of newly generated neurons in the adult brain. *Nature* 2006;439(7076):589–93.

13. Dayer A, Ford A, Cleaver K, Yassaee M, Cameron H. Short-term and long-term survival of new neurons in the rat dentate gyrus. *J Comp Neurol* 2003;460(4):563–72.
14. Markakis EA, Gage FH. Adult-generated neurons in the dentate gyrus send axonal projections to field CA3 and are surrounded by synaptic vesicles. *J Comp Neurol* 1999;406(4):449–60.
15. Hastings NB, Gould E. Rapid extension of axons into the CA3 region by adult-generated granule cells. *J Comp Neurol* 1999; 413(1):146–54.
16. van Praag H, Schinder AF, Christie BR, Toni N, Palmer TD, Gage FH. Functional neurogenesis in the adult hippocampus. *Nature* 2002;415(6875):1030–4.
17. Saxe MD, Battaglia F, Wang J-W, et al. Ablation of hippocampal neurogenesis impairs contextual fear conditioning and synaptic plasticity in the dentate gyrus. 2006;103(46):17501–6.
18. Helmstaedter C. Effects of chronic epilepsy on declarative memory systems. *Prog Brain Res* 2002;135:439–53.
19. Stefan H, Pauli E. Progressive cognitive decline in epilepsy: An indication of ongoing plasticity. *Prog Brain Res* 2002;135:409–17.
20. Doetsch F, Hen R. Young and excitable: The function of new neurons in the adult mammalian brain. *Curr Opin Neurobiol* 2005;15(1):121–8.
21. Ming GL, Song H. Adult neurogenesis in the mammalian central nervous system. *Annu Rev Neurosci* 2005;28(1):223–50.
22. Engel Jr J, Williamson PD, H-G W. Mesial temporal lobe epilepsy. In: J EJ, TA P, eds. *Epilepsy:* A *Comprehensive Textbook.* Philadelphia: Lippincott-Raven; 1997:2417–26.
23. Engel J. A Proposed Diagnostic Scheme for People with Epileptic Seizures and with Epilepsy: Report of the ILAE Task Force on Classification and Terminology. *Epilepsia* 2001;42 (6):796–803.
24. Helmstaedter C, Kurthen M, Lux S, Reuber M, Elger C. Chronic epilepsy and cognition: A longitudinal study in temporal lobe epilepsy. *Ann Neurol* 2003;54(4):425–32.
25. Elger CE, Helmstaedter C, Kurthen M. Chronic epilepsy and cognition. *Lancet Neurol* 2004;3(11):663–72.
26. Helmstaedter C, Sonntag-Dillender M, Hoppe C, Elger CE. Depressed mood and memory impairment in temporal lobe epilepsy as a function of focus lateralization and localization. *Epilepsy Behav* 2004;5(5):696–701.
27. Mazza M, Orsucci F, De Risio S, Bria P, Mazza S. Epilepsy and depression: Risk factors for suicide? *Clin Ter* 2004;155 (10):425–7.
28. Tauck DL, Nadler JV. Evidence of functional mossy fiber sprouting in hippocampal formation of kainic acid-treated rats. *J Neurosci* 1985;5(4):1016–22.
29. Laurberg S, Zimmer J. Lesion-induced sprouting of hippocampal mossy fiber collaterals to the fascia dentata in developing and adult rats. *J Comp Neurol* 1981; 200(3):433–59.
30. Buckmaster PS. Laboratory animal models of temporal lobe epilepsy. *Comp Med* 2004;54(5):473–85.
31. Parent JM, Lowenstein DH. Mossy fiber reorganization in the epileptic hippocampus. *Curr Opin Neurol* 1997;10(2):103–9.
32. Gray WP, Sundstrom LE. Kainic acid increases the proliferation of granule cell progenitors in the dentate gyrus of the adult rat. *Brain Res* 1998;790(1–2):52–9.
33. Scott BW, Wang S, Burnham WM, De Boni U, Wojtowicz JM. Kindling-induced neurogenesis in the dentate gyrus of the rat. *Neurosci Lett* 1998;248(2):73–6.
34. Nadler JV, Tu B, Timofeeva O, Jiao Y, Herzog H. Neuropeptide Y in the recurrent mossy fiber pathway. *Peptides* 2007; 28(2):357–64.
35. de Lanerolle NC, Kim JH, Robbins RJ, Spencer DD. Hippocampal interneuron loss and plasticity in human temporal lobe epilepsy. *Brain Res* 1989;495(2):387–95.
36. Houser CR, Miyashiro JE, Swartz BE, Walsh GO, Rich JR, Delgado-Escueta AV. Altered patterns of dynorphin immunoreactivity suggest mossy fiber reorganization in human hippocampal epilepsy. *J Neurosci* 1990;10(1):267–82.
37. Babb TL, Kupfer WR, Pretorius JK, Crandall PH, Levesque MF. Synaptic reorganization by mossy fibers in human epileptic fascia dentata. *Neuroscience* 1991;42(2): 351–63.
38. Mello LEAM, Cavalheiro EA, Tan AM, et al. Circuit mechanisms of seizures in the pilocarpine model of chronic epilepsy: Cell loss and mossy fiber sprouting. *Epilepsia* 1993;34 (6):985–95.
39. Cavalheiro EA, Leite JP, Bortolotto ZA, Turski WA, Ikonomidou C, Turski L. Long-term effects of pilocarpine in rats: Structural damage of the brain triggers kindling and spontaneous recurrent seizures. *Epilepsia* 1991;32(6):778–82.

40. Williams PA, Wuarin JP, Dou P, Ferraro DJ, Dudek FE. Reassessment of the effects of cycloheximide on mossy fiber sprouting and epileptogenesis in the pilocarpine model of temporal lobe epilepsy. *J Neurophysiol* 2002;88(4):2075–87.

41. Jessberger S, Zhao C, Toni N, Clemenson GD, Jr., Li Y, Gage FH. Seizure-associated, aberrant neurogenesis in adult rats characterized with retrovirus-mediated cell labeling. *J Neurosci* 2007;27(35):9400–7.

42. Parent JM, Elliott RC, Pleasure SJ, Barbaro NM, Lowenstein DH. Aberrant seizure-induced neurogenesis in experimental temporal lobe epilepsy. *Ann Neurol* 2006; 59(1):81–91.

43. Parent JM, Yu TW, Leibowitz RT, Geschwind DH, Sloviter RS, Lowenstein DH. Dentate granule cell neurogenesis is increased by seizures and contributes to aberrant network reorganization in the adult rat hippocampus. *J Neurosci* 1997; 17(10):3727–38.

44. Hattiangady B, Rao MS, Shetty AK. Chronic temporal lobe epilepsy is associated with severely declined dentate neurogenesis in the adult hippocampus. *Neurobiol Dis* 2004;17(3):473–90.

45. Dashtipour K, Tran PH, Okazaki MM, Nadler JV, Ribak CE. Ultrastructural features and synaptic connections of hilar ectopic granule cells in the rat dentate gyrus are different from those of granule cells in the granule cell layer. *Brain Res* 2001;890(2):261–71.

46. Scharfman HE, Goodman JH, Sollas AL. Granule-like neurons at the hilar/CA3 border after status epilepticus and their synchrony with area CA3 pyramidal cells: Functional implications of seizure-induced neurogenesis. *J Neurosci* 2000;20(16): 6144–58.

47. Shapiro LA, Ribak CE. Integration of newly born dentate granule cells into adult brains: Hypotheses based on normal and epileptic rodents. *Brain Res Brain Res Rev* 2005;48 (1):43–56.

48. Scharfman HE, Sollas AE, Berger RE, Goodman JH, Pierce JP. Perforant path activation of ectopic granule cells that are born after pilocarpine-induced seizures. *Neuroscience* 2003;121(4):1017–29.

49. Parent JM, von dem Bussche N, Lowenstein DH. Prolonged seizures recruit caudal subventricular zone glial progenitors into the injured hippocampus. *Hippocampus* 2006;16(3):321–8.

50. Gong C, Wang TW, Huang HS, Parent JM. Reelin regulates neuronal progenitor migration in intact and epileptic hippocampus. *J Neurosci* 2007; 27 (8):1803–11.

51. Scharfman H, Goodman J, Macleod A, Phani S, Antonelli C, Croll S. Increased neurogenesis and the ectopic granule cells after intrahippocampal BDNF infusion in adult rats. *Exp Neurol* 2005;192(2): 348–56.

52. Amaral DG. A Golgi study of cell types in the hilar region of the hippocampus in the rat. *J Comp Neurol* 1978;182(4 Pt 2): 851–914.

53. Seress L, Pokorny J. Structure of the granular layer of the rat dentate gyrus. A light microscopic and Golgi study. *J Anat* 1981;133(Pt 2):181–95.

54. Buckmaster PS, Dudek FE. In vivo intracellular analysis of granule cell axon reorganization in epileptic rats. *J Neurosci* 1999;81 (2):712–21.

55. Ribak CE, Tran PH, Spigelman I, Okazaki MM, Nadler JV. Status epilepticus-induced hilar basal dendrites on rodent granule cells contribute to recurrent excitatory circuitry. *J Comp Neurol* 2000;428(2):240–53.

56. Spigelman I, Yan XX, Obenaus A, Lee EY, Wasterlain CG, Ribak CE. Dentate granule cells form novel basal dendrites in a rat model of temporal lobe epilepsy. *Neuroscience* 1998;86(1):109–20.

57. Shapiro LA, Ribak CE. Newly born dentate granule neurons after pilocarpine-induced epilepsy have hilar basal dendrites with immature synapses. *Epilepsy Res* 2006;69 (1):53–66.

58. Walter C, Murphy BL, Pun RY, Spieles-Engemann AL, Danzer SC. Pilocarpine-induced seizures cause selective time-dependent changes to adult-generated hippocampal dentate granule cells. *J Neurosci* 2007;27(28):7541–52.

59. Dashtipour K, Wong AM, Obenaus A, Spigelman I, Ribak CE. Temporal profile of hilar basal dendrite formation on dentate granule cells after status epilepticus. *Epilepsy Res* 2003;54(2–3):141–51.

60. Jung K-H, Chu K, Kim M, et al. Continuous cytosine-b-D-arabinofuranoside infusion reduces ectopic granule cells in adult rat hippocampus with attenuation of spontaneous recurrent seizures following pilocarpine-induced status epilepticus. *Eur J Neurosci* 2004;19(12):3219–26.

61. Jung KH, Chu K, Lee ST, et al. Cyclooxygenase-2 inhibitor, celecoxib, inhibits

the altered hippocampal neurogenesis with attenuation of spontaneous recurrent seizures following pilocarpine-induced status epilepticus. *Neurobiol Dis* 2006;23(2):237–46.

62. Jessberger S, Nakashima K, Clemenson GD, Jr., et al. Epigenetic modulation of seizure-induced neurogenesis and cognitive decline. *J Neurosci* 2007;27(22):5967–75.
63. Parent JM, Lowenstein DH. Seizure-induced neurogenesis: Are more new neurons good for an adult brain? *Prog Brain Res* 2002;135:121–31.
64. Jakubs K, Nanobashvili A, Bonde S, et al. Environment matters: Synaptic properties of neurons born in the epileptic adult brain develop to reduce excitability. *Neuron* 2006;52(6):1047–59.
65. Walter C, Murphy BL, Pun RYK, Spieles-Engemann AL, Danzer SC. Pilocarpine-induced seizures cause selective time-dependent changes to adult-generated hippocampal dentate granule cells. *J Neurosci* 2007; 27(28):7541–52.
66. Yan XX, Spigelman I, Tran PH, Ribak CE. Atypical features of rat dentate granule cells: Recurrent basal dendrites and apical axons. *Anat Embryol (Berl)* 2001;203(3):203–9.
67. Mathern GW, Leiphart JL, De Vera A, et al. Seizures decrease postnatal neurogenesis and granule cell development in the human fascia dentata. *Epilepsia* 2002;43(Suppl)5: 68–73.
68. Fahrner A, Kann G, Flubacher A, et al. Granule cell dispersion is not accompanied by enhanced neurogenesis in temporal lobe epilepsy patients. *Exp Neurol* 2007;203 (2):320–32.
69. Shetty AK, Hattiangady B. Concise review: Prospects of stem cell therapy for temporal lobe epilepsy. *Stem Cells* 2007;25(10): 396–407.
70. Hattiangady B, Rao MS, Zaman V, Shetty AK. Incorporation of embryonic CA3 cell grafts into the adult hippocampus at 4-months after injury: Effects of combined neurotrophic supplementation and caspase inhibition. *Neuroscience* 2006;139(4):1369–83.
71. Rao MS, Hattiangady B, Rai KS, Shetty AK. Strategies for promoting anti-seizure effects of hippocampal fetal cells grafted into the hippocampus of rats exhibiting chronic temporal lobe epilepsy. *Neurobiol Dis* 2007;27(2):117–32.
72. Shetty AK, Turner DA. Glutamic acid decarboxylase-67-positive hippocampal interneurons undergo a permanent reduction in number following kainic acid-induced degeneration of ca3 pyramidal neurons. *Exp Neurol* 2001;169(2):276–97.
73. Shetty AK, Turner DA. Fetal hippocampal grafts containing CA3 cells restore host hippocampal glutamate decarboxylase-positive interneuron numbers in a rat model of temporal lobe epilepsy. *J Neurosci* 2000;20 (23):8788–801.
74. Kobayashi M, Buckmaster PS. Reduced inhibition of dentate granule cells in a model of temporal lobe epilepsy. *J Neurosci* 2003;23(6):2440–52.
75. Remy S, Beck H. Molecular and cellular mechanisms of pharmacoresistance in epilepsy. *Brain* 2006:18–35.
76. Alvarez-Dolado M, Calcagnotto ME, Karkar KM, et al. Cortical inhibition modified by embryonic neural precursors grafted into the postnatal brain. *J Neurosci* 2006; 26(28):7380–9.
77. Wichterle H, Garcia-Verdugo JM, Herrera DG, Alvarez-Buylla A. Young neurons from medial ganglionic eminence disperse in adult and embryonic brain. *Nat Neurosci* 1999;2(5):461–6.
78. Li T, Steinbeck JA, Lusardi T, et al. Suppression of kindling epileptogenesis by adenosine releasing stem cell-derived brain implants. *Brain* 2007;130(5):1276–88.

INDEX

J

K

L

M

N

O

P

R

S

T

Printed in the United States of America